Chitin and Chitosan - Physicochemical Properties and Industrial Applications

Edited by Mohammed Berrada

Published in London, United Kingdom

IntechOpen

Supporting open minds since 2005

Chitin and Chitosan - Physicochemical Properties and Industrial Applications
http://dx.doi.org/10.5772/intechopen.91553
Edited by Mohammed Berrada

Contributors
Khuriah Abdul Hamid, Siti Zuhairah Zainuddin, Erol Pehlivan, Şerife Parlayıcı, Sadik Buyukyoruk, Ranju Kandra, Sunil Bajpai, Nashiru Billa Billa, Rayan Sabra, Mohammed Berrada, Damiri Fouad, Yahya Bachra, Grouli Ayoub, Amine Ouaket, Noureddine Knouzi, Ahmed Bennamara, Fouad Dabbarh, Noureddin Elbakali Kassimi, Takashi Kawabata, Beata Kolesinska Kolesinska, Piotr Rosiak, Paulina Paul, Ilona Latanska, Witold Sujka, Keita Kashima, Tomoki Takahashi, Ryo-ichi Nakayama, Masanao Imai, Marouene Bejaoui, Haykel Galai, Hanène Oueslati, Ella Kusumastuti, Fadila Mauliani, Lukman Atmaja, F. Widhi Mahatmanti, Jumaeri, Nurul Widiastuti, Brahim El Ibrahimi, Lei Guo, Jéssica Verger Nardeli, Oukhrib Rachid

First published in London, United Kingdom, 2021 by IntechOpen
IntechOpen is the global imprint of INTECHOPEN LIMITED, registered in England and Wales, registration number: 11086078, 5 Princes Gate Court, London, SW7 2QJ, United Kingdom
Printed in Croatia

British Library Cataloguing-in-Publication Data
A catalogue record for this book is available from the British Library

Additional hard and PDF copies can be obtained from orders@intechopen.com

Chitin and Chitosan - Physicochemical Properties and Industrial Applications
Edited by Mohammed Berrada
p. cm.
Print ISBN 978-1-78984-424-5
Online ISBN 978-1-78984-425-2
eBook (PDF) ISBN 978-1-83968-695-5

For EU product safety concerns:
IN TECH d.o.o., Prolaz Marije Krucifikse Kozulić 3, 51000 Rijeka, Croatia, info@intechopen.com or visit our website at intechopen.com.

Meet the editor

Dr. Mohammed Berrada is a pioneering researcher in polymer science involved in numerous university-industry transfer technology projects. He is head of the Innovation and Technology Platform at Hassan II University, Casablanca, Morocco, where he is also a professor of Organic Chemistry and Polymer Science. He received a Ph.D. in Polymer Science from the Macromolecular Sciences Division of Pierre and Marie Curie University, Paris, France, in 1992 and a Ph.D. in Organic Chemistry from Faculty of Sciences Ben M'Sik in 1998. From 2006 to 2009, Dr. Berrada was director of the Canadian Research and Development Department at Intersand Inc., where he patented the innovative OdourLock Litter Technology, which was a breakthrough and a huge success for the company.

Contents

Preface

This book is an introduction to the physicochemical properties and industrial applications of chitosan. It discusses the synthesis, purification, analysis, and properties of chitin-chitosan polymers as a renewable source of natural biodegradable polysaccharides and presents various practical and potential new applications of these materials. Each chapter deals with a very specific research topic, updates the latest results obtained by researchers, and offers readers interesting academic research perspectives. This book is highly recommended for industrial personnel involved in bioprocesses, biochemistry, microbiology, bioengineering, food engineering, and medical and pharmaceutical industries as well as readers interested in environmental protection and who wish to specialize in chitin and chitosan derivative technologies.

I am grateful to all persons that contributed to the editorial process in any manner or capacity, particularly Dr. Yahya Bachra for his contribution to the revision of the chapters and Author Service Managers Mr. Mateo Pulko and Mrs. Maja Bozicevic at IntechOpen for their assistance during the preparation of this book.

Mohammed Berrada
Professor,
Faculty of Sciences Ben M'Sick,
Department of Chemistry,
University Hassan II of Casablanca,
Casablanca, Morocco

Section 1

Drug Delivery Systems

Chapter 1

A Novel Drug Delivery System Based on Nanoparticles of Magnetite Fe_3O_4 Embedded in an Auto Cross-Linked Chitosan

Damiri Fouad, Yahya Bachra, Grouli Ayoub, Amine Ouaket, Ahmed Bennamara, Noureddine Knouzi and Mohammed Berrada

Abstract

Recently, chitosan (CS) was given much attention as a functional biopolymer for designing various hydrogels for industrial, environmental and biomedical applications, but their biomedical use is limited due to the toxicity of the crosslinker agents. To overcome this inconvenience, we developed an auto cross-linked material based on a chitosan backbone that carries an amino and aldehyde moieties. This new drug delivery system (DDS) was designed by using oxidized chitosan (OCS) that crosslinks chitosan (CS). In the first part, a simple, rapid, low-cost and eco-friendly green method was introduced to synthesize magnetite nanoparticles (Fe_3O_4-NPs) successfully. These nanoparticles Fe_3O_4 have received a great deal of attention in the biomedical field. Especially in a targeted drug delivery system, drug-loaded Fe_3O_4-NPs can accumulate at the tumor site by the aid of an external magnetic field and increase the effectiveness of drug release to the tumor site. In the second part, we have incorporated the Fe_3O_4-NPs into chitosan/oxidized chitosan solution because of their unique magnetic properties, outstanding magnetism, biocompatibility, lower toxicity, biodegradability, and other features. Three drugs (5-Fluorouracil (5-FU), Caffeine and Ascorbic acid)) were embedded into the magnetite solution that became quickly a hydrogel. The successful fabrication of the hydrogels and ferrogels was confirmed by (FT-IR), (TGA), (SEM), (VSM) analysis at room temperature. Finally, results showed that our hydrogels and ferrogels may be technologically used as devices for drug delivery in a controllable manner.

Keywords: magnetic nanoparticles, ferrogel, oxidized chitosan (OCS), chitosan (CS), controlled drug delivery

1. Introduction

Nowadays, the global increase in the number of people with chronic diseases as "cancer, diabetes, etc." have affected the health and quality of life of many citizens around the world. For example, cancer is considered a public health problem because of its high incidence and mortality. The World Health Organization (WHO) estimates 27 million cases of cancer and 17 million deaths disease for the year 2030 [1]. Therefore, study on cancer treatment has attracted many scientists. Among therapies, cancer chemotherapy is widely used despite its disadvantages. Chemotherapy usually causes serious side effects because of the low selectivity of anti-cancer drugs, which affects not only cancer cells but also normal cells [2]. Thus, increasing attention is being paid to targeted drug delivery systems, which have been used to increase the efficiency of drug delivery to specific tissues and to decrease the associated side effects.

The most efficient solution is to use nanoparticles embedded in the hydrogel; this innovative-targeted drug delivery strategy involves coupling the drug to magnetic nanoparticles (NP_s) that can be guided to the target using external magnetic fields [3]. Once they reach the target, the nanoparticles release the drugs under the influence of an alternating magnetic field [4]. Nanoparticle (NP) targeting has shown great potential for cancer drug delivery applications over the past decade. From nanoparticle targeting, magneto-particle have been widely studied because of their ability to target when an external magnetic field is applied [5].

An increasing population causes serious environmental pollution, waste production is also increasing and major proportion of by-products generated by contemporary food remains underused which may often contain precious substances. The crucial problem confronting by industries and society in food processing is the elimination of food waste. Chitin is an important natural resource and the world's annual production of it is approximately 10^{10}–10^{12} tons [6]. This latter is principally produced by mollusks, arthropods (crustaceans and insects) and fungi [7]. However, chitin and its derivatives have a high economic impact due to their numerous applications in the pharmaceutical [8], food [9, 10], cosmetics [11], textile [12], wastewater treatment and agricultural sectors [13].

In the past few decades, drug delivery systems (DDS) have been of great interest and resulted in many efforts to realize the effectiveness and targeted drug delivery tendency as well as to reduce the associated side effects [14]. However, DDS provide several advantages as compared to conventional dosages in terms of improved efficacy, reduced toxicity, improved patient compliance, and convenience [15]. Thus, the carriers used for drug release are generally biodegradable polymers which are extensively used for designing the control drug delivery systems [16].

On the other hand, with the rapid development of technology much attention has been given to use biopolymer based hydrogels in many applications including pharmaceuticals [17, 18], cosmetics [19], agriculture [20] and biotechnology [21, 22]. However, porous biomaterials fabricated from natural polymers "chitosan" were given significant attention for years. Chitosan is a unique natural cationic biopolymer produced by N-deacetylation of chitin and is the second most abundant natural polymer after cellulose [23]. Chitosan has been widely used in the biomedical field due to its superior properties including good biodegradability [24], biocompatibility [24], low toxicity [25, 26], mucoadhesive properties [27], antibacterial activity [28] and low cost [24, 29]. Chitosan is an excellent candidate for different applications particularly it has been employed in various FDA (Food and Drug Administration) approved biomedical applications. The -NH_2 group of the (CS) chains is a pH-sensitive polymer with pK_a around 6.5 due to variation of charge density at the pH range of 6–6.5, which is useful for wide range of pharmaceuticals applications. The pH value of the soluble-insoluble transition in the range 6–6.5 [30]. At pH levels beneath the pK_a, high charge density of chitosan results in polyelectrolyte formation, in contrast at neutral pH, the low charge density of chitosan eases the intracellular release of biomolecules and contributes to its low cytotoxicity [31]. Chitosan has increasingly been used in the pharmaceutical field as it is one of the excellent choices for the Schiff's base linkages to form an injectable hydrogel due to the nature of abundant amino groups on its backbone. Hydrogels from chitosan are usually prepared through physical interchain interaction or chemical reaction of the free amine groups with crosslinker agents (e.g. glutaraldehyde, glyoxal [32]). The disadvantage of these crosslinking agents, especially glutaraldehyde [33], benzaldehyde [34] and glyoxal is their toxicity to human tissues even at small traces [35], which has limited the use of chitosan hydrogels as scaffolds for drug delivery.

However, this paper discusses the recent trends in drug delivery systems (DDS) applications using macroscopic hydrogels. Hydrogel has received extensive attention due to its interconnected cross-linked porous hydrophilic polymer networks which can absorb large amounts of water or biological fluids. Additionally, hydrogels can be divided into three categories according to their size: macroscopic gels, nanogels (<200 nm), and microgels (0.5–10 μm). Hydrogels are promising, fashionable, intelligent and "smart" drug delivery vehicles that meet specific requirements for targeting drugs to specific sites and controlling drug release. The hydrogel-based drug carrier loaded with 5-fluorouracil (5-FU) drug for an up to 36 days sustained delivery has been studied by Xueyun Chen et al. [36].

Hydrogels are trendy, promising, intelligent and 'smart' drug delivery vehicles have become a great search field for targeting drugs to the specific sites and controlling drug release. Among several hardware platforms, ferrogels (FG) have a high potential for use in drug delivery. Ferrogels (FG) are consisting mainly of a polymer matrix embedded with magnetic micro and nano-particles (Fe_3O_4) [37–39]. Upon the application of an external magnetic field, the polymer matrix of the ferrogel can deform due to the magnetic force generated by the embedded magnetic particles, which would allow actuation and magnetically driven drug release on demand. The main advantage of ferrogels is that a larger quantity of drug can be loaded, compared to that transported by simple magnetic dispersion nanoparticles.

Ferrogel (FG) are typically prepared by incorporating magnetic particles into hydrogels [40]. Magnetic nanoparticles (Fe_3O_4) shown in **Figure 1**, have attracted much attention in the last few years as carriers for drug delivery systems. The potential use of nanoparticles as drug carriers has been presented in recent years as a major challenge, as nanoparticles have been designed to improve pharmacological and therapeutic effects in terms of reducing their toxic side effects. Besides, magnetite (Fe_3O_4) is considered as an important type of magnetic material with cubic inverse spinel structure. This property makes it very interesting because of its wide field of use, including magnetic recording, ferrofluid [41], catalyst [42] and some biomedical applications like magnetic resonance imaging (MRI) [43], bio separation [44], in addition to drug delivery system and therapeutic hyperthermia as well, to treat cancer and tumors [45]. Several methods have been developed recently

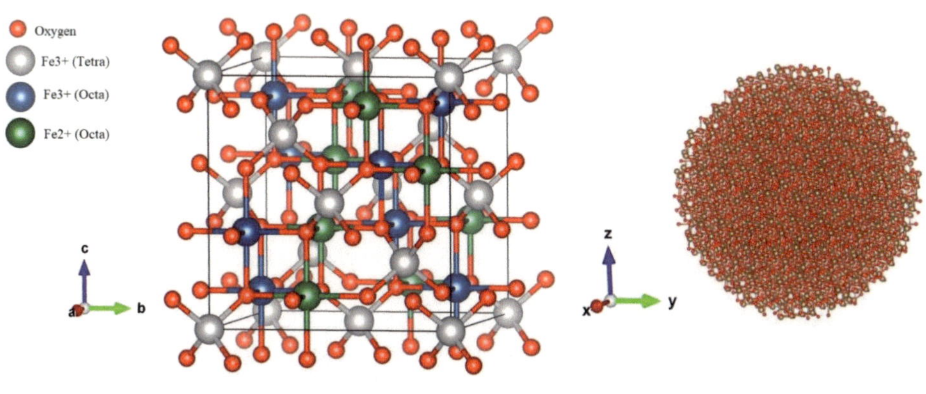

Figure 1.
Model of crystal structure of magnetite (Fe_3O_4) and spherical Fe_3O_4 nanoparticles.

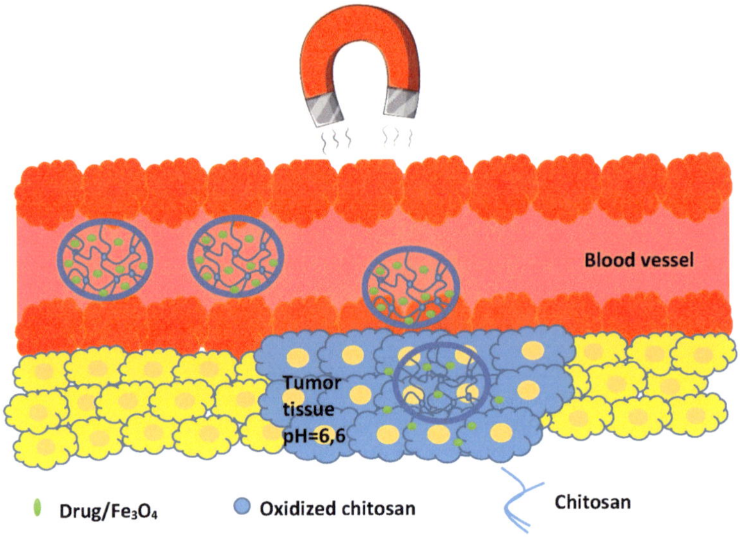

Figure 2.
Schematic representation of magnetic drug delivery system under the influence of external magnetic field.

for preparing magnetic nanoparticles, such as co-precipitation [46], sol–gel [47], solvothermal [48], sonochemical and chemical vapor deposition phase (CVD) [49]. Among these methods, co-precipitation is considered as the simplest, most efficient, and most economic method.

Ferrogels characterized by the presence of magnetic particles incorporated in polymer gels, are the subject of extensive research due to those magnetic particles and magnetic fields which have an extended application and clinical acceptance [50, 51]. Recent studies have shown controlled release of many drugs from ferrogels subject to magnetic fields [52, 53]. Ferrogels (FG) have made also injectable and biodegradable. However, typical drug delivery ferrogels have a disadvantage due to the cross-linking agent toxicity, which limits the macroporous biomaterials synthesis [37].

There are only a few reports in the literature on the synthesis of Fe_3O_4 by co-precipitation method. In this paper, a macroporous ferrogel which is sensitive to magnetic field was studied. Furthermore, we are probably the first scientific team that reports the preparation of novel hydrogels and ferrogels based on chitosan and oxidized chitosan as cross-linking agent embedded Fe_3O_4/drug (5-FU, caffeine and ascorbic acid). **Figure 2** shows the magnetic drug delivery system under the influence of external magnetic field. The kinetics and in-vitro drug release profile of the drugs were studied in PBS pH (7.4) buffered solution at 37°C.

2. Materials and methods

2.1 Materials

Chitosan (in powder form was prepared in our laboratory from exoskeletons of shrimp waste and purified, with degree of deacetylation >90% was determined by conductimetric titration), iron (II) sulfate heptahydrate ($FeSO_4.7H_2O$, sigma-Aldrich), iron (III) chloride ($FeCl_3.6H_2O$, Sigma-Aldrich), Caffeine (sigma-Aldrich), 5-Fluorouracil (sigma-Aldrich), Ascorbic acid (Fluka), Sodium metaperiodate ($NaIO_4$, sigma-Aldrich), Hydrochloric acid (sigma-Aldrich), Ethylene glycol (sigma-Aldrich), Acetic acid (sigma-Aldrich), Sodium bicarbonate (Panreac), Sodium hydroxide (Sigma-Aldrich).

2.1.1 Synthesis of oxidized chitosan (OCS)

Oxidized chitosan (**Figure 3**) was prepared according to a previously reported method [54, 55]. The purified chitosan (1 g) ([GlcN] = 5.34 mM) was dispersed in 50 ml Hydrochloric acid solution HCl (10^{-3} M) at pH ranging from 4 to 5, and kept under magnetic stirring at 4°C for 30 min. Then 1 ml aqueous solution of sodium periodate 0.534 mM, P_0 = 0.1 (P_0 = moles of $NaIO_4$ x moles of GlcN) was added to the mixture. The reaction system was covered with aluminum foil to prevent photo induced decomposition of periodate ion. The reaction lasted for 30 min at 4°C and it was interrupted by the addition of 1 ml ethylene glycol to inactivate any unreacted periodate in a molar ratio of 1:1. The oxidized derivative was washed by distilled water for 5 h and the dry product was obtained by freeze-drying.

2.1.2 Preparation of oxidized chitosan (OCS)/chitosan (CS) hydrogels

The procedure for preparation of the hydrogel was as follows: chitosan (0.1 g) was dispersed in 10 ml of (1% acetic acid) at pH = 4.8 in an ice bath under magnetic stirring until a clear solution was obtained. Afterwards, the drug (5-FU 10 mg, Caffeine 10 mg or Ascorbic acid 10 mg) was added into the solution and gently stirred until the complete dissolution for 30 min. The oxidized chitosan (60 mg) was added to the solution of chitosan soluble drug under continuous stirring to facilitate crosslinking between amino of chitosan (NH_2) and aldehyde groups of chitosan (OCS). Then the 6% (w/v) $NaHCO_3$ solution was slowly added to the solution in an ice bath under magnetic stirring to get homogeneous mixture at pH = 7 [56], and a transparent gel was obtained as shown in **Figure 4**. The hydrogel was washed with ethanol, and filtered to remove traces of unreacted reagents.

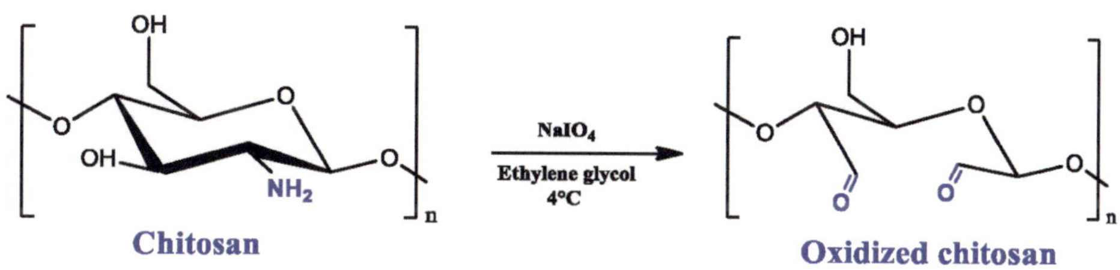

Figure 3.
Schematic representation of reaction between chitosan (CS) by sodium periodate.

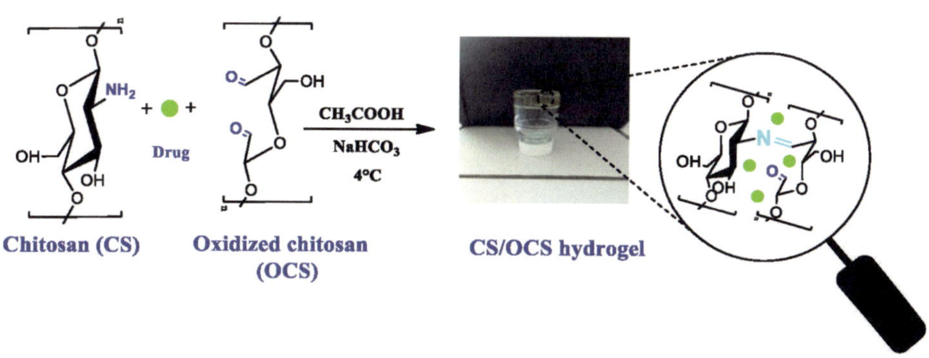

Figure 4.
Schematic representation of the chemical (CS)/(OCS) cross-linking mechanism and the formation of the hydrogel.

2.1.3 Synthesis of magnetite nanoparticles (Fe3O4) by co-precipitation

According to the previous work, magnetite (Fe_3O_4) nanoparticles were synthesized by chemical co-precipitation method. Magnetic Fe_3O_4 NPs were synthesized by dissolving 50 mL of $FeCl_3.6H_2O$ and 25 mL of $FeSO_4.7H_2O$ in 350 mL of distilled water under nitrogen atmosphere under vigorous stirring. Upon addition of (35 ml) NaOH, the pH was adjusted to about 10, the solution turned black, indicating the formation of magnetite nanoparticles. Further stirring is continued for 1 h to uniformly disperse the magnetic nanoparticles. After raising the reaction temperature to 80°C. Then, a formed black precipitate was collected with an external magnet, washed several times with ethanol and distilled water, and dried in vacuum at 60°C. The entire reaction is given by the equation as shown in **Figure 5**.

2.1.4 Synthesis of ferrogel

Firstly, chitosan solution was prepared by dissolving 0.1 g chitosan (CS) in 10 mL acetic acid solution (1%). 50 mg iron oxide nanoparticles were mixed with chitosan solution and stirred for 2 h at 40°C to give chitosan coated magnetic nanoparticles (CS-Fe_3O_4). Then, the drug (5-FU 10 mg, Caffeine 10 mg or Ascorbic acid 10 mg) was added into the solution (CS-Fe_3O_4) and gently stirred for 30 min. 60 mg of oxidized chitosan (OCS) was added as crosslinking agent and were mixed with (CS-Fe_3O_4) solution. The reaction mixture was stirred at 0°C for 3 hours. The pH of the solution was adjusted to pH = 7 with 6% (w/v) $NaHCO_3$. The product (CS-Fe_3O_4-OCS) was washed with deionized water. **Figure 6** shows a representation of the prepared drug delivery ferrogel.

2.1.5 Fourier transform infrared (FTIR) spectra of hydrogel

Fourier transform infrared (FTIR) spectra of the chitosan, oxidized chitosan, CS/OCS hydrogel, CS/drug/OCS lyophilized hydrogel, Fe_3O_4 NPs and the

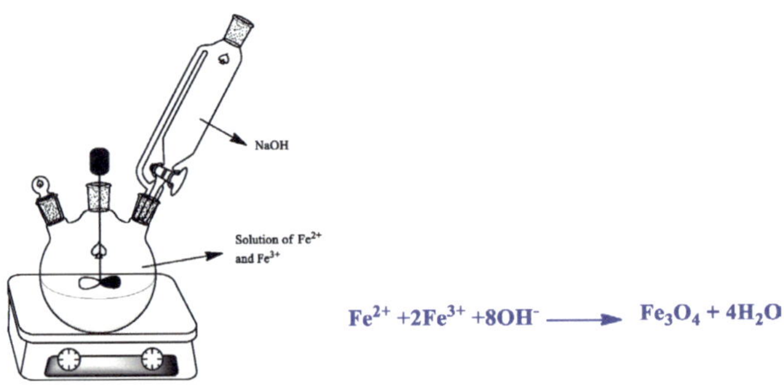

Figure 5.
Representation of superparamagnetic magnetite nanoparticles synthesis technique.

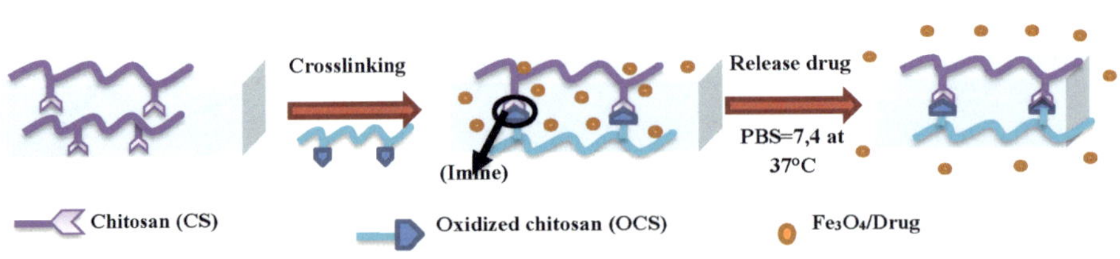

Figure 6.
Representation of superparamagnetic drug delivery hydrogel.

freeze-dried CS/Fe_3O_4/OCS ferrogel were obtained from discs containing 2.0 mg dry sample in approximately 198 mg potassium bromide (KBr). The measurements were recorded by a Perkin–Elmer FTIR spectrophotometer at the resolution of 4 cm^{-1} in the wave number region 400–4000 cm^{-1}.

2.1.6 Thermogravimetric (TGA) analysis

The thermal properties of lyophilized hydrogel, pure chitosan, oxidized chitosan, Fe_3O_4 NPs and CS/Fe_3O_4/OCS lyophilized ferrogel were investigated by thermogravimetric analysis (TGA). Samples were placed in the balance system and heated from 40 to 800°C at a heating rate of 10°C/min using a TA Instruments TGA (Q_{500}) device.

2.1.7 Magnetization studies using (SQUID) analysis

The saturation magnetization of Fe_3O_4 NPs, CS-Fe_3O_4-OCS ferrogel was measured by vibrating sample magnetometer VSM (SQUID model MPMS XL 7) from Quantum at room temperature between magnetic fields of −14,000 (Oe) to 14,000 (Oe).

2.1.8 Scanning electron microscopy (SEM)

The chitosan-based hydrogels cross-linked with oxidized chitosan was frozen at −75°C for 24 h and then lyophilized (by Alpha 1–2 LD_{plus}) for 48 h. The lyophilized sample was obtained and then examined by a scanning electron microscopy (SEM) (Mini SEM Hirox SH-4000).

2.1.9 Hydrogel swelling

Freeze-dried CS/OCS hydrogel were immersed in phosphate-buffered saline (PBS) at 37°C (pH = 1.2, pH = 5.8 and pH = 7.4). The samples were weighed before being put into PBS (W_0). After the vial was sealed and held at 37°C for 24 h, and the excess solution on the surface of the hydrogels was quickly absorbed with filter paper [55]. The equilibrium-swelling ratio (SR) was calculated using the following equation:

$$\mathbf{SR\%} = \frac{\mathbf{W_s - W_d}}{\mathbf{W_d}}, \tag{1}$$

where, W_d is the weight of dry hydrogel after lyophilization and W_s is the weight of swollen hydrogel.

2.1.10 In vitro drug release from the hydrogels and ferrogels

(5-FU, caffeine or ascorbic acid) was selected and used as a model drug in the release experiments. In vitro drug release test was performed in a phosphate buffer solution PBS (pH 7.4 at 37°C) under shaking. The hydrogels and ferrogels (m = 1.14 g) (3 cm x 4 cm) were placed in a cartouche before immersing in 1000 mL of phosphate-buffered saline (PBS). At predetermined time intervals, 5 mL of release medium was withdrawn. Then 5 mL of fresh buffer was added to the original to maintain the total volume. The drug release was determined by

UV–Vis spectrophotometry at λ_{max} (5-Fluorouracil (266 nm), caffeine (273 nm) and ascorbic acid (265 nm)). The concentration of the active ingredient in the (PBS, pH = 7.4 at 37°C) has been achieved from the calibration curve, and the amount of drug released at time t (M_t) was calculated by accumulating the total active ingredient release up to that time. In vitro drug release tests were performed in triplicate (n = 3). There are a few steps, which mainly control drug release phenomena from the polymer matrix, dissolution of the drug, liquid penetration into the matrix and diffusion of the drug from the drug encapsulated in the matrix. In order to understand the release kinetics and the mechanism of the active ingredient release, release kinetics data obtained in vitro using ferrogels and hydrogels are fitted with kinetics model. The release data are best fitted with the Korsmeyer–Peppas (KP) model. The (KP) model deal with Eq [57]:

$$\%\textbf{Cumulative release} = \frac{\mathbf{M_t}}{\mathbf{M_0}} \times \mathbf{100} \tag{2}$$

where "M_t" is the amount of drug released at time (t), "M_0" is the maximal amount of the drug released at maximum interval. It is interesting to note that three drugs (5-FU, caffeine and ascorbic acid) in hydrogels exhibit a Fickian nature of drug diffusion. However, the interaction of the drug molecules with the matrix play an important role in the drug release kinetics occurring through a diffusion mechanism.

2.2 Statistical analysis

The experimental data are expressed as the mean values of at least three replicates ± standard deviation (SD). The results were analyzed and showed usage Kaleida graph.

3. Results and discussion

3.1 Hydrogel and ferrogel formation

The concept of this study is depicted in (**Figure 7**). Oxidized chitosan (OCS) was prepared following a well-known method where chitosan is oxidized with sodium periodate ($NaIO_4$). Oxidization of chitosan created multiple aldehyde groups all along the polymeric chain using the method described in literature [54, 58]. The hydrogels and ferrogels (magnetic Fe_3O_4 embedded in novel hydrogel) were prepared by crosslinking chitosan (CS) with oxidized chitosan (OCS). The crosslinking of hydrogel and ferrogel was achieved by (–C=N-) bonds of Schiff-base reaction. Our results indicate that the process synthesis of the hydrogel and ferrogel was embedded with three drugs (5-FU, caffeine and ascorbic acid) has successfully loaded in the carrier polymeric. The schematic representation of smart ferrogel "magnetic hydrogel" was shown in (**Figure 7**).

The **Figure 8** shows the procedure for preparation of the hydrogel and a photograph of (CS-drug-OCS) hydrogel.

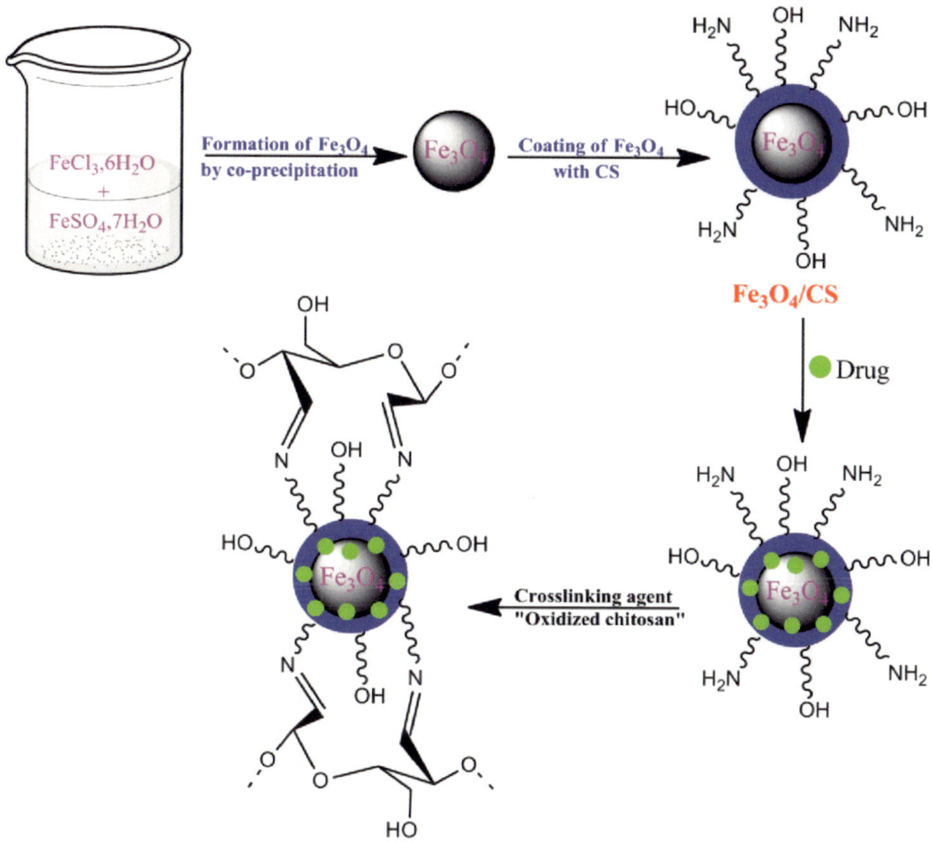

Figure 7.
Schematic representation of (CS-Fe_3O_4-OCS) ferrogel.

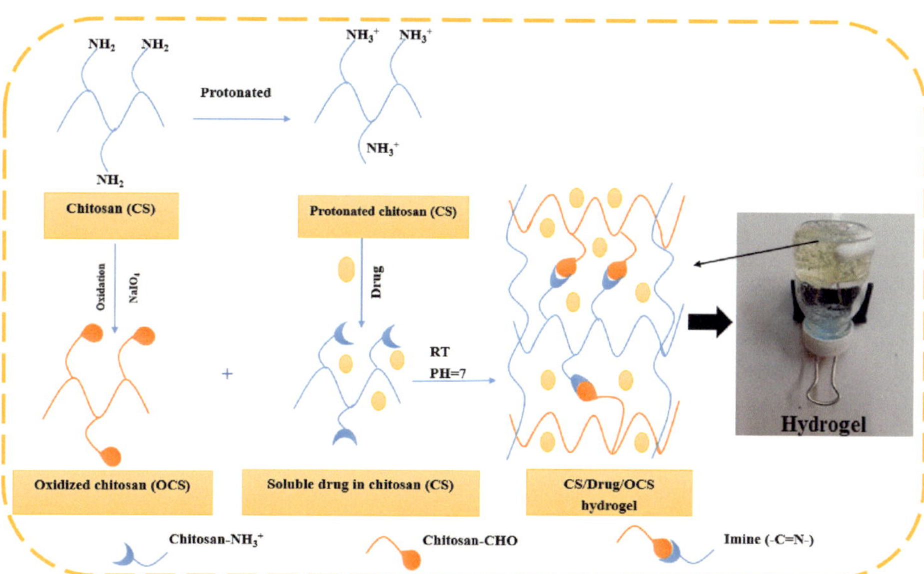

Figure 8.
Schematic representation of the synthesis process of CS/drug/OCS hydrogel.

3.2 Fourier transform infrared (FTIR) spectra of hydrogel

The IR spectrum of CS, OCS, (CS/OCS) and (CS/drug/OCS) lyophilized hydrogel are listed in (**Figures 9** and **10**) and major functional moieties are labeled with wavenumbers was recorded in the region of 4000–40 cm^{-1}. The FTIR spectrum of pure chitosan (**Figure 9(a)**) shows wide band around 3450 cm^{-1} corresponding to amine N–H symmetrical vibration and H bonded O–H group. The peak observed between 3400 and 3700 cm^{-1} corresponding to combination of the band O-H, NH_2 intra and intermolecular hydrogen bonding. The peaks at 2920 and 2320 cm^{-1} are assigned to the symmetric and asymmetric may be attributed to –CH vibrations of carbohydrate ring [59]. The bands at 1650 cm^{-1} and 1545 cm^{-1} may be attributed to C=O stretching (amide I vibration) and N-H bending (–NH_2 bending of amide II) in amide group, respectively and 1390 cm^{-1} (N–H stretching or C–N bond stretching vibrations, amide III vibration) [60]. The peak observed at 1050 cm^{-1} has the contribution to the symmetric stretching of C–O–C groups. The absorption peaks in the range 900–1200 cm^{-1} are due to the antisymmetric C–O stretching of saccharide structure of chitosan. In order to understand the oxidation of oxidized chitosan (OCS), the FTIR spectra results for (OCS) in (**Figure 9(b)**) verified successful oxidation, while a new absorption peak appeared around 1725 cm^{-1} [55], which was assigned to an aldehyde group (-C=O) bond, indicating that the CS has been successfully oxidized by the $NaIO_4$ [61]. Furthermore, the peak (**Figure 9(c)** at 1637 cm^{-1} caused by C=O and C=N is reduced significantly. These differences indicate that the aldehyde groups of OCS reacted with the amino groups of CS to generate a Schiff base"imine" [62]. FTIR analysis was performed (**Figure 9(d)**) exhibits the IR spectra of the prepared nanoparticles. The spectrum of Fe_3O_4 magnetic nanoparticles shows the formation of two strong absorption bands between 636 cm^{-1} and 592 cm^{-1}. Furthermore, the band at 592 cm^{-1} was confirmed as the Fe-O stretching vibration of tetrahedral sites of spinel structure. The absorption bands at 459 cm^{-1}, assigned to tetrahedral and octahedral sites, peaks at 3400 cm^{-1} due to the O-H stretching model adsorbed on the surface of the Fe_3O_4 nanoparticles [63].

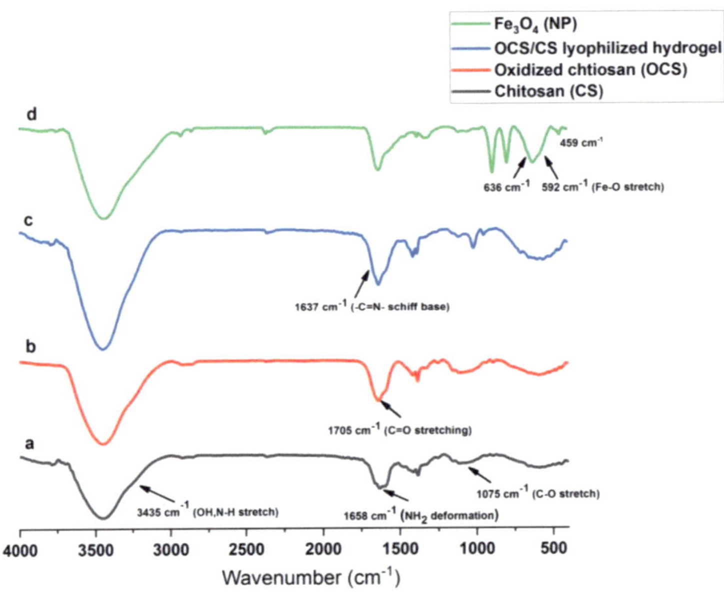

Figure 9.
FTIR spectra of (a) chitosan, (b) oxidized chitosan, (c) CS/OCS lyophilized hydrogel, (d) Fe_3O_4 NP.

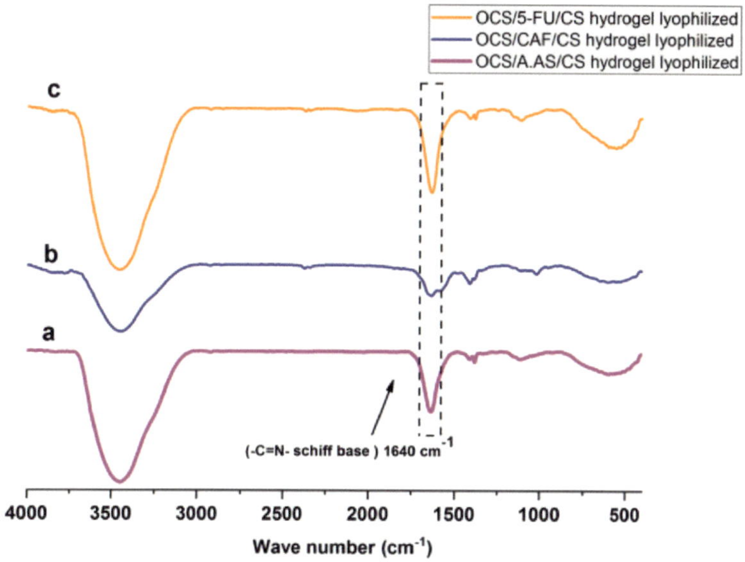

Figure 10.
FTIR spectra of (a) CS/a.AS/OCS lyophilized hydrogel, (b) CS/CAF/OCS lyophilized hydrogel; (c) CS/5FU/OCS lyophilized hydrogel.

For the spectra of three drugs (5-FU, caffeine and ascorbic acid) loaded (CS/OCS) in hydrogels (**Figure 10**); exhibited the characteristic absorption of imine stretching vibration (-C=N–) at 1637 cm^{-1} [54, 64], suggests that the coupling reaction was occurred between –CHO of OCS and $–NH_2$ of CS.

3.3 Magnetization studies using (SQUID) analysis

The measurements of the magnetic field-dependence of the magnetization of the uncoated and coated magnetite nanoparticles at 25°C are presented in (**Figure 11**). The plots indicate that both samples exhibit superparamagnetic behavior with zero remanence and coercivity. (**Figure 11**) shows the magnetic curves as a function of applied field at room temperature obtained for Fe_3O_4 and CS-Fe_3O_4-OCS ferrogel, respectively. The magnetization saturations were found to be 60.57 emu/g for Fe_3O_4, 17.25 emu/g for CS-Fe_3O_4-OCS ferrogel [65]. The magnetization value decreased after coating due to the existence of oxidized chitosan and chitosan, which formed polymerized multilayers. It can be concluded that the Ms. value of CS/Fe_3O_4/OCS ferrogel is less than the Fe_3O_4 (NPs) that can be attributed to the creation of a non-magnetic polymer layer around Fe_3O_4 (NPs) in the hydrogel [66]. Taking into account the magnetic properties of the prepared by ferrogel, it may be able to deliver the drug to the target area in the presence of an external magnetic field.

3.4 Scanning electron microscopy (SEM)

A perfect injectable hydrogel must have pores in the range of 50–100 μm and a high degree of interconnectivity to facilitate nutrient and oxygen transport, as well as cell adhesion and migration. By using SEM, we studied the pore size distribution in hydrogels (**Figure 12**). In this study, morphology of freeze-dried hydrogel (CS/OCS) were observed with scanning electron microscope (SEM). As can be seen in (**Figure 12**), the (CS/OCS) lyophilized hydrogel had continuous and porous structures with interconnecting pores, pores diameter ranging from several tens to

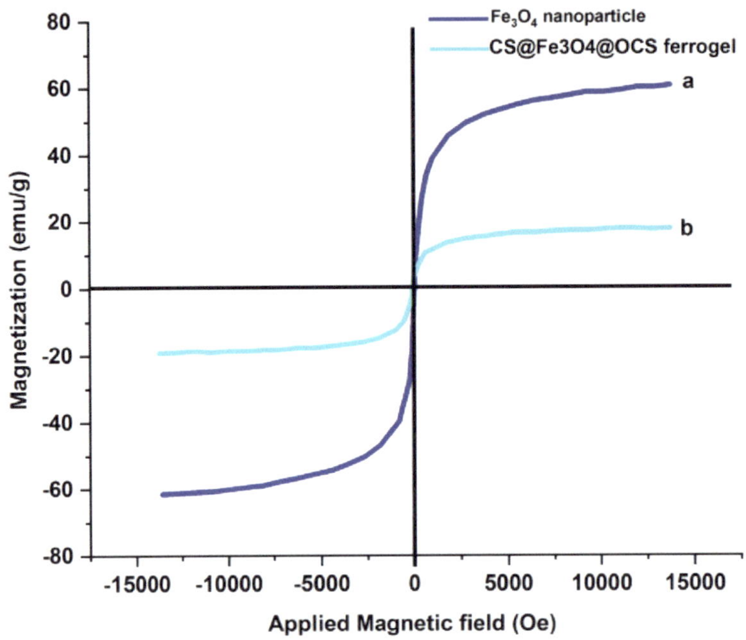

Figure 11.
The magnetic hysteresis loops for Fe_3O_4 NPs and CS@Fe_3O_4@OCS hydrogel measured by SQUID at room temperature.

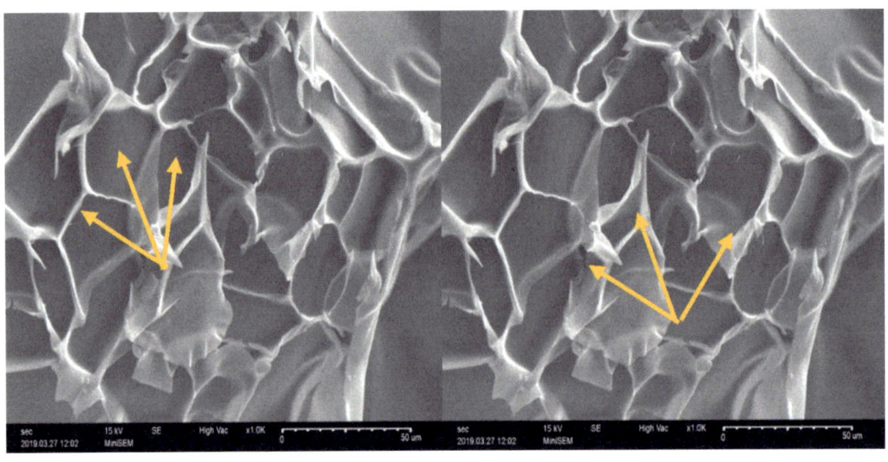

Figure 12.
Micrographs of chitosan cross-linked oxidized chitosan hydrogel at low and high magnification.

several thousands of micrometers [55]. Moreover, the more crosslinking between amino group and aldehyde group, the diameter of the pores are decreased and the compactness of pores increased. The micrographs of CS/OCS hydrogel it was clearly observed that the hydrogel had a three-dimensional porous structure, which was beneficial for drug delivery systems.

3.5 Thermogravimetric (TGA) analysis

To evaluate the thermal stability of the chitosan (CS) oxidized chitosan (OCS) and (CS/OCS) lyophilized hydrogel, TGA thermograms were obtained as shown in (**Figure 13**). The TGA of pure chitosan shows two-stage weight loss in the range 40

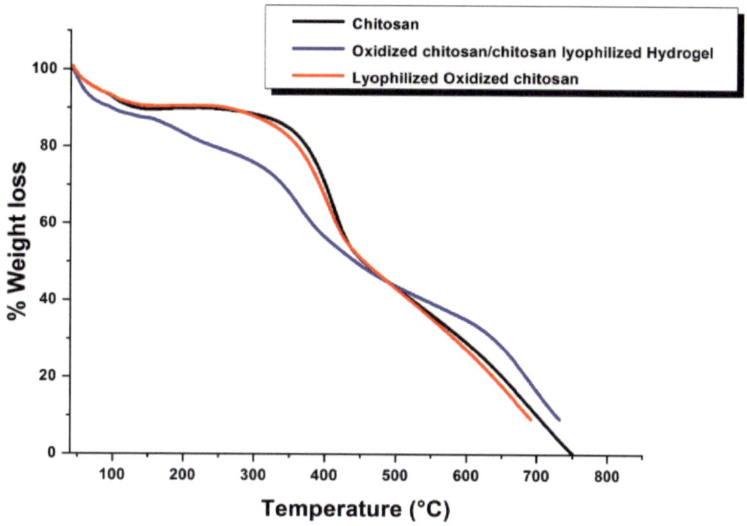

Figure 13.
TGA graphs of chitosan (CS), oxidized chitosan (OCS) and (CS/OCS) lyophilized hydrogel.

to 750°C (**Figure 13**), it is clear that chitosan started to degrade at 40–130°C with 8% weight is due to the loss of water molecules. Initial decomposition around 130°C for pure chitosan can be attributed to the strong water adsorptive nature of chitosan. The second stage of degradation occurred at 340°C and continued up to 460°C [67]. There was 38.43% weight loss occurring in the second stage due to degradation of pure chitosan biopolymer and the temperature at which maximum degradation observed was 274.74°C. At the end of 750°C, the total weight loss of sample was 100% [54, 67]. TGA of oxidized chitosan (OCS) showed two steps of degradation (**Figure 13**), the first stage ranges between 40 and 130°C and shows about 8.2% loss in weight corresponded to water release for the initial step. The second stage decomposition was observed from 300°C and continued up to 460°C, during this time there was 39% weight loss due to the degradation of chitosan. At the end of 700°C, the total weight loss of sample was 90% [10, 54]. For CS/OCS lyophilized hydrogel, the degradation starts at a lower temperature compared to chitosan and oxidized chitosan (**Figure 13**). For the CS/OCS hydrogel, shows two-stage weight loss in three stages. The first stages of degradation takes place from 40 to 160°C with a weight loss of 13% could be due to the loss of both the loosely bound water and tightly bound water [68]. The free water and hydrogen-bonded water are released at a temperature between 40 and 100°C. The hydrogel contains many hydrophilic groups that retain water more tightly in the hydrogel skeleton by polar interaction. As a result, it is harder to lose. Thus, this tightly bound water is released in the temperature region 100–156°C. The second stage of hydrogel degradation started between 160° C and 385° C with a 41% weight loss. The three-stages of degradation biopolymers, at the end of 700°C, the total weight loss of sample was 90%. This phase of the weight loss mainly could be caused by a series of thermal and oxidative decomposition in the process including dehydration of the sugar cycle, degradation, N-deacetylation of the molecular chain of the chitosan cracking unit and vaporization and removal of volatile products. It can be concluded the TGA curve shows at about 222°C. This is probably due to the formation of (-C=N-) and this proves that biopolymer-based Schiff base is thermally less stable.

The TGA curves of uncoated Fe_3O_4 and the lyophilized ferrogel (CS-Fe_3O_4-OCS) are shown in **Figure 14**. For uncoated Fe_3O_4 NPs (**Figure 14(a)**), the TGA curve showed that the weight loss over the temperature range 40–750°C was about 2.2%.

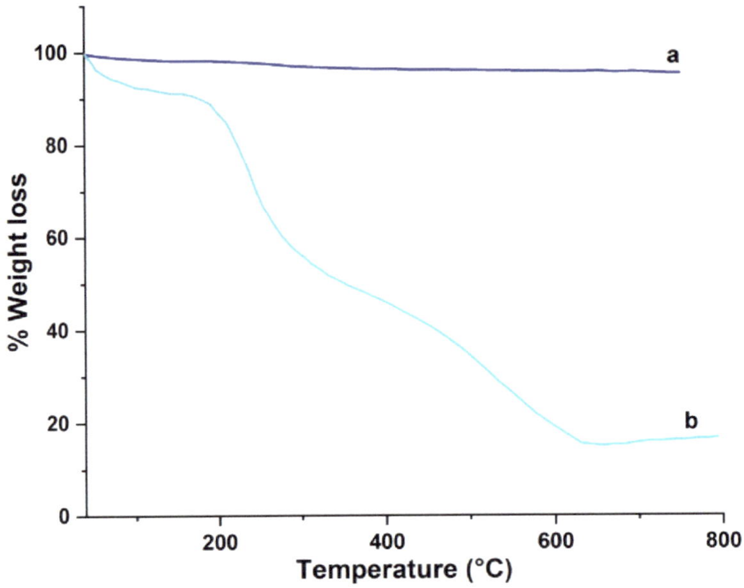

Figure 14.
TGA curves for a) uncoated Fe_3O_4 b) CS-Fe_3O_4-OCS lyophilized ferrogel.

Hence, this weight loss is related to removal of the physically adsorbed water and/or hydroxyl groups on the surface of Fe_3O_4 nanoparticles. For coated nanoparticles Fe_3O_4. Similarly, the weight curve of CS/Fe_3O_4/OCS (**Figure 14(b)**) showed a progressive decrease of 85%. This is due to the degradation of the polymer and hydrogen-bound water in the temperature range of 40–166°C, which forms the polysaccharide structure of OCS and CS. In addition, a uniform and steady decrease in weight loss at 166–630°C is CS/Fe_3O_4/OCS was observed. This may be partly attributed to the degradation and decomposition of organic skeletal structure, amino groups, and other functional groups. By comparing the curves of CS/Fe_3O_4/OCS and Fe_3O_4, it was observed that Fe_3O_4 particles are wrapped into CS and OCS can enhance the thermal stability of the whole system. A temperature of 630°C or higher, the remaining material was carbonized completely. The indicated that chitosan and oxidized chitosan coated Fe_3O_4 successfully and penetrated deeply into in the matrix CS/OCS.

3.6 Hydrogel swelling

To investigate the pH dependent swelling behavior of the CS/OCS hydrogels for drug delivery, PBS with (pH = 1.2; pH = 5.8 and pH = 7.4) at T = 37°C were used to simulate the physiological medium and were used for testing swelling of the hydrogels. The results of the equilibrium-swelling ratio are presented in (**Figure 15**). The CS/OCS hydrogels showed large differences in swelling behavior at different pH values. The pK_a value of the D-glucosamine residue in chitosan was approximately 6.2 to 7.0. Therefore, the amino groups in the chitosan were protonated and positively charged in acidic PBS (pH = 1.2 and pH = 5.8), and the electrostatic repulsion between positively charged -NH^{3+} groups would lead to swelling of the hydrogels. The % equilibrium swelling values were found to be higher at pH = 1.2, than at pH = 7.4. This can be explained by protonation of the unreacted NH_2 groups of chitosan at acidic pH, leading to dissociation of the hydrogen bonding involving the amino groups, and thus facilitating the entry of the solvent into the material. The swelling process of hydrogels involves the ionization [69]. Furthermore, the Schiff base bonds

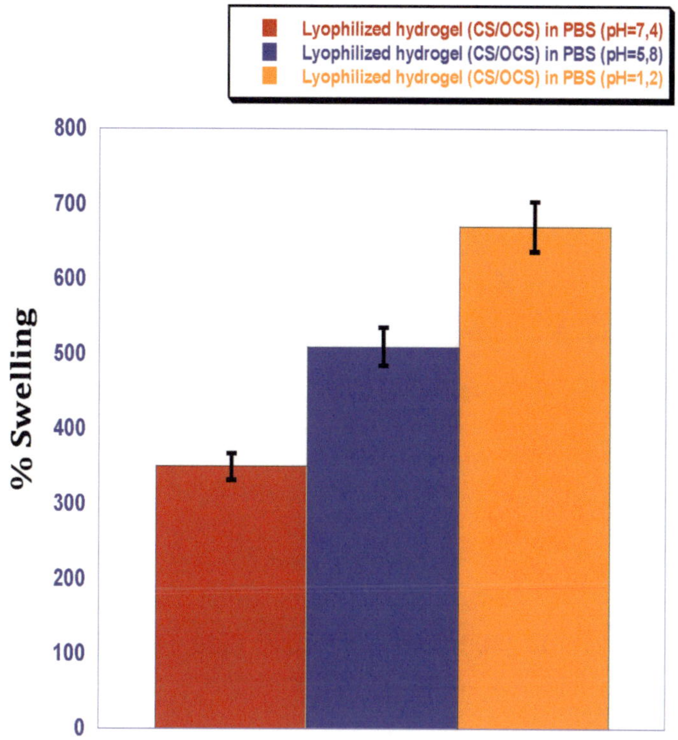

Figure 15.
% Swelling of CS/OCS hydrogel in buffer solutions of pH = 1,2, pH = 5,8 and pH = 7,4 at T = 37°C.

between -NH_2 and -CHO as a cross-linker became weak in PBS (pH = 7.4 at 37°C); the swelling ratio of CS/OCS was 350%, resulting in swelling of the hydrogel. Therefore, the CS/OCS hydrogel in PBS (pH = 1.2 at 37°C) exhibited the largest swelling ratio of 670%; the swelling ratio (pH = 5.8 at 37°C) was 509%. The decreased swelling ratio occurred due to increased crosslinking density in the hydrogels. Meanwhile, the equilibrium swelling ratios of the hydrogels in a pH = 7.4 solution dramatically decreased. This is because hydrogels exhibited pH-responsive swelling behavior, and the hydrogels showed a much higher swelling ratio in an acidic medium than that in a pH 7.4 medium.

3.7 In vitro release from the hydrogels

The model drugs (5-FU, caffeine and ascorbic acid) was encapsulated in the hydrogel matrix used for the release kinetics in PBS (pH = 7.4 at 37°C) are depicted in (**Figure 16**). The purpose here was to study whether the release of the drug trapped in the hydrogels, as well as by the simple diffusion. In a system, where drug is entrapped in a biodegradable matrix, the release rate depends on three parameters: the size of the drug molecule, the drug solubility (soluble-sparingly soluble-insoluble), the cross-linking density and the degradation rate. The ability of the chitosan-imine-oxidized chitosan hydrogels to act as matrix for controlled release was investigated in vitro by monitoring the release profile of the (5-FU, caffeine and ascorbic acid) three-model drug from system in phosphate buffer of physiological pH (7.4), at the human body temperature of 37°C (**Figure 16**). In order to study the release behavior of 5-FU, caffeine, ascorbic acid incorporated in the chitosan-based hydrogel cross-linked by oxidized chitosan, they were incubated in release media (phosphate buffer pH = 7.4 at 37°C) and evaluated by UV spectrophotometry.

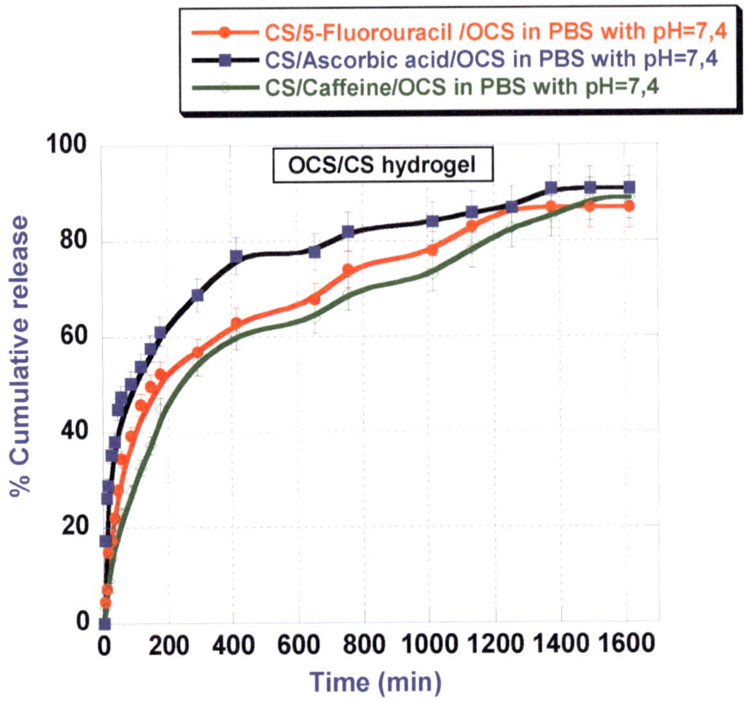

Figure 16.
In vitro release of 5-FU, caffeine and ascorbic acid embedded in hydrogels (pH = 7,4 at 37°C). Values reported are an average of n = 3 ± standard deviation.

(**Figure 16**) shows release profiles of 5-FU, caffeine and ascorbic acid up to 26 h of incubation period. As shown in (**Figure 16**), the chitosan hydrogels showed an initial burst release of 5-FU, caffeine and ascorbic acid over a period of 3 h for all incubation media, which was of the order of (5 -FU "52%", caffeine "43%" and ascorbic acid "60%"). This initial rapid release, characterized by a "burst effect», because certain quantities of 5-FU, caffeine and ascorbic acid were localized on the surface of the hydrogels by adsorption which could be released easily by diffusion. After 3h the release percentages of (5-FU "86%", caffeine "89%" and ascorbic acid "91%"), respectively [70]. The remaining part of the drug can be trapped in hydrogels because the amino functions of chitosan can enter into the Schiff reaction with the aldehyde groups of oxidized chitosan. Indeed, the hydrogel contained large pore size, which is beneficial for the diffusion of the drug, this initial bursting effect, a slower sustained and controlled release occurred throughout the incubation period and the amount of release. The release profiles confirmed that the (5-FU, caffeine and ascorbic acid) drugs were encapsulated in hydrogels.

As shown in **Figure 17**, magneto-hydrogel showed an initial burst release of (5-FU, caffeine and ascorbic acid) in a period of 5 h, which was in the range of (37%, 34%, and 42%). The initial burst phase is caused by drug adsorbed on the surface of the nanoparticles Fe_3O_4 embedded in hydrogel. The release kinetics at pH = 7.4 at 37°C within 45 h clearly indicated that Fe_3O_4 embedded in hydrogel influenced drugs release. At pH = 7.4 at 37°C, about (95%, 93% and 97%) amounts of 5-FU, caffeine and ascorbic acid were release after 45 h, respectively. Due to the slower swelling rate of nanotransporters in solution, the rate of drug release is also slower. It is well known that there are a large number of amino and hydroxyl groups on the surface of chitosan molecules, which provide functional groups and favorable characteristics for biological molecules. In PBS (pH = 7.4 at 37°C), amino

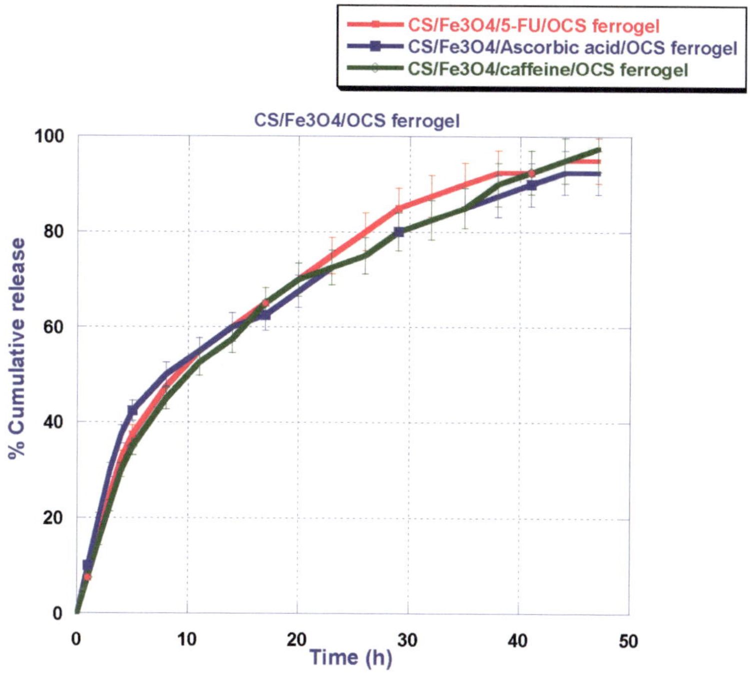

Figure 17.
Cumulative release profiles of 5-FU-gel, caffeine-gel, ascorbic acid-gel with external magnetic field. Values reported are an average of n = 3 ± standard deviation.

groups of chitosan mainly attach to the surface of the nanoparticles, so as to reduce the surface voids and render the pore blockage, lower penetration, and thus slow down the release rate of the drug [36]. However, due to the magnetic orientation role of MNPs, magnetic nano-drug carriers could be transported by applying an external magnetic field and maintain drug concentration for extended periods. Such rapid transport and slow release of the nanocarriers to the target site may be desirable for many biomedical applications, minimizing drug leakage to undesirable sites and reducing the risk of heart attack due to high dose in a short period of time. These results clearly illustrated that the chitosan hydrogel containing Fe_3O_4 resulted in a barrier system for the sustained release of 5-Fu, caffeine and ascorbic acid. We speculate that this barrier structure would block the drug loss in the early burst release, which is benefit to reduce the toxic side effects.

4. Conclusion

Many conclusions can be made from the present work, the ferrogels (FG) are cross-linked polymer networks containing magnetic nanoparticles: a) magnetic magnetite (Fe_3O_4) were synthesized successfully by chemical co-precipitation and has been confirmed using FT-IR, VSM analysis, TGA. The advantage of the co-precipitation method are low cost, rapidity, ease, reproducibility and high-yield synthesis. b) The hydrogel was formulated by cross-linking chitosan (CS) and oxidized chitosan (OCS) via the Schiff-base (-C=N-) reaction. Obviously, these results indicate that this exhibit non-toxic, biodegradable, good injectability, less expensive and respect the environment, quick gelation time, in vitro pH-dependent equilibrated swelling ratios, interconnected porosity. c) Magneto-responsive hydrogels are typically prepared by incorporating magnetic particles into hydrogels and has been confirmed using (FT-IR), (TGA), (VSM) analysis at room temperature. d) The role of Fe_3O_4 embedded in hydrogel, which allows reducing the surface voids and rendering the pore blockage, lower penetration, and thus slowing down the release rate of the drug. e) A 5-Fluorouracil (5-FU), caffeine and ascorbic acid release tests were used to demonstrate the excellent in vitro drug release behavior of these hydrogels and ferrogels. However, all these results indicate that this type of biomaterial based on chitosan with oxidized chitosan in the presence of the three drugs with different solubility for the preparation of hydrogels effective for the controlled release of the drug. Compared to other hydrogels based on chitosan, the present study brings to the attention of researchers a novel strategy to design a non-toxic and biodegradable matrix for drug delivery systems, by the simple use of appropriate oxidized chitosan without incorporating any crosslinking agents. This design expands the variety of hydrogel matrices, guiding additional efforts in the development of the new ideas for pharmaceuticals applications. As a perspective and future challenges, we will test this type of ferrogels for cancer treatment by hyperthermia.

Author details

Damiri Fouad*, Yahya Bachra, Grouli Ayoub, Amine Ouaket, Ahmed Bennamara, Noureddine Knouzi and Mohammed Berrada
University Hassan II of Casablanca, Faculty of Sciences Ben M'Sick, Department of Chemistry, Laboratory of Biomolecules and Organic Synthesis (BIOSYNTHO), Casablanca, Morocco

*Address all correspondence to: fouad.damiri@outlook.fr

References

[1] L. Boyle, P. & Bernard, World Cancer Report 2008, (IARC Press. Lyon). (2008) (IARC Press. Lyon).

[2] X.-J. Liang, C. Chen, Y. Zhao, P.C. Wang, Circumventing Tumor Resistance to Chemotherapy by Nanotechnology, in: 2010: pp. 467-488. https://doi.org/10.1007/978-1-60761-416-6_21.

[3] B. GAIHRE, M. KHIL, D. LEE, H. KIM, Gelatin-coated magnetic iron oxide nanoparticles as carrier system: Drug loading and in vitro drug release study, Int. J. Pharm. 365 (2009) 180-189. https://doi.org/10.1016/j.ijpharm.2008.08.020.

[4] J. Chomoucka, J. Drbohlavova, D. Huska, V. Adam, R. Kizek, J. Hubalek, Magnetic nanoparticles and targeted drug delivering, Pharmacol. Res. 62 (2010) 144-149. https://doi.org/10.1016/j.phrs.2010.01.014.

[5] R. Sensenig, Y. Sapir, C. MacDonald, S. Cohen, B. Polyak, Magnetic nanoparticle-based approaches to locally target therapy and enhance tissue regeneration in vivo, Nanomedicine. 7 (2012) 1425-1442. https://doi.org/10.2217/nnm.12.109.

[6] G.S. Dhillon, S. Kaur, S.K. Brar, M. Verma, Green synthesis approach: extraction of chitosan from fungus mycelia, Crit. Rev. Biotechnol. 33 (2013) 379-403. https://doi.org/10.3109/07388551.2012.717217.

[7] R.N. Tharanathan, F.S. Kittur, Chitin — The Undisputed Biomolecule of Great Potential, Crit. Rev. Food Sci. Nutr. 43 (2003) 61-87. https://doi.org/10.1080/10408690390826455.

[8] B.K. Park, M.-M. Kim, Applications of Chitin and Its Derivatives in Biological Medicine, Int. J. Mol. Sci. 11 (2010) 5152-5164. https://doi.org/10.3390/ijms11125152.

[9] V. Manigandan, R. Karthik, S. Ramachandran, S. Rajagopal, Chitosan Applications in Food Industry, in: Biopolym. Food Des., Elsevier, 2018: pp. 469-491. https://doi.org/10.1016/B978-0-12-811449-0.00015-3.

[10] A. Laaraibi, F. Moughaoui, F. Damiri, A. Ouakit, I. Charhouf, S. Hamdouch, A. Jaafari, A. Abourriche, N. Knouzi, A. Bennamara, M. Berrada, Chitosan-Clay Based (CS-NaBNT) Biodegradable Nanocomposite Films for Potential Utility in Food and Environment, in: Chitin-Chitosan - Myriad Funct. Sci. Technol., InTech, 2018. https://doi.org/10.5772/intechopen.76498.

[11] I. Aranaz, N. Acosta, C. Civera, B. Elorza, J. Mingo, C. Castro, M. Gandía, A. Heras Caballero, Cosmetics and Cosmeceutical Applications of Chitin, Chitosan and Their Derivatives, Polymers (Basel). 10 (2018) 213. https://doi.org/10.3390/polym10020213.

[12] J. Roy, F. Salaün, S. Giraud, A. Ferri, J. Guan, Chitosan-Based Sustainable Textile Technology: Process, Mechanism, Innovation, and Safety, in: Biol. Act. Appl. Mar. Polysaccharides, InTech, 2017. https://doi.org/10.5772/65259.

[13] R. Sharp, A Review of the Applications of Chitin and Its Derivatives in Agriculture to Modify Plant-Microbial Interactions and Improve Crop Yields, Agronomy. 3 (2013) 757-793. https://doi.org/10.3390/agronomy3040757.

[14] M.C. Gortari, R.A. Hours, Biotechnological processes for chitin recovery out of crustacean waste: A mini-review, Electron. J. Biotechnol. 16 (2013). https://doi.org/10.2225/vol16-issue3-fulltext-10.

[15] G. Tiwari, R. Tiwari, S. Bannerjee, L. Bhati, S. Pandey, P. Pandey, B.

Sriwastawa, Drug delivery systems: An updated review, Int. J. Pharm. Investig. 2 (2012) 2. https://doi.org/10.4103/2230-973X.96920.

[16] N. Kamaly, B. Yameen, J. Wu, O.C. Farokhzad, Degradable Controlled-Release Polymers and Polymeric Nanoparticles: Mechanisms of Controlling Drug Release, Chem. Rev. 116 (2016) 2602-2663. https://doi.org/10.1021/acs.chemrev.5b00346.

[17] P. Baldrick, The safety of chitosan as a pharmaceutical excipient, Regul. Toxicol. Pharmacol. 56 (2010) 290-299. https://doi.org/10.1016/j.yrtph.2009.09.015.

[18] E. Ruel-Gariépy, M. Shive, A. Bichara, M. Berrada, D. Le Garrec, A. Chenite, J.-C. Leroux, A thermosensitive chitosan-based hydrogel for the local delivery of paclitaxel, Eur. J. Pharm. Biopharm. 57 (2004) 53-63. https://doi.org/10.1016/S0939-6411(03)00095-X.

[19] M.E. Parente, A. Ochoa Andrade, G. Ares, F. Russo, Á. Jiménez-Kairuz, Bioadhesive hydrogels for cosmetic applications, Int. J. Cosmet. Sci. 37 (2015) 511-518. https://doi.org/10.1111/ics.12227.

[20] B. Cheng, B. Pei, Z. Wang, Q. Hu, Advances in chitosan-based superabsorbent hydrogels, RSC Adv. 7 (2017) 42036-42046. https://doi.org/10.1039/C7RA07104C.

[21] D. Huber, G. Tegl, M. Baumann, E. Sommer, E.G. Gorji, N. Borth, G. Schleining, G.S. Nyanhongo, G.M. Guebitz, Chitosan hydrogel formation using laccase activated phenolics as cross-linkers, Carbohydr. Polym. 157 (2017) 814-822. https://doi.org/10.1016/j.carbpol.2016.10.012.

[22] M. Periayah, A. Halim, A.M. Saad, Chitosan: A promising marine polysaccharide for biomedical research, Pharmacogn. Rev. 10 (2016) 39. https://doi.org/10.4103/0973-7847.176545.

[23] M.A. Elgadir, M.S. Uddin, S. Ferdosh, A. Adam, A.J.K. Chowdhury, M.Z.I. Sarker, Impact of chitosan composites and chitosan nanoparticle composites on various drug delivery systems: A review, J. Food Drug Anal. 23 (2015) 619-629. https://doi.org/10.1016/j.jfda.2014.10.008.

[24] T. Kean, M. Thanou, Biodegradation, biodistribution and toxicity of chitosan, Adv. Drug Deliv. Rev. 62 (2010) 3-11. https://doi.org/10.1016/j.addr.2009.09.004.

[25] J.-Q. Gao, Hu, Wang, Han, Toxicity evaluation of biodegradable chitosan nanoparticles using a zebrafish embryo model, Int. J. Nanomedicine. (2011) 3351. https://doi.org/10.2147/IJN.S25853.

[26] A.S. Kritchenkov, S. Andranovitš, Y.A. Skorik, Chitosan and its derivatives: vectors in gene therapy, Russ. Chem. Rev. 86 (2017) 231-239. https://doi.org/10.1070/RCR4636.

[27] H. Sashiwa, Y. Shigemasa, Chemical modification of chitin and chitosan 2: preparation and water soluble property of N-acylated or N-alkylated partially deacetylated chitins, Carbohydr. Polym. 39 (1999) 127-138. https://doi.org/10.1016/S0144-8617(98)00167-2.

[28] M.S. Benhabiles, R. Salah, H. Lounici, N. Drouiche, M.F.A. Goosen, N. Mameri, Antibacterial activity of chitin, chitosan and its oligomers prepared from shrimp shell waste, Food Hydrocoll. 29 (2012) 48-56. https://doi.org/10.1016/j.foodhyd.2012.02.013.

[29] J. Cao, L. Xiao, X. Shi, Injectable drug-loaded polysaccharide hybrid hydrogels for hemostasis, RSC Adv. 9 (2019) 36858-36866. https://doi.org/10.1039/C9RA07116D.

[30] M. Dash, F. Chiellini, R.M. Ottenbrite, E. Chiellini, Chitosan—A versatile semi-synthetic polymer in biomedical applications, Prog. Polym. Sci. 36 (2011) 981-1014. https://doi.org/10.1016/j.progpolymsci.2011.02.001.

[31] M. Huang, E. Khor, L.-Y. Lim, Uptake and Cytotoxicity of Chitosan Molecules and Nanoparticles: Effects of Molecular Weight and Degree of Deacetylation, Pharm. Res. 21 (2004) 344-353. https://doi.org/10.1023/B:PHAM.0000016249.52831.a5.

[32] R. Spanneberg, F. Osswald, I. Kolesov, W. Anton, H.-J. Radusch, M.A. Glomb, Model Studies on Chemical and Textural Modifications in Gelatin Films by Reaction with Glyoxal and Glycolaldehyde, J. Agric. Food Chem. 58 (2010) 3580-3585. https://doi.org/10.1021/jf9039827.

[33] M.P. Shaffer, D. V. Belsito, Allergic contact dermatitis from glutaraldehyde in health-care workers, Contact Dermatitis. 43 (2000) 150-156. https://doi.org/10.1034/j.1600-0536.2000.043003150.x.

[34] F. Damiri, Y. Bachra, C. Bounacir, A. Laaraibi, M. Berrada, Synthesis and Characterization of Lyophilized Chitosan-Based Hydrogels Cross-Linked with Benzaldehyde for Controlled Drug Release, J. Chem. 2020 (2020) 1-10. https://doi.org/10.1155/2020/8747639.

[35] H.W. Sung, R.N. Huang, L.L. Huang, C.C. Tsai, In vitro evaluation of cytotoxicity of a naturally occurring cross-linking reagent for biological tissue fixation., J. Biomater. Sci. Polym. Ed. 10 (1999) 63-78. https://doi.org/10.1163/156856299x00289.

[36] X. Chen, M. Fan, H. Tan, B. Ren, G. Yuan, Y. Jia, J. Li, D. Xiong, X. Xing, X. Niu, X. Hu, Magnetic and self-healing chitosan-alginate hydrogel encapsulated gelatin microspheres via covalent cross-linking for drug delivery, Mater. Sci. Eng. C. 101 (2019) 619-629. https://doi.org/10.1016/j.msec.2019.04.012.

[37] B.Y. Shin, B.G. Cha, J.H. Jeong, J. Kim, Injectable Macroporous Ferrogel Microbeads with a High Structural Stability for Magnetically Actuated Drug Delivery, ACS Appl. Mater. Interfaces. 9 (2017) 31372-31380. https://doi.org/10.1021/acsami.7b06444.

[38] R. Lamouri, L. Fkhar, E. Salmani, O. Mounkachi, M. Hamedoun, M. Ait Ali, A. Benyoussef, H. Ez-Zahraouy, A combined experimental and theoretical study of the magnetic properties of bulk CoFe2O4, Appl. Phys. A. 126 (2020) 325. https://doi.org/10.1007/s00339-020-03510-9.

[39] R. Lamouri, O. Mounkachi, E. Salmani, M. Hamedoun, A. Benyoussef, H. Ez-Zahraouy, Size effect on the magnetic properties of CoFe2O4 nanoparticles: A Monte Carlo study, Ceram. Int. 46 (2020) 8092-8096. https://doi.org/10.1016/j.ceramint.2019.12.035.

[40] R. Muzzalupo, L. Tavano, C.O. Rossi, N. Picci, G.A. Ranieri, Novel pH sensitive ferrogels as new approach in cancer treatment: Effect of the magnetic field on swelling and drug delivery, Colloids Surfaces B Biointerfaces. 134 (2015) 273-278. https://doi.org/10.1016/j.colsurfb.2015.06.065.

[41] T. Guo, X. Bian, C. Yang, A new method to prepare water based Fe3O4 ferrofluid with high stabilization, Phys. A Stat. Mech. Its Appl. 438 (2015) 560-567. https://doi.org/10.1016/j.physa.2015.06.035.

[42] Z.-X. Li, D. Luo, M.-M. Li, X.-F. Xing, Z.-Z. Ma, H. Xu, Recyclable Fe3O4 Nanoparticles Catalysts for Aza-Michael Addition of Acryl Amides by

Magnetic Field, Catalysts. 7 (2017) 219. https://doi.org/10.3390/catal7070219.

[43] Z.R. Stephen, F.M. Kievit, M. Zhang, Magnetite nanoparticles for medical MR imaging, Mater. Today. 14 (2011) 330-338. https://doi.org/10.1016/S1369-7021(11)70163-8.

[44] J. Kim, V.T. Tran, S. Oh, C.-S. Kim, J.C. Hong, S. Kim, Y.-S. Joo, S. Mun, M.-H. Kim, J.-W. Jung, J. Lee, Y.S. Kang, J.-W. Koo, J. Lee, Scalable Solvothermal Synthesis of Superparamagnetic Fe 3 O 4 Nanoclusters for Bioseparation and Theragnostic Probes, ACS Appl. Mater. Interfaces. 10 (2018) 41935-41946. https://doi.org/10.1021/acsami.8b14156.

[45] J. Jose, R. Kumar, S. Harilal, G.E. Mathew, D.G.T. Parambi, A. Prabhu, M.S. Uddin, L. Aleya, H. Kim, B. Mathew, Magnetic nanoparticles for hyperthermia in cancer treatment: an emerging tool, Environ. Sci. Pollut. Res. 27 (2020) 19214-19225. https://doi.org/10.1007/s11356-019-07231-2.

[46] A. Radoń, A. Drygała, Ł. Hawełek, D. Łukowiec, Structure and optical properties of Fe 3 O 4 nanoparticles synthesized by co-precipitation method with different organic modifiers, Mater. Charact. 131 (2017) 148-156. https://doi.org/10.1016/j.matchar.2017.06.034.

[47] O.M. Lemine, K. Omri, B. Zhang, L. El Mir, M. Sajieddine, A. Alyamani, M. Bououdina, Sol–gel synthesis of 8nm magnetite (Fe3O4) nanoparticles and their magnetic properties, Superlattices Microstruct. 52 (2012) 793-799. https://doi.org/10.1016/j.spmi.2012.07.009.

[48] W. Zhang, F. Shen, R. Hong, Solvothermal synthesis of magnetic Fe3O4 microparticles via self-assembly of Fe3O4 nanoparticles, Particuology. 9 (2011) 179-186. https://doi.org/10.1016/j.partic.2010.07.025.

[49] D. Stadler, D.N. Mueller, T. Brede, T. Duchoň, T. Fischer, A. Sarkar, M. Giesen, C.M. Schneider, C.A. Volkert, S. Mathur, Magnetic Field-Assisted Chemical Vapor Deposition of Iron Oxide Thin Films: Influence of Field–Matter Interactions on Phase Composition and Morphology, J. Phys. Chem. Lett. 10 (2019) 6253-6259. https://doi.org/10.1021/acs.jpclett.9b02381.

[50] R. Weissleder, A. Bogdanov, E.A. Neuwelt, M. Papisov, Long-circulating iron oxides for MR imaging, Adv. Drug Deliv. Rev. 16 (1995) 321-334. https://doi.org/10.1016/0169-409X(95)00033-4.

[51] S. Mitragotri, J. Lahann, Physical approaches to biomaterial design, Nat. Mater. 8 (2009) 15-23. https://doi.org/10.1038/nmat2344.

[52] Z. Lu, M.D. Prouty, Z. Guo, V.O. Golub, C.S.S.R. Kumar, Y.M. Lvov, Magnetic Switch of Permeability for Polyelectrolyte Microcapsules Embedded with Co@Au Nanoparticles, Langmuir. 21 (2005) 2042-2050. https://doi.org/10.1021/la047629q.

[53] P.J. Reséndiz-Hernández, O.S. Rodríguez-Fernández, L.A. García-Cerda, Synthesis of poly(vinyl alcohol)–magnetite ferrogel obtained by freezing–thawing technique, J. Magn. Magn. Mater. 320 (2008) e373–e376. https://doi.org/10.1016/j.jmmm.2008.02.073.

[54] B.M. Charhouf I, Bennamara A, Abourriche A, Chenite A, Zhu J, Characterization of a Dialdehyde Chitosan Generated by Periodate Oxidation, Int. J. Sci. Basic Appl. Res. 16(2) (2014) 336-348.

[55] H. Tan, C.R. Chu, K.A. Payne, K.G. Marra, Injectable in situ forming biodegradable chitosan–hyaluronic acid based hydrogels for cartilage tissue engineering, Biomaterials. 30 (2009) 2499-2506. https://doi.org/10.1016/j.biomaterials.2008.12.080.

[56] L. Liu, X. Tang, Y. Wang, S. Guo, Smart gelation of chitosan solution in the presence of NaHCO3 for injectable drug delivery system, Int. J. Pharm. 414 (2011) 6-15. https://doi.org/10.1016/j.ijpharm.2011.04.052.

[57] R.W. Korsmeyer, R. Gurny, E. Doelker, P. Buri, N.A. Peppas, Mechanisms of solute release from porous hydrophilic polymers, Int. J. Pharm. 15 (1983) 25-35. https://doi.org/10.1016/0378-5173(83)90064-9.

[58] C. Müller, D. Rahmat, F. Sarti, K. Leithner, A. Bernkop-Schnürch, Immobilization of 2-mercaptoethylamine on oxidized chitosan: A substantially mucoadhesive and permeation enhancing polymer, J. Mater. Chem. 22 (2012) 3899-3908. https://doi.org/10.1039/c2jm15164b.

[59] M. Fernandes Queiroz, K. Melo, D. Sabry, G. Sassaki, H. Rocha, Does the Use of Chitosan Contribute to Oxalate Kidney Stone Formation?, Mar. Drugs. 13 (2014) 141-158. https://doi.org/10.3390/md13010141.

[60] M. KASAAI, A review of several reported procedures to determine the degree of N-acetylation for chitin and chitosan using infrared spectroscopy, Carbohydr. Polym. 71 (2008) 497-508. https://doi.org/10.1016/j.carbpol.2007.07.009.

[61] Y. Jia, Y. Hu, Y. Zhu, L. Che, Q. Shen, J. Zhang, X. Li, Oligoamines conjugated chitosan derivatives: Synthesis, characterization, in vitro and in vivo biocompatibility evaluations, Carbohydr. Polym. 83 (2011) 1153-1161. https://doi.org/10.1016/j.carbpol.2010.09.046.

[62] M. Mehedi Hasan, M. Nuruzzaman Khan, P. Haque, M.M. Rahman, Novel alginate-di-aldehyde cross-linked gelatin/nano-hydroxyapatite bioscaffolds for soft tissue regeneration, Int. J. Biol. Macromol. 117 (2018) 1110-1117. https://doi.org/10.1016/j.ijbiomac.2018.06.020.

[63] Y. Wei, B. Han, X. Hu, Y. Lin, X. Wang, X. Deng, Synthesis of Fe3O4 Nanoparticles and their Magnetic Properties, Procedia Eng. 27 (2012) 632-637. https://doi.org/10.1016/j.proeng.2011.12.498.

[64] S. Lü, M. Liu, B. Ni, An injectable oxidized carboxymethylcellulose/N-succinyl-chitosan hydrogel system for protein delivery, Chem. Eng. J. 160 (2010) 779-787. https://doi.org/10.1016/j.cej.2010.03.072.

[65] Z. Zarnegar, J. Safari, Fe 3 O 4 @ chitosan nanoparticles: a valuable heterogeneous nanocatalyst for the synthesis of 2,4,5-trisubstituted imidazoles, RSC Adv. 4 (2014) 20932-20939. https://doi.org/10.1039/C4RA03176H.

[66] J. Wu, W. Jiang, R. Tian, Y. Shen, W. Jiang, Facile synthesis of magnetic-/pH-responsive hydrogel beads based on Fe 3 O 4 nanoparticles and chitosan hydrogel as MTX carriers for controlled drug release, J. Biomater. Sci. Polym. Ed. 27 (2016) 1553-1568. https://doi.org/10.1080/09205063.2016.1218212.

[67] H. Ghasemzadeh, M. Sheikhahmadi, F. Nasrollah, Full polysaccharide crosslinked-chitosan and silver nano composites, for use as an antibacterial membrane, Chinese J. Polym. Sci. 34 (2016) 949-964. https://doi.org/10.1007/s10118-016-1807-3.

[68] E.F.S. Vieira, A.R. Cestari, C. Airoldi, W. Loh, Polysaccharide-Based Hydrogels: Preparation, Characterization, and Drug Interaction Behaviour, Biomacromolecules. 9 (2008) 1195-1199. https://doi.org/10.1021/bm7011983.

[69] S.J. Kim, S.J. Park, S.I. Kim, Swelling behavior of interpenetrating polymer network hydrogels composed

of poly(vinyl alcohol) and chitosan, React. Funct. Polym. 55 (2003) 53-59. https://doi.org/10.1016/S1381-5148(02)00214-6.

[70] Y. Liang, X. Zhao, P.X. Ma, B. Guo, Y. Du, X. Han, pH-responsive injectable hydrogels with mucosal adhesiveness based on chitosan-grafted-dihydrocaffeic acid and oxidized pullulan for localized drug delivery, J. Colloid Interface Sci. 536 (2019) 224-234. https://doi.org/10.1016/j.jcis.2018.10.056.

Chapter 2

Gastrointestinal Delivery of APIs from Chitosan Nanoparticles

Rayan Sabra and Nashiru Billa

Abstract

Successful clinical treatment outcomes rely on achieving optimal systemic delivery of therapeutics. The oral route of administering Active Pharmaceutical Ingredients (API) remains formidable because of ease to the patient and convenience. Yet, the gastrointestinal tract (GIT) poses several barriers that need to be surmounted prior to systemic availability, especially for Class IV type drugs. Drug delivery systems in the form of nanoparticles (NP), can be appropriately formulated to alter the physicochemical properties of APIs, thereby addressing constraints related to absorption from the GIT. Polymers offer amenability in the fabrication of NP due to their diversity. Chitosan has emerged as a strong contender in orally deliverable NP because it is biocompatible, biodegradable and muco-adhesive. Due to the positively charged amine moieties within chitosan (NH_3^+), interactions with the negatively charged sialic acid of mucin within the mucosa is possible, which favors delayed GI transit and epithelial uptake. This ultimately results in improved systemic bioavailability. Thus, we expect research in the use of chitosan in oral NP delivery to intensify as we transcend the frontier toward clinical testing of viable formulations.

Keywords: chitosan, gastrointestinal, cellular uptake, nanoparticles, drug delivery, formulation

1. Introduction

According to the US National Institute of Health, drug delivery is a process that permit the influx of therapeutic substances in to the body. Drug delivery systems are designed to enhance the efficiency and safety of therapeutics by regulating the rate, time and place of release in the body [1, 2]. Drug delivery technology has emerged as an essential tool for the improvement of drug bioavailability, reduction in the side effects of medication, all of which generate remarkable clinical outcomes [3]. Drugs may be administered to the body via local application, enterally or parenterally. The parenteral route typically relates to administration that excludes absorption from the gastrointestinal tract (GIT). It consists of administration by injection, inhalation and via transdermal routes. The enteral route is associated with the absorption of the drug via the GIT, this includes oral, sublingual, and rectal administration. Aptly, the mode of drug administration depends on the disease, the desired therapeutic effect and the nature of the product available [4]. Moreover, each delivery route has inherent benefits and constraints. Nevertheless, the majority of manufactured medicines in the pharmaceutical industry are delivered orally, owing to the distinctive advantages offered by this route, including versatility in accommodating various types of drugs, simplicity of administration and accessibility, patient compliance and safety profiles [5–7]. Additionally, the intestinal epithelium is an ideal platform for drug absorption due to the viscous mucosal layer lined with abundant enterocytes, goblet cells, and Peyer's patches that trap drug molecules within the mucus as they transit the GIT. [8, 9].

In comparison to other routes of administration, the oral route is exceptionally complex in expression of anatomical features physiology throughout the GIT [10]. Furthermore, these expressions vary along the GIT in both intensity structure. For example, the mucus layer varies in composition and physical properties along the GIT and the pH varies significantly in the main sections of the GIT. The gastrointestinal motility also varies in intensity and form along the GIT and also depends on food status [11]. Even though, these features can impede drug delivery across the GIT, through careful interplay between formulation science and GIT physiology, scientists have been able to exploit this variance for improved drug delivery. In this regard, nanoparticle formulations have immerged as strong contenders able to surmount some of the constraints associated with GIT absorption. Nanoparticles have gained great interest by researchers in recent years as they can be used to improve drug solubility and bioavailability in the harsh GIT environment due to increased surface area to volume ratio, thus provide a rapid onset of therapeutic action [12]. They can also be used to targeting specific sites within the GIT and hence reduce the effects of enzymatic degradation, all of which can improve the safety and effectiveness of drugs [12, 13].

Nanoparticle formulations may be presented in various forms however, polymeric nanoparticles present the versatility of polymers and can be tailored to achieve superior drug stability, enhanced drug payload capacity, longer circulation times and controlled drug release capabilities, when compared with other their colloidal counterparts [14, 15]. In this regard, chitosan-based nanoparticle formulation have been shown to present several of the desirable attributes listed above in addition to being biodegradable, having low toxicity, amenable to tuneable physical properties and bio-adhesive properties [16, 17].

In this chapter we will be discussing the interplay between the GIT physiology/anatomy and drug physicochemical/biopharmaceutical factors in the absorption process that influence oral therapeutics. We, will also review the physicochemical properties of chitosan relevant for effective GIT delivery, including methods of formulation. The most utilised nanoparticle formulation methods used for chitosan-based nanoparticles are also examined. Finally, we will highlight the recent developments on chitosan-based nanoparticles used in the oral delivery of different drugs.

2. The GIT

The GIT, also known as the digestive tract or alimentary canal, is approximately 9 meters long and can be functionally divided into two parts, the upper and the lower GIT (**Figure 1**). The upper GIT; consisting of mouth, pharynx, oesophagus, stomach and small intestine, play a major role in the transport of the swallowed food bolus, enzymatic digestion and absorption of nutrients [18]. The lower GIT is usually referred to the large intestine and is responsible for the adsorption of water, fermentation of undigested sugars and the storage and evacuation of stool [19]. Following oral dosing, the drug traverses several semipermeable cell membranes through its trajectory to absorption and eventually enters the general circulatory system. Drugs cross cell membranes, which comprise of bimolecular lipid matrix, either by passive diffusion or active transport.

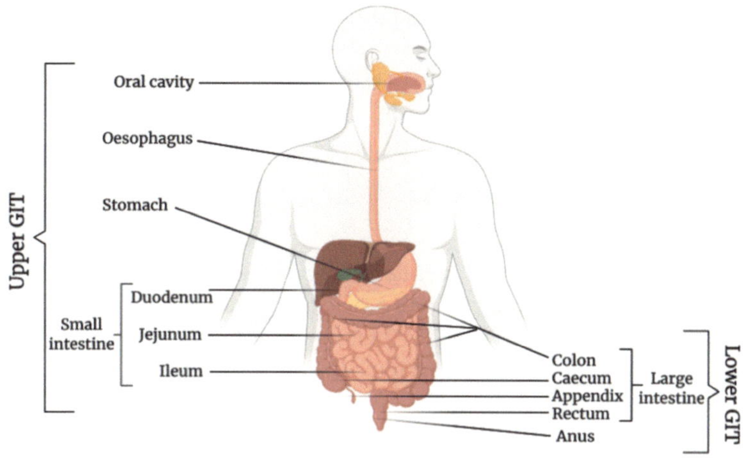

Figure 1.
The anatomy of the human gastrointestinal tract.

2.1 Passive diffusion

The most prevalent form of absorption of the majority of orally administered drugs is by passive diffusion across cell membranes. This process comprises of a three-step process, whereby the permeant first transverses into the membrane, disperses across it and then is released into the cytosol [20]. Typically, drug molecules move down a concentration gradient, from a region of high concentration (e.g., GI fluids) to one of low concentration (e.g., blood), without the expenditure of energy [21]. Usually, a concentration gradient is manifested as a disparity in concentration of a substance within an area and is linearly related to the diffusion rate. The latter is also governed by the lipid solubility, size and polarity of the drug species.

Most drugs are either weak acids or bases and occur either in the unionized or ionized form as a function of pH [22]. For lipophilic drugs, the unionized form of drug, may penetrate cell membranes easily as the membrane is lipoidal. On the other hand, hydrophilic drugs, present an ionized form of the drug, which has high electrical resistance and thus cannot traverse the cell membranes easily but may diffuse through the para-cellular spaces. However, it is worth noting that the para-cellular junctions contribute to less than 0.01% of the entire GIT surface area and furthermore, the permeability of these junctures diminishes down the GIT [23]. Additionally, the capability of drugs to traverse a membrane also relies on the acid–base dissociation constant (pKa) of the drug in question. The pKa is the pH at which concentrations of ionized and unionized forms are equivalent [24]. So, if the pH is less than the pKa, the unionized form of a weak acid prevails, and *vice versa* for weak bases. Henceforth, when a weak acid is administered orally, nearly all the drug in the stomach remains unionized, preferring diffusion via the gastric mucosa. On the other hand, for a weak bases with a pKa = 4.4, majority of the drug in the stomach will be ionized [21].

Overall, the process by which molecules traverse cell membranes is by passive diffusion, down the concentration gradient. However large hydrophilic ionic molecules and charged molecules cannot freely traverse the phospholipid bilayer cell membrane passively. Their transport may be confined to protein channels and distinct transport mechanisms present within the membrane [25]. Such drugs gain access through the membrane by facilitated diffusion whereby molecules integrate with embedded protein carriers to shuttle them across the membrane. This process does not expend energy and is also down the concentration gradient though quicker than would be anticipated by diffusion alone [26]. A frequent case of facilitated diffusion is the migration of glucose into cells during the production of adenosine triphosphate (ATP). Glucose is both large and polar thereby unable to pass the lipid bilayer via simple diffusion. Hence, glucose molecules are delivered into the cell via a unique carrier protein (glucose transporter) to promote its internalisation in cells [27].

2.2 Active Transport

Active transport is an energy-dependent process that translocates drug molecules against their concentration gradient by a molecular pump [20]. Carrier-mediated active transport demand energy via ATP hydrolysis or by accompanying the co-transport of counter ions down its electrochemical gradient (e.g., Na^+, H^+, Cl^-) [28]. The most common active transport system is the sodium-potassium pump and receptor-mediated endocytosis. Energy can either be directly provided to the ion pump or indirectly by connecting a pump-action to an activated ionic gradient. It is often encountered in the gut mucosa, the liver, renal tubules and the blood–brain barrier [22]. Active transport is typically restricted to drugs that structurally resemble endogenous substances; e.g., vitamins and amino acids, and that are absorbed via specific sites in the small intestine. Targeting drugs to these transporters can enhance their bioavailability and distribution [21].

The sodium-potassium pump system (Na^+/K^+ ATPase), utilises ATP to move Na^+ and K^+ in and out of the cell. It is a vital ion pump located in the membranes of various cell types, such as the Na+/amino acid symport in the mucosal cells of the small bowel [22, 29].

Cells control the endocytosis of certain substances via receptor-mediated endocytosis. The use of this form of endocytosis in the GIT is crucial for oral delivery of drugs because it delays the transit of drugs in GIT. Receptor-mediator endocytosis involves the internalisation of macromolecules by binding the latter to receptors considered as membrane-associated protein [30]. There are more than 20 different receptors involved in the internalisation of macromolecules [31]. Following binding to the receptor on the cell surface, the cell will endocytose the portion of the cell membrane enclosing the receptor-ligand complex via a clathrin-dependent endocytic process [28]. Clathrin plays a significant role in the formation of clathrin-coated pits; invaginated regions of the plasma membrane, and pinch off to form clathrin-coated vesicles that transport molecules within cells [31].

In summary, drug adsorption may occur passively or via active transport. In either case, absorption occurs predominantly in the small intestine due to its more permeable membrane and larger surface area provided by the microvilli. Even though, the stomach has a relatively broad epithelial surface, yet the dense mucus layer and transient transit times expended by dosage forms contribute to an impeded absorption. Moreover, the colon with an absorptive surface area of about $5m^2$ has negligible contribution to drug absorption in GIT, due to slow caecal arrival times of dosage forms, the presence of numerous gut bacteria and solid stool that impede lateral diffusion. All in all, absorption of oral drugs is interlinked and controlled by various intrinsic factors; like drug solubility, dissolution and permeability across the mucosal barriers, and physiological factors; such as gastrointestinal transit time, pH and gut microbiome [13, 32].

2.3 Drug dissolution, solubility and permeability

Drug dissolution, solubility and permeability are the three fundamental parameters used in the Biopharmaceutics Classification System (BCS) to predict the factors limiting drug absorption from GIT [33]. The BCS is recognised as a useful tool for designing drug delivery systems and is adopted by the US Food and Drug Administration (FDA), the European Medicines Agency (EMA) and the World Health Organization (WHO) [34]. According to the BCS, all drug substances are classified into four categories: class I—high soluble and high permeable, class II—low soluble and high permeable, class III—low soluble and high permeable and class IV—low soluble and low permeable [35].

Drug solubility is crucial outcome in pharmaceutical dosage form. In the BCS system, a drug is deemed highly soluble when the maximal dose strength is soluble in 250 mL of aqueous media across the pH range of 1 to 7.5 [35]. However, more than 40% of the established new chemical entities in the pharmaceutical sector are considered insoluble in water, causing inadequate bioavailability [36]. This makes solubility amongst the most important rate limiting parameters in GIT absorption.

Drug dissolution reflects a dynamic consequence to drug absorption [33], whereby drug is released, dissolved and made accessible for absorption. With the exception of enteric formulations and drugs with low acid solubility, the dissolution process for majority of drugs starts in the stomach where the volume of gastric fluid is sufficient to attain effective drug dissolution [37]. Thus, the gastric fluid containing the disintegrated immediate-release dosage forms brings the solubilized drug into contact with the absorptive surface of the small intestine as absorption in the stomach is generally minimal.

Drug permeability represents the final frontier in the sequence of rate-liming steps to systemic drug availability. It is a measure of the ease of permeation of the drug across the intestinal wall. There is a positive association between the intestinal permeability and drug solubility GI milieu, which in turn depends on the physicochemical characteristics of the drug [38], including the pKa, particle size, lipophilicity, as discussed in the sections below. The ultimate amount of drug absorbed from the GIT also bears dependence on its transit time in the GIT [39].

2.4 Gastrointestinal pH

The GI pH influences the extent of ionization of drug molecules and thereby impacts on its absorption across the epithelium. Variations in pH across the GIT can be exploited for delayed drug release in desired section of the GIT in order to achieve efficient absorption. The fasted stomach is acidic, with pH range of 1–3, which increases upon food or liquid intake. Food is known to buffer the acidic content of the stomach. A rise in pH resumes in response to the continual gastric secretion and then finally, the pH reverts to the original levels due to gastric emptying of content; [40]. The gastric emptying rate significantly affects the rate of drug absorption because it regulates arrival in the duodenum, where the epithelial surface is suited for absorption [41]. Moreover, the disparity in gastric pH conditions affects the drug delivery behaviour of modified release dosage forms such as enteric coated products, where the onset of release along with the overall release kinetics may be changed [42].

The arrival of orally administered dosage forms into the small intestine is met by a pH of about 6 in the duodenum through to pH 7.4 at the terminal ileum [43]. This high pH variability is due to duodenal secretion of alkaline bicarbonate. During postprandial state, the initial intestinal pH drops due to the influx of acidic chyme, which is buffered by bicarbonate secretion as it travels distally [13]. Besides, the mean pH in proximal small intestine during the first hour of transit is usually 6.6, which is further decreased to 5–6 in the distal duodenum [44].

Typically, the pH in the caecum drops to just below pH 6 owing to the fermentation processes of the colonic microbiota and then rises to pH 7 at the rectum [42]. The drop in the amount of short chain fatty acids at the distal colon causes the secretion of colonic mucosal bicarbonate that leads to a neutral pH. Short chain fatty acids are the end products of fermentation of dietary fibres by the anaerobic intestinal microbiota [45]. As a consequence of the neutral pH of the colonic luminal fluid, the solubilisation of drug is the rate-limiting factor in colonic drug absorption [46]. The unspecific interactions of drugs with colonic content (e.g. dietary residues, intestinal secretions or faecal matter) all adds to the odds of effective adsorption across the colon [47].

2.5 GIT transit time

Generally, the GIT transit time of most orally administered doses through buccal cavity and oesophagus is transient. The stomach is naturally the first segment of the GIT, wherein disintegration and dissolution of solids such as drugs and formulations occur [42]. The period required for a dosage form to exit the stomach is inconstant and relies on several physiological factors, such as age, body posture, gender and food intake [48]. Gastric transit can span from 0 to 2 h in the fasted state and can be extended up to 6 h after food intake [47]. The small intestine is the region of choice for drug absorption with a transit time ranging from 2 to 6 h in healthy individuals. The dissolution of poorly soluble, weakly acidic compounds and lipophilic compounds is greatly enhanced in this region [13]. In colon-specific drug delivery, the drug has to cross the whole GIT prior to arrival at the colon. Thus, the transit time across the colon can be highly variable, and ranges from 20 to 56 h in healthy humans, although higher variations are also reported in literature amounting up to 72 h [42, 49, 50]. Variations in colonic transit time are affected by dosing time, bowel movements as well as gender, whereby females generally have longer colonic transit times than males [51, 52].

2.6 Gut microbiome

Enzymatic and microbial degradation of GIT content affects the amount ultimately made available for absorption. The active sites for most endogenous enzymes are the stomach and small intestine. Even though these enzymes may affect the stability of orally administered drugs, it is possible to exploit this property for regional drug delivery of formulations in the GIT [47]. On the other hand, the intestinal microbiome which includes 500–1000 bacterial species is also important for the digestion of food and the metabolism of drugs [53]. Gastrointestinal microbiome is found in both upper and lower GIT, whereby, a lower bacterial number (10^{13}–10^{14} bacteria mL^{-1} of intestinal content) is in the upper GIT because of the fast luminal flow, intestinal fluid volume, and the secretion of bactericidal compounds in this part of the GIT, and highest bacterial community (10^{10}–10^{11} bacteria mL^{-1} of intestinal content) is in the colon, in which the redox potential is low and the transit time is long [54, 55]. Therefore, greater number of the intestinal microbiome exists in the anaerobic colon, in which the fermentation of carbohydrates contributes to their nourishment. Usually, orally administered drugs are transformed to bioactive, bio-inactive, or toxic metabolites by the gut microbial population, all of which can impede the bioavailability of drug. However, gut microflora can improve drug bioavailability by eliminating polar moiety from derived conjugates and thereby promoting biliary recycling of compounds [13].

Thus, formulation scientist must be cognizant of the interplay between drug and physiological and anatomical manifestations within the GIT when designing orally administered dosage forms. For example, enteric coating can be applied to dosage forms to delay the release of the API in the acidic gastric fluid until pH above 5.0 [56]. Enteric coating may also be used to shield acid-labile drugs from gastric distress, and upon arrival to the alkaline pH milieu, the enteric polymer coating disintegrates within the intestinal fluid, releasing the drug [57]. Despite employing such coatings and other conventional interventions, numerous pharmaceuticals still display insufficient bioavailability through the oral route of administration. This necessitates the use of alternate strategies. One area of research that is gaining traction more recently is the employment of nanoparticles.

3. Nanoparticle technology

Nanoparticle technology is a multidisciplinary field that utilizes principles from chemistry, biology, physics and engineering to design and fabricate submicronic (< 1 μm) colloidal systems [58]. Nanotechnology has several pharmaceutical and medical applications wherein nanoparticles (NPs), with sizes comparable to large biological molecules such as enzymes can be employed in the delivery of therapeutic agents [59]. The effectiveness of the nanoscale drug delivery vehicles lies on their ability to attain the following key attributes [60]:

- The NP must be able to bind or contain the appropriate drug.
- The nanocarrier must stay stable in the serum to allow systemic delivery of the therapeutics and only release the drug once at the required site.
- The NP-drug complex has to reach the required site either via receptor-mediated interactions or by the enhanced permeability and retention (EPR) effect.
- The residual NP carrier should ideally be made of a biological or biologically inert material with a limited lifespan to allow safe degradation.

There are several types of NP drug delivery systems, which may be broadly divided as organic and inorganic NPs [61]. Their particle size, surface charge (ζ potential), hydrophilicity/hydrophobicity, composition, etc. can be tailored for a diverse applications [62]. The primary consideration when designing orally administered NP drug delivery system is to maximise drug concentration in the GI therapeutic window.

3.1 Oral organic NP

Organic NP (**Figure 2**) are solid particles comprised of organic compounds (usually lipidic or polymeric) ranging from 10 nm to 1 μm [63]. They can be formulated by simple techniques to encapsulate therapeutic agents. Preferably, compounds used in formulation of organic NPs should be biodegradable and biocompatible [61]. Manifestations of organic NP include liposomal, polymeric and solid lipid NP, each system possessing requisite features that addresses physiological and anatomic constraints addressed in sections above. In addition, others systems such as micelles, dendrimers etc. have been also explored as effective nanocarriers for effective deployment of APIs in the GIT [14, 64].

3.2 Oral inorganic NP

Inorganic NP represent a wide spectrum of systems synthesized from metals, metal oxides, and metal sulphides [65]. Gold, silica and superparamagnetic oxide NP are among the long list of inorganic NP (**Figure 3**). They have been studied for

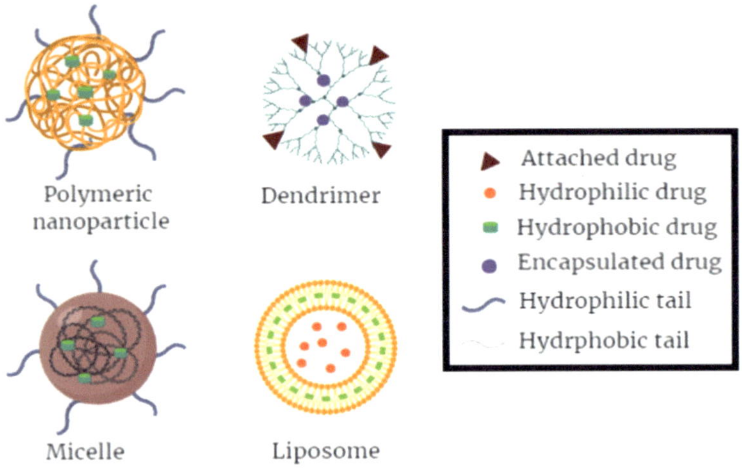

Figure 2.
Examples of organic nanoparticle platforms for drug delivery.

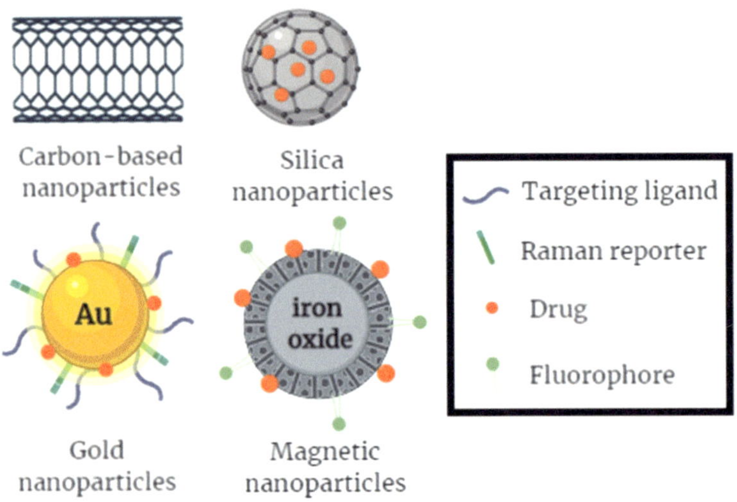

Figure 3.
Examples of inorganic NP platforms for drug delivery.

use in imaging on nuclear magnetic resonance and high-resolution superconducting quantum interference devices, and their intrinsic properties have been utilised for therapy [66]. Inorganic NP can easily be conjugated to ligands for tumour targeting and/or with chemotherapeutics for tumour therapy. Additionally, their surface composition can be feasibly manipulated to create NP that can escape the reticulo-endothelial system [67]. Even though inorganic NP present good stability characteristics, they have not been the focus of attention in oral NP research, possibly due to concerns on the degradation and elimination end products, which can be potentially toxic [68].

Generally, inorganic NPs differ conceptually from organic NPs in terms of fabrication principles. Inorganic NPs can be formed by the precipitation of inorganic salts, which are linked within a matrix, whilst, most organic NPs are formed by several organic molecules through self-organization or chemical binding [61]. Notwithstanding, both types of NP are very promising in the formulation of oral delivery system and forms part of the evolutional success in several clinical applications. Polymeric NP arguably presents more desirable attributes as orally delivered NP because of their higher stability, enhanced drug payload and controlled drug release capabilities compared with their colloidal counterparts [14, 69].

3.3 Polymeric NP

According to Alexis F. et al., polymeric NP represent the most effective nanocarrier system for prolonged drug delivery [70]. 'Polymeric NPs' include any type of polymer formed as NP. Nanospheres are solid spherical NP with molecules attached or adsorbed to their surface, whilst nanocapsules are vesicular systems with substances confined within a cavity consisting of a liquid core (either water or oil) surrounded by a solid shell [71]. Characteristic properties of polymers such as molecular weight, hydrophobicity and crystallinity can be explored to manifest controlled drug release kinetics and entrapment of therapeutic agents [72]. Polymers also provide significant flexibility in the design of oral NP and many exhibit biodegradability [73]. In this regard, synthetic and natural variants have been studied. For example poly-lactic-co-glycolic-acid (PLGA) and poly-lactic-acid (PLA) are synthetic whilst natural polymers include gelatine, dextran, and chitosan [74]. The use of natural polymers is preferred over the synthetic ones as the former usually exhibit less toxicity, widely available and have lower production costs [75]. Chitosan is arguable one of the most studied polymer in NP formulation in view of its distinctive properties. In orally administered NP, chitosan offers added desirability including muco-adhesiveness, augmenting the dissolution rate of poorly water-soluble drugs; useful in drug targeting in the GIT [76].

4. Chitosan polymer

Chitosan is a hydrophilic, cationic polysaccharide soluble in dilute acids such as acetic acid and formic acid, due to protonated amine groups ($NH3^+$) [75]. It is an N-acetylated derivative of chitin, a natural polysaccharide found in the shells of marine crustaceans. Chitin is chemically inert and thus has fewer applications that chitosan [77]. The acetamido group of chitin, (C_2H_4NO) can be turned into amino group to yield chitosan by the alkaline deacetylation of chitin. Chitosan is approved as safe by the United States Food and Drug Administration (US-FDA) for dietary use and wound dressing applications, but its toxicity increases with electrical charge and degree of deacetylation [17]. Chemically, it comprises of β- [1–4] -linked D-glucosamine and N-acetylated units (**Figure 4**).

The amine group has pKa of 6.2–6.5 [78]. At slightly acidic pH values, the amine groups ($NH3^+$) become protonated, hence possessing the ability to effectively form electrostatic interactions with negatively charged species within mucin in the GIT [75]. Positively charged moieties of chitosan also interact with the tight junctions of the intestinal epithelial cells and thus modulate drug permeation and absorption through the interstitial space between epithelial cells [79]. Moreover, the existence of both hydroxyl and amino groups offers various possibilities for chemical modification. Chemical modifications give rise to different functional derivatives of chitosan like carboxylation, thiolation, alkylation, acylation etc. that further imparts desirable physiochemical and biopharmaceutical properties, such as solubility, adsorption and pH sensitivity in oral drug delivery [80]. For example, N-trimethyl chitosan chloride is developed to amplify the intestinal solubility of chitosan; thiolated chitosan is produced to augment the mucoadhesiveness of chitosan;

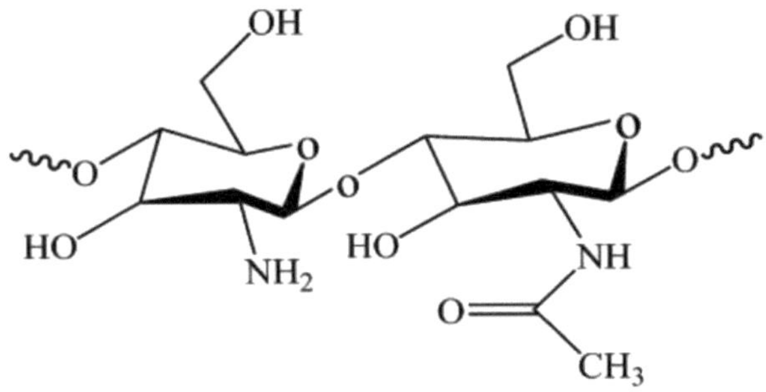

Figure 4.
Chemical structure of chitosan, comprising N-acetyl-D-glucosamine (right) and D-glucosamine (left) units.

quaternization of chitosan reinforces its impact on the tight junctions of the GIT epithelium whilst grafting carboxylated chitosan with poly(methyl methacrylate) imparts increased pH sensitivity [81]. Physical modification through blending with other polymers may be used to enhance desirable physical properties. For example, blending of chitosan with polyethylene glycol (PEG) and polyvinyl alcohol (PVA) ameliorate the hydrophilic property of chitosan, while blending of chitosan with cellulose improves its antibacterial properties [82].

4.1 Mucoadhesion from chitosan

Some of the key desirable features in orally administered dosage forms is delayed GI transit in the duodenum and ability to traverse the epithelium effectively. In this regard, chitosan-based NP have been shown to possess these attributes. Mucoadhesion refers to the adhesion between two materials, one of which is mucosal [83]. It can be utilised to prolong the GI transit of dosage forms in the duodenum, thereby improving bioavailability. Delayed transit results from interactions of positively charged moieties in chitosan with negatively charged moieties in sialic acid within mucin [81]. Chitosan is also capable of physically penetrating the mucous network. Prolonged GI residence results in higher net drug flux across the GIT membrane. Drug flux is a combination of passive diffusion and uptake of whole NP by Peyer's patches [84]. Moreover, chitosan offers controlled drug release capabilities via diffusion from the matrix. Yin et al. prepared thiolated trimethyl chitosan NP for the oral delivery of insulin, where increase in the mucoadhesion resulted in increased insulin transport through rat intestine and uptake by Peyer's patches compared to controls. They attributed these results to the disulfide bond formation between the NP and mucin [85]. Overall, to achieve the desired properties of interest such as particle size, particle size distribution and area of application, the mode of preparation of chitosan NP plays an essential role.

4.2 Fabrication methods for chitosan NP

The preparation of chitosan NP is principally divided into two approaches. The first approach is based on a two-step procedure, where an emulsification system is carried out to generate nanodroplets in which organic compounds (polymer, monomer, and lipid) are solubilized, followed by precipitation or polymerisation into NP [61]. The second approach involves a one-step procedure where the NP are directly generated via different mechanisms such as nanoprecipitation or ionic gelation [86]. An example of each of the two general approaches is summarized in the following.

4.2.1 Chitosan NP by ionic gelation

Ionic gelation, also known as ionotropic gelation or polyelectrolyte complexation involves the gradual addition of a cross-linking agent (tripolyphosphate, glutardehyde etc.) into an aqueous solution of chitosan under continuous stirring to form hydrogels [87]. The polyanions from the cross-linker forms a meshwork of structures by interacting with the polyvalent cations within chitosan, leading to gelation [88]. APIs can be loaded into these hydrogels during the production where it becomes encapsulated or added to the formed NP, where it can be adsorbed into the matrix. The choice of the cross-linker should be matched to the desired physical characteristics of the NP, such as mechanical strength, as well as safety profiles. For example, glutardehyde reported to be toxic when used in high concentrations and results in NP with low mechanical strength. This has been attributed to its double bond (–C=N–) association with the amine group in chitosan [89]. Genipin is a natural cross-linker obtained from iridoid glucoside (geniposide) and present in gardenia fruits that can be cross-linked with chitosan. It displays slower degradation rate than glutaraldehyde and possess higher biocompatibility. Sodium tripolyphosphate (STPP) displays better crosslinker characteristics than each of the above because of its inorganic nature and consequently, results in production of chitosan NP with better mechanical stability. The size dimension derived from STPP gelled chitosan NP is of lower order as well. Another attractive feature of STPP is that it is nontoxic, relatively inexpensive, multivalent, has quick gelling property and thus, widely utilised as a crosslinker in chitosan-based NP [90–92].

4.2.2 Emulsion evaporation

Polymeric nano-emulsions are formulated whereby organic solvent is added to a solution of chitosan with surfactant and mixed via sonication [93]. Basically, the emulsion droplets are converted into NP suspension as the organic solvent evaporates by continuous magnetic stirring at room temperature. The NP suspensions are then centrifuged, washed with distilled water to remove additives such as surfactants and finally lyophilized [94]. Poovi et al. encapsulated the poorly water-soluble drug, repaglinide, into chitosan NP using the emulsion evaporation for sustained release. They proved that the NP exhibited a controlled release of repaglinide and obtained a high drug loading (11.22% w/w) and encapsulation efficiency (97.0%) [95]. In another study, Lee et al. employed solvent evaporation method to formulate polymeric NP from chitosan derivatives fluorescein isothiocyanate (FITC) - conjugated glycol CSs (FGCs) using diluted chloroform as the solvent. Size range of 150–500 nm were obtained and the NP remained stable in phosphate buffered saline for 20 days at 37°C [96].

4.3 In vitro drug release from chitosan NP

In vitro drug release studies give us insights on the response of formulated delivery systems to challenges in *in vivo*. The rate and extent of *in vitro* drug release from chitosan-based NP is influenced by a host of factors, notably, shape and size of the of the delivery system, physicochemical properties of the drug and external media [97]. Three primary mechanisms govern the drug release from chitosan NP, which includes desorption (release of drug from surface), diffusion, and erosion/degradation of the particle matrix [98]. Erosion or degradation of polymers lead to successive physical depletion of the polymer as chains and bonds break [99]. Drug release from the chitosan NP matrix is often pH dependent because of the solubility of chitosan in acidic media [100]. In acidic media, the matrix swells or disentangles and may act as an effective barrier to drug diffusion. The extent of drug diffusion through this gelled matrix depends on the diffusivity of the drug [99]. In alkaline media, the polymer matrix does not swell and drug release is controlled mainly by passive diffusion into the medium and the polymer plays an insignificant role in the drug release profile. If the drug is weakly bound to the surface of the NP, an initial burst release occurs [97]. In vitro drug release from chitosan NP usually show a two-step pattern with an initial rapid release followed by sustained release [101]. Patel et al. observed that rifampicin- chitosan NP presents a burst effect in the early stages followed by slow sustained drug release in which 90% of rifampicin was released in the range of 28–34 h. A good correlation fit was obtained between the cumulative drug released and square root of time, signifying that the drug release from the NP is diffusion-controlled as described by the Higuchi model. They concluded that rifampicin release from chitosan NP is pH dependent, i.e., faster at a lower pH than around neutral pH [102]. Similarly, Avadi et al. observed that insulin-loaded gum arabic/chitosan NP present a burst effect release in acidic medium, relating it to high solubility of both chitosan and insulin. No burst release was observed at higher pH values of 6.5 and 7.2. They concluded that the release followed a non-Fickian transport, governed by diffusion and/or swelling of the chitosan chains [103]. The performance of chitosan NP in the GIT depends on its response to the external milieu as discussed above. Equally important is how the GIT responds to the presence of NP. The following section describes the consequence of NP deployment in the GIT in the management of selected diseases and expected responses.

4.4 Chitosan as an oral delivery vehicle for therapeutics

As mentioned in sections 4.1 and 4.2, extensive research presented the potential of chitosan as an oral absorption enhancer owing to its mucoadhesive properties and ability to loosen tight junctions within the GI epithelia, hence permitting the passage of macromolecular therapeutics across a "well-organised" epithelia [100]. Moreover, due to various characteristics; i.e. non-toxic, biodegradable, biocompatible, antimicrobial property etc. [104], chitosan NP hold promise as a suitable oral delivery vehicle for a wide spectrum of therapeutics including, anti-cancer drugs, antibacterial agents, polyphenolic compounds and protein drugs.

4.4.1 Anti-cancer drug delivery

Chemotherapeutic APIs usually exhibit low bioavailability following oral administration. Several studies have investigated chitosan-based NP as a possible delivery system to address this issue. For example, doxorubicin (Dox), broadly employed to treat breast, bladder and other cancers, is typically delivered intravenously. The oral bioavailability of Dox is low due to efflux transporter P-glycoprotein, which identifies Dox as a substrate, restraining its cellular uptake [105]. In 2013, Feng et al. developed chitosan/o-carboxymethyl chitosan (CS/CMCS) NP as a pH responsive carrier for the oral delivery of Dox. They investigated the bioavailability of orally administered Dox-CS/CMCS NP and free Dox drug on Sprague–Dawley rats. Negligible Dox was detected in plasma after the oral dosage of free Dox, representing its poor absorption. On the other hand, 2.3-folds increase in plasma concentration of Dox was registered after an oral dose of Dox-CS/CMCS NP. Moreover, accumulation of Dox in the liver, spleen and lungs were demonstrated in rats treated with oral Dox- CS/CMCS NP, as opposed to DoX solution which was more concentrated in the kidneys. They concluded that the NP matrix improved the intestinal absorption of Dox and thus improved oral bioavailability [106].

Gemcitabine (Gem) is a widely prescribed anticancer agent used in pancreatic, lung and advanced colon cancer. Oral administration of Gem results in low oral bioavailability, high first-pass clearance gastrointestinal toxicity, such as nausea, vomiting and diarrhoea [107]. Hosseinzadeh et al. synthesised and characterised chitosan/Pluronic® F-127 (Gem-Chi/PF) NP in oral delivery of Gem for the treatment of colon cancer. *In vitro* studies showed that the NP presented enhanced cytotoxicity effects against HT-29 cell line and concluded that Gem-Chi/PF NP is a potential candidate for colon cancer treatment [108].

4.4.2 Anti-bacterial agent delivery

Chitosan impedes the growth of bacteria, fungi, and yeast [109]. It exhibits potential antimicrobial properties at pH below 6.0 because of the positively charged – NH_3^+ at the C-2 position within the glucosamine. Low molecular weight chitosan derived NP integrate with bacterial DNA, impeding mRNA synthesis. Conversely, the NH_3^+ in high molecular chitosan derived NP interact with the negatively-charged cell wall in microorganisms and subsequently amend cell permeability [110]. Alqahtani et al. formulated chitosan NP from high and low molecular weight variants to encapsulate the non-antibiotic diclofenac sodium (DIC). The antibacterial properties of NP from low and high molecular weight of chitosan on *Staphylococcus aureus* and *Bacillus subtilis* was significantly higher than from DIC alone. The antibacterial activity of chitosan was higher from the high molecular weight chitosan at pH = 5.5 [111]. In another *in vitro* study, Qi et al., investigated the antibacterial activity of chitosan NP and copper-loaded chitosan NP against various microorganisms (*E. coli, S. choleraesuis, S. typhimurium and S. aureus*). The antibacterial activity of chitosan NP and copper-loaded chitosan NP were significantly higher than from chitosan and doxycycline alone. Furthermore, copper-loaded NP indicated higher antibacterial activity against microorganism compared to chitosan NP void of copper. They concluded that this is due to the higher surface charge density of copper-loaded NP that improves the affinity of the cargo with the negatively charged bacteria membrane. Clearly, the antimicrobial property of chitosan is demonstrable and may augment the antibacterial effects of therapeutic antimicrobial when administered orally.

4.4.3 Polyphenolic compounds delivery

Secondary plant metabolites in the form of polyphenolic compounds have gained wide attention by scientists due to their wide spectrum of pharmacological activities, including antioxidant, antimicrobial and anticancer properties. Most however suffer from poor systemic bioavailability following oral administration due to low solubility and susceptibility to GI degradation. To overcome this constraint, chitosan-based NP have been proposed as a possible delivery intervention, which not only protect these APIs from GI degradation but also improves bioavailability [112]. Curcumin (CUR) is a polyphenol that has been studied extensively. It is derived from the rhizomes of *Curcuma longa* and active against a range of cancers in *in vitro* setups [113, 114]. However, preclinical and clinical data indicate that oral administration of CUR results in poor systemic bioavailability and high susceptibility to metabolic degradation [115]. In a study by AlKhader et al., the pharmacokinetic and anti-colon cancer properties of curcumin-containing chitosan-pectinate NP (CUR-CS-PEC-NPs) were evaluated. The cellular uptake and subsequent anti-proliferative effects of the CUR-CS-PEC-NPs were boosted at low CUR concentration after 48 and 72 hours of treatment compared to free CUR at equivalent dose. Besides, the carrier provided protection to CUR from acidic degradation. After oral administration of CUR-CS-PEC-NPs and free CUR at

10 mg/ml in rats a 4-fold increase in CUR concentration was detected compared to that of free CUR. Their findings indicated a null release of CUR in the upper GIT and a successful delivery of CUR to the colon with increased bioavailability of delivered CUR with time from CUR-CS-PEC-NPs for 24 h. Hence, rapid degradation metabolism of free CUR was noticed at the same duration. They concluded that this formulation may serve as a suitable delivery system for CUR to the colon in which CUR will be available on site for its chemotherapeutic activity against tumour [116].

4.4.4 Protein drug delivery

Proteins are the building blocks of life and required in replicating organisms. Their high molecular weight, chemical and enzymatic susceptibility in the GIT, low diffusion rate through the mucosa barrier and fast systemic clearance, limit their delivery via oral route. As a result, most proteins are administered parenterally. Fortunately, chitosan-based NP are emerging as promising means for the delivery of protein drugs by the oral route through a combination of shielding GI pH, enzymatic degradation and facilitation of epithelial uptake [117]. In a study by He et al., chitosan-STPP insulin NP (CS/STPP/insulin) were orally administered to Type I diabetic rat models in comparison to free insulin solution. Free insulin solution failed to elicit any difference in the blood glucose level, whilst CS/STPP/insulin NP distinctly reduced the blood glucose levels by up to 59% within 8 hours. Crucially, CS/TPP/insulin NP allowed for a fast recovery of blood sugar level when fasting was halted. Moreover, the CS/TPP/insulin NP exhibited negligible toxicity to liver enzymes, confirming the safety profile of the orally delivered CS/TPP/insulin NP. They concluded that CS/TPP NP are an effective oral delivery vehicle for insulin [118]. In another study, Tan et al. demonstrated better *in vitro* uptake and safety profile from amphotericin B-containing chitosan coated nanostructured lipid carrier (ChiAmpB NLC) than from uncoated NLC [119]. The same authors later demonstrated better *in vivo* uptake from ChiAmpB NLC in rats than from uncoated NCL [120]. They attributed the observed increase in systemic bioavailability to increased mean absorption and mean residence times (MAT and MRT) from ChiAmpB NLC than from naked NLC. This was prompted by the mucoadhesive effect imposed by chitosan.

5. Conclusion

The oral route of administration remains formidable in the systemic delivery of therapeutics. It affords patient compliance, ease of administration and flexibility and remains the favourite choice for administration by patients. However, orally administered therapeutics may undergo premature release in the upper GIT which may render them to enzymatic or pH degradation. Therapeutics that are delivered to the absorptive window are susceptible to efflux pump and metabolic enzymes (e.g., cytochrome P450 enzymes) within the GIT epithelia, which is itself a structural barrier. Scientist involved in the design of therapeutics intended for GI delivery must be cognizant of the above constraints and balance these with the physicochemical properties of the therapeutic. Recent evidence attest to the fact that appropriately formulated NP may be fit for this pursuit. In this regard, chitosan NP is the subject of intense interest because it is readily available, biocompatible, biodegradable, mucoadhesive and influences traversing of therapeutics across the GI epithelia. We expect to see more evidence on the application of chitosan in the oral delivery of therapeutics, especially in the form of NP. Further studies on toxicity related issues *in vivo* will assist in discerning any unanticipated effects in humans. These will pave the way for running clinical trials in humans in near future.

Author details

Rayan Sabra[1] and Nashiru Billa[2]*

1 The University of Nottingham Malaysia Campus, Malaysia

2 College of Pharmacy, QU Health, Qatar University, Doha, Qatar

*Address all correspondence to: nbilla@qu.edu.qa

References

[1] M. Rashid, Q. Zaid Ahmad, and Tajuddin, "Trends in Nanotechnology for Practical Applications," in *Applications of Targeted Nano Drugs and Delivery Systems*, Elsevier, 2019, pp. 297-325.

[2] C. Witharana and J. Wanigasekara, "Drug Delivery Systems: A New Frontier in Nano-technology," *Int. J. Med. Res. Heal. Sci.*, vol. 6, no. 9, pp. 11-14, 2017, [Online]. Available: http://www.ijmrhs.com/medical-research/drug-delivery-systems-a-new-frontier-in-nanotechnology.pdf.

[3] L. Kulinsky and M. J. Madou, *BioMEMs for drug delivery applications*, vol. 2015. Woodhead Publishing Limited, 2012.

[4] K. K. Jain, "Drug Delivery Systems - An Overview," in *Methods in Molecular Biology*™, vol. 437, Humana Press, 2008, pp. 1-50.

[5] P. Shrivastava *et al.*, "Nanotechnology for oral drug delivery and targeting," in *Nanoengineered Biomaterials for Advanced Drug Delivery*, Elsevier, 2020, pp. 473-498.

[6] A. El-Kattan and M. Varm, "Oral Absorption, Intestinal Metabolism and Human Oral Bioavailability," in *Topics on Drug Metabolism*, InTech, 2012, pp. 1-37.

[7] J. Zhang, Z. Xie, N. Zhang, and J. Zhong, "Nanosuspension drug delivery system: preparation, characterization, postproduction processing, dosage form, and application," in *Nanostructures for Drug Delivery*, Elsevier, 2017, pp. 413-443.

[8] G. Kaur, M. Arora, and M. N. V. R. Kumar, "Oral Drug Delivery Technologies-A Decade of Developments," J. Pharmacol. Exp. Ther., vol. 370, pp. 529-543, 2019, doi: 10.1124/jpet.118.255828.

[9] B. Homayun, X. Lin, and H.-J. Choi, "Challenges and Recent Progress in Oral Drug Delivery Systems for Biopharmaceuticals," Pharmaceutics, vol. 11, no. 3, pp. 1-29, 2019, doi: 10.3390/pharmaceutics11030129.

[10] L. S. Schanker, "Absorption of Drugs from the Gastrointestinal Tract," in *Concepts in Biochemical Pharmacology*, Berlin, Heidelberg: Springer Berlin Heidelberg, 1971, pp. 9-24.

[11] A. A. Date, J. Hanes, and L. M. Ensign, "Nanoparticles for oral delivery: Design, evaluation and state-of-the-art," J. Control. Release, vol. 240, pp. 504-526, 2016, doi: 10.1016/j.jconrel.2016.06.016.

[12] W. Cui *et al.*, "Application of the Nano-Drug Delivery System in Treatment of Cardiovascular Diseases," *Front. Bioeng. Biotechnol.*, vol. 7, no. 489, pp. 1-18, 2020, doi: 10.3389/fbioe.2019.00489.

[13] P. Viswanathan, Y. Muralidaran, and G. Ragavan, "Challenges in oral drug delivery: A nano-based strategy to overcome," in *Nanostructures for Oral Medicine*, Andronescu Ecaterina and G. A. Miha, Eds. Elsevier Inc., 2017, pp. 173-201.

[14] F. Din *et al.*, "Effective use of nanocarriers as drug delivery systems for the treatment of selected tumors," Int. J. Nanomedicine, vol. 12, pp. 7291-7309, 2017, doi: 10.2147/IJN.S146315.

[15] J. K. Patra *et al.*, "Nano based drug delivery systems: recent developments and future prospects," J Nanobiotechnol, vol. 16, no. 71, pp. 1-33, 2018, doi: 10.1186/s12951-018-0392-8.

[16] A. Rampino, M. Borgogna, P. Blasi, B. Bellich, and A. Cesàro, "Pharmaceutical nanotechnology

Chitosan nanoparticles: Preparation, size evolution and stability," *Int. J. Pharm.*, vol. 455, pp. 219-228, 2013, doi: 10.1016/j.ijpharm.2013.07.034.

[17] M. A. Mohammed, J. T. M. Syeda, K. M. Wasan, and E. K. Wasan, "An overview of chitosan nanoparticles and its application in non-parenteral drug delivery," Pharmaceutics, vol. 9, no. 4, pp. 1-26, Dec. 2017, doi: 10.3390/pharmaceutics9040053.

[18] P. M. Treuting, M. J. Arends, and S. M. Dintzis, "Upper Gastrointestinal Tract," in *Comparative Anatomy and Histology*, Elsevier, 2018, pp. 191-211.

[19] P. M. Treuting, M. J. Arends, and S. M. Dintzis, "Lower Gastrointestinal Tract," in *Comparative Anatomy and Histology*, Elsevier, 2018, pp. 213-228.

[20] N. J. Yang and M. J. Hinner, "Getting across the cell membrane: an overview for small molecules, peptides, and proteins," *Methods Mol. Biol.*, vol. 1266, pp. 29-53, 2015, doi: 10.1007/978-1-4939-2272-7_3.

[21] Le Jennifer, "Drug Absorption ," *MSD Manual Professional Edition*, May 2019. https://www.msdmanuals.com/professional/clinical-pharmacology/pharmacokinetics/drug-absorption# (accessed Sep. 26, 2020).

[22] T. E. Peck, S. A. Hill, T. E. Peck, and S. A. Hill, "Drug passage across the cell membrane," in *Pharmacology for Anaesthesia and Intensive Care*, Cambridge University Press, 2014, pp. 1-8.

[23] H. F. Helander and L. Fändriks, "Surface area of the digestive tract-revisited," Scand. J. Gastroenterol., vol. 49, no. 6, pp. 681-689, 2014, doi: 10.3109/00365521.2014.898326.

[24] S. Kumar, "Local anesthetics," in *Anesthesia Secrets*, Elsevier Inc., 2011, pp. 105-111.

[25] OpenStax, "The Cellular Level of Organization," in *Anatomy and Physiology*, OpenStax, 2013, pp. 1-18.

[26] K. S. Ramos, "Introduction to Molecular Toxicology," in *Comprehensive Toxicology: Second Edition*, Second., vol. 2, Elsevier Inc., 2010, pp. 1-8.

[27] A. M. Navale and A. N. Paranjape, "Glucose transporters: physiological and pathological roles," Biophys. Rev., vol. 8, no. 1, pp. 5-9, Mar. 2016, doi: 10.1007/s12551-015-0186-2.

[28] J. H. Hamman, P. H. Demana, and E. I. Olivier, "Targeting receptors, transporters and site of absorption to improve oral drug delivery.," *Drug Target Insights*, vol. 2, pp. 71-81, 2007, Accessed: Oct. 07, 2020. [Online]. Available: http://www.ncbi.nlm.nih.gov/pubmed/21901064.

[29] J. F. Collawn, "Unlocking the mysteries of Na+-K+-ATPase endocytosis: Phosphorylation is the key," Am. J. Respir. Cell Mol. Biol., vol. 35, no. 1, pp. 1-2, Jul. 2006, doi: 10.1165/rcmb.f317.

[30] W. Stillwell, "Moving Components Through the Cell," in *An Introduction to Biological Membranes*, Elsevier, 2016, pp. 369-379.

[31] R. J. Kaufman and L. Popolo, "Protein Synthesis, Processing, and Trafficking," in *Hematology: Basic Principles and Practice*, Elsevier Inc., 2018, pp. 45-58.

[32] P. B. Shekhawat and V. B. Pokharkar, "Understanding peroral absorption: regulatory aspects and contemporary approaches to tackling solubility and permeability hurdles," Acta Pharm. Sin. B, vol. 7, no. 3, pp. 260-280, May 2017, doi: 10.1016/j.apsb.2016.09.005.

[33] J. Alsenz, "The Impact of Solubility and Dissolution Assessment on Formulation Strategy and Implications

for Oral Drug Disposition," in *Encyclopedia of Drug Metabolism and Interactions*, Hoboken, NJ, USA: John Wiley & Sons, Inc., 2012, pp. 1-70.

[34] A. Beig, M. Markovic, and A. Dahan, "Solubility, Permeability, and Their Interplay," in *Methods and Principles in Medicinal Chemistry*, John Wiley & Sons, Ltd, 2018, pp. 171-202.

[35] K. T. Savjani, A. K. Gajjar, and J. K. Savjani, "Drug Solubility: Importance and Enhancement Techniques," ISRN Pharm., vol. 2012, pp. 1-10, 2012, doi: 10.5402/2012/195727.

[36] S. Sareen, L. Joseph, and G. Mathew, "Improvement in solubility of poor water-soluble drugs by solid dispersion," *Int. J. Pharm. Investig.*, vol. 2, no. 1, p. 12, 2012, doi: 10.4103/2230-973x.96921.

[37] A. L. Golub, R. W. Frost, C. I. Betlach, and M. A. Gonzalez, "Physiologic considerations in drug absorption from the gastrointestinal tract," J Allergy Clin Immunol., vol. 78, pp. 689-94, 1986.

[38] A. Dahan and J. M. Miller, "The solubility-permeability interplay and its implications in formulation design and development for poorly soluble drugs," AAPS J., vol. 14, no. 2, pp. 244-251, Jun. 2012, doi: 10.1208/s12248-012-9337-6.

[39] T. Kimura and K. Higaki, "Gastrointestinal Transit and Drug Absorption," Biol. Pharm. Bull., vol. 25, no. 2, pp. 149-164, Feb. 2002, doi: 10.1248/bpb.25.149.

[40] D. M. Mudie, G. L. Amidon, and G. E. Amidon, "Physiological Parameters for Oral Delivery and In vitro Testing," Mol. Pharm., vol. 7, no. 5, pp. 1388-1405, 2010, doi: 10.1021/mp100149j.

[41] S. Riley, F. Sutcliffe, M. Kim, M. Kapas, M. Rowland, and L. Turnberg, "The influence of gastrointestinal transit on drug absorption in healthy volunteeers.," Br. J. Clin. Pharmacol., vol. 34, no. 1, pp. 32-39, 1992, doi: 10.1111/j.1365-2125.1992.tb04104.x.

[42] M. Koziolek *et al.*, "Investigation of pH and Temperature Profiles in the GI Tract of Fasted Human Subjects Using the Intellicap® System," J. Pharm. Sci., vol. 104, no. 9, pp. 2855-2863, Sep. 2015, doi: 10.1002/jps.24274.

[43] J. Bratten and M. P. Jones, "New Directions in the Assessment of Gastric Function: Clinical Applications of Physiologic Measurements," Dig. Dis., vol. 24, no. 3-4, pp. 252-259, Jul. 2006, doi: 10.1159/000092878.

[44] D. F. Evans, G. Pye, R. Bramley, A. G. Clark, J. Dyson, and J. D. Hardcastle, "Measurement of gastrointestinal pH profiles in normal ambulant human subjects," Gut, vol. 29, pp. 1035-1041, 1988, doi: 10.1136/gut.29.8.1035.

[45] G. Den Besten, K. Van Eunen, A. K. Groen, K. Venema, D. J. Reijngoud, and B. M. Bakker, "The role of short-chain fatty acids in the interplay between diet, gut microbiota, and host energy metabolism," J. Lipid Res., vol. 54, no. 9, pp. 2325-2340, Sep. 2013, doi: 10.1194/jlr.R036012.

[46] S. Amidon, J. E. Brown, and V. S. Dave, "Colon-Targeted Oral Drug Delivery Systems: Design Trends and Approaches," AAPS PharmSciTech, vol. 16, no. 4, pp. 731-741, Aug. 2015, doi: 10.1208/s12249-015-0350-9.

[47] S. Hua, "Advances in Oral Drug Delivery for Regional Targeting in the Gastrointestinal Tract - Influence of Physiological, Pathophysiological and Pharmaceutical Factors," *Frontiers in Pharmacology*, vol. 11. Frontiers Media S.A., p. 524, Apr. 28, 2020, doi: 10.3389/fphar.2020.00524.

[48] L. Kagan and A. Hoffman, "Systems for region selective drug delivery in the gastrointestinal tract:

Biopharmaceutical considerations," Expert Opin. Drug Deliv., vol. 5, no. 6, pp. 681-692, Jun. 2008, doi: 10.1517/17425247.5.6.681.

[49] B. R. Southwell, M. C. C. Clarke, J. Sutcliffe, and J. M. Hutson, "Colonic transit studies: Normal values for adults and children with comparison of radiological and scintigraphic methods," Pediatr. Surg. Int., vol. 25, no. 7, pp. 559-572, Jul. 2009, doi: 10.1007/s00383-009-2387-x.

[50] A. H. Maurer *et al.*, "The SNMMI and EANM Practice Guideline for Small-Bowel and Colon Transit 1.0*," *J. Nucl. Med.*, vol. 54, no. 11, pp. 2004-2013, Nov. 2013, doi: 10.2967/jnumed.113.129973.

[51] G. Sathyan, S. Hwang, and S. K. Gupta, "Effect of dosing time on the total intestinal transit time of non-disintegrating systems," Int. J. Pharm., vol. 204, no. 1-2, pp. 47-51, Jun. 2000, doi: 10.1016/S0378-5173(00)00472-5.

[52] S. Buhmann, C. Kirchhoff, R. Ladurner, T. Mussack, M. F. Reiser, and A. Lienemann, "Assessment of colonic transit time using MRI: A feasibility study," Eur. Radiol., vol. 17, no. 3, pp. 669-674, Mar. 2007, doi: 10.1007/s00330-006-0414-z.

[53] S. Tuddenham and C. L. Sears, "The intestinal microbiome and health," Curr. Opin. Infect. Dis., vol. 28, no. 5, pp. 464-470, Jan. 2015, doi: 10.1097/QCO.0000000000000196.

[54] E. T. Hillman, H. Lu, T. Yao, and C. H. Nakatsu, "Minireview Microbial Ecology along the Gastrointestinal Tract," Microbes Env., vol. 32, no. 4, pp. 300-313, 2017, doi: 10.1264/jsme2.ME17017.

[55] S. E. Aidy, B. van den Bogert, and M. Kleerebezem, "The small intestine microbiota, nutritional modulation and relevance for health," Curr. Opin. Biotechnol., vol. 32, pp. 14-20, Apr. 2015, doi: 10.1016/j.copbio.2014.09.005.

[56] D. Kapoor, R. Maheshwari, K. Verma, S. Sharma, P. Ghode, and R. K. Tekade, "Coating technologies in pharmaceutical product development," in *Drug Delivery Systems*, Elsevier, 2019, pp. 665-719.

[57] M. Long and Y. Chen, "Dissolution Testing of Solid Products," in *Developing Solid Oral Dosage Forms*, Elsevier Inc., 2009, pp. 319-340.

[58] I. Brigger, C. Dubernet, P. Couvreur, and C. Couvreur, "N anoparticles in cancer therapy and diagnosis," *Adv. Drug Deliv. Rev.*, vol. 54, pp. 631-651, 2002, [Online]. Available: www.elsevier.com/locate/drugdeliv.

[59] Y. Xing, J. Zhao, P. S. Conti, and K. Chen, "Radiolabeled Nanoparticles for Multimodality Tumor Imaging," Theranostics, vol. 4, no. 3, pp. 290-306, 2014, doi: 10.7150/thno.7341.

[60] A. S. Thakor and S. S. Gambhir, "Nanooncology: The future of cancer diagnosis and therapy," Cancer J. Clin., vol. 63, no. 6, pp. 395-418, Nov. 2013, doi: 10.3322/caac.21199.

[61] G. Romero and S. E. Moya, "Synthesis of organic nanoparticles," in *Frontiers of Nanoscience*, vol. 4, no. 1, Elsevier Ltd, 2012, pp. 115-141.

[62] N. Ma *et al.*, "Influence of nanoparticle shape, size, and surface functionalization on cellular uptake," J. Nanosci. Nanotechnol., vol. 13, no. 10, pp. 6485-6498, Oct. 2013, doi: 10.1166/jnn.2013.7525.

[63] J. Allouche, "Synthesis of organic and bioorganic nanoparticles: An overview of the preparation methods," in *Nanomaterials: A Danger or a Promise? A Chemical and Biological Perspective*, Brayner R., Fiévet F., and Coradin T., Eds. Springer, London, 2013, pp. 27-74.

[64] S. Gurunathan, M.-H. Kang, M. Qasim, and J.-H. Kim, "Nanoparticle-Mediated Combination Therapy: Two-in-One Approach for Cancer," Int. J. Mol. Sci., vol. 19, pp. 1-37, 2018, doi: 10.3390/ijms19103264.

[65] S. Bhattacharyya, R. A. Kudgus, R. Bhattacharya, and P. Mukherjee, "Inorganic Nanoparticles in Cancer Therapy," Pharm. Res., vol. 28, no. 2, pp. 237-259, 2011, doi: 10.1007/s11095-010-0318-0.

[66] H. C. Huang, S. Barua, G. Sharma, S. K. Dey, and K. Rege, "Inorganic nanoparticles for cancer imaging and therapy," J. Control. Release, vol. 155, no. 3, pp. 344-357, Nov. 2011, doi: 10.1016/j.jconrel.2011.06.004.

[67] Z. P. Xu, Q. H. Zeng, G. Q. Lu, and A. B. Yu, "Inorganic nanoparticles as carriers for efficient cellular delivery," Chem. Eng. Sci., vol. 61, no. 3, pp. 1027-1040, Feb. 2006, doi: 10.1016/j.ces.2005.06.019.

[68] S. J. Soenen, P. Rivera-Gil, J. M. Montenegro, W. J. Parak, S. C. De Smedt, and K. Braeckmans, "Cellular toxicity of inorganic nanoparticles: Common aspects and guidelines for improved nanotoxicity evaluation," Nano Today, vol. 6, no. 5, pp. 446-465, Oct. 2011, doi: 10.1016/j.nantod.2011.08.001.

[69] W. Rao *et al.*, "Thermally responsive nanoparticle-encapsulated curcumin and its combination with mild hyperthermia for enhanced cancer cell destruction," Acta Biomater., vol. 10, no. 2, pp. 831-842, Feb. 2014, doi: 10.1016/j.actbio.2013.10.020.

[70] F. Alexis, E. M. Pridgen, R. Langer, and O. C. Farokhzad, "Nanoparticle technologies for cancer therapy," in *Handbook of Experimental Pharmacology*, vol. 197, Springer Berlin Heidelberg, 2010, pp. 55-86.

[71] S. S. Guterres, M. P. Alves, and A. R. Pohlmann, "Polymeric Nanoparticles, Nanospheres and Nanocapsules, for Cutaneous Applications," *Drug Target Insights*, vol. 2, pp. 147-157, Jul. 2007, doi: 10.1177/117739280700200002.

[72] J. M. Chan, P. M. Valencia, L. Zhang, R. Langer, and O. C. Farokhzad, "Polymeric nanoparticles for drug delivery.," in *Cancer Nanotechnology. Methods in molecular biology (Methods and Protocols)*, vol. 624, Grobmyer S. and Moudgil B., Eds. Clifton, N.J.: Humana Press, 2010, pp. 163-175.

[73] J. P. Rao and K. E. Geckeler, "Polymer nanoparticles: Preparation techniques and size-control parameters," Prog. Polym. Sci., vol. 36, no. 7, pp. 887-913, Jul. 2011, doi: 10.1016/j.progpolymsci.2011.01.001.

[74] A. Sionkowska, "Current research on the blends of natural and synthetic polymers as new biomaterials: Review," Prog. Polym. Sci., vol. 36, no. 9, pp. 1254-1276, Sep. 2011, doi: 10.1016/j.progpolymsci.2011.05.003.

[75] K. Kavitha, M. Rupesh Kumar, and S. Jagadeesh Singh, "Novel Mucoadhesive Polymers-A Review," J. Appl. Pharm. Sci., vol. 01, no. 08, pp. 37-42, 2011.

[76] T. Ahmed and B. Aljaeid, "Preparation, characterization, and potential application of chitosan, chitosan derivatives, and chitosan metal nanoparticles in pharmaceutical drug delivery," Drug Des. Devel. Ther., vol. 10, pp. 483-507, Jan. 2016, doi: 10.2147/DDDT.S99651.

[77] M. W. Anthonsen, K. M. Vårum, and O. Smidsrød, "Solution properties of chitosans: conformation and chain stiffness of chitosans with different degrees of N-acetylation," Carbohydr. Polym., vol. 22, no.

3, pp. 193-201, Jan. 1993, doi: 10.1016/0144-8617(93)90140-Y.

[78] Q. Z. Wang *et al.*, "Protonation constants of chitosan with different molecular weight and degree of deacetylation," Carbohydr. Polym., vol. 65, no. 2, pp. 194-201, Jul. 2006, doi: 10.1016/j.carbpol.2006.01.001.

[79] A. Muheem *et al.*, "A review on the strategies for oral delivery of proteins and peptides and their clinical perspectives," *Saudi Pharmaceutical Journal*, vol. 24, no. 4. Elsevier B.V., pp. 413-428, Jul. 01, 2016, doi: 10.1016/j.jsps.2014.06.004.

[80] N. Islam, I. Dmour, and M. O. Taha, "Degradability of chitosan micro/nanoparticles for pulmonary drug delivery," *Heliyon*, vol. e01684, pp. 1-9, May 2019, doi: 10.1016/j.heliyon.2019.e01684.

[81] U. I. Garg, S. I. Chauhan, U. I. Nagaich, and N. Jain, "Current Advances in Chitosan Nanoparticles Based Drug Delivery and Targeting," Tabriz Univ. Med. Sci., vol. 7, no. 2, pp. 113-117, 2019, doi: 10.15171/apb.2019.023.

[82] E. A. Elhefian, M. Nasef, and A. H. Yahaya, "Chitosan-Based Polymer Blends: Current Status and Applications," J. Chem. Soc. Pakistan, vol. 36, no. 1, pp. 11-27, 2014.

[83] R. Shaikh, T. Raj Singh, M. Garland, A. Woolfson, and R. Donnelly, "Mucoadhesive drug delivery systems," J. Pharm. Bioallied Sci., vol. 3, no. 1, pp. 89-100, Jan. 2011, doi: 10.4103/0975-7406.76478.

[84] B. M. Boddupalli, Z. N. K. Mohammed, R. Nath A., and D. Banji, "Mucoadhesive drug delivery system: An overview," *J. Adv. Pharm. Technol. Res.*, vol. 1, no. 4, pp. 381-387, Oct. 2010, doi: 10.4103/0110-5558.76436.

[85] L. Yin, J. Ding, C. He, L. Cui, C. Tang, and C. Yin, "Drug permeability and mucoadhesion properties of thiolated trimethyl chitosan nanoparticles in oral insulin delivery," Biomaterials, vol. 30, no. 29, pp. 5691-5700, Oct. 2009, doi: 10.1016/j.biomaterials.2009.06.055.

[86] R. K. Shatrohan Lal, "Synthesis of Organic Nanoparticles and their Applications in Drug Delivery and Food Nanotechnology: A Review," *J. Nanomater. Mol. Nanotechnol.*, vol. 03, no. 04, Aug. 2014, doi: 10.4172/2324-8777.1000150.

[87] R. R, M. I. Y, A. U, V. V, R. V, and K. moorthy S.B, "CHITOSAN NANOPARTICLES - AN EMERGING TREND IN NANOTECHNOLOGY," *Int. J. Drug Deliv.*, vol. 6, no. 3, pp. 204-229, Sep. 2014, Accessed: Oct. 13, 2020. [Online]. Available: http://www.arjournals.org/index.php/ijdd/article/view/1452.

[88] B. Sarmento, D. Ferreira, F. Veiga, and A. Ribeiro, "Characterization of insulin-loaded alginate nanoparticles produced by ionotropic pre-gelation through DSC and FTIR studies," Carbohydr. Polym., vol. 66, no. 1, pp. 1-7, Oct. 2006, doi: 10.1016/j.carbpol.2006.02.008.

[89] T. Mitra, G. Sailakshmi, and A. Gnanamani, "Could glutaric acid (GA) replace glutaraldehyde in the preparation of biocompatible biopolymers with high mechanical and thermal properties?," *J. Chem. Sci*, vol. 126, no. 1, pp. 127-140, 2014, doi: https://doi.org/10.1007/s12039-013-0543-2.

[90] R. Sabra, N. Billa, and C. J. Roberts, "An augmented delivery of the anticancer agent, curcumin, to the colon," React. Funct. Polym., vol. 123, pp. 54-60, 2018, doi: 10.1016/j.reactfunctpolym.2017.12.012.

[91] A. Lange de Pinho Neves, C. Cardoso Milioli, L. Müller, H. Gracher Riella, N. Cabral Kuhnen, and H. Karine Stulzer, "Factorial design as tool in chitosan nanoparticles development by ionic gelation technique," *Physicochem. Eng. Asp.*, vol. 445, pp. 34-39, 2014, doi: 10.1016/j.colsurfa.2013.12.058.

[92] L. M. Zhao *et al.*, "Preparation and application of chitosan nanoparticles and nanofibers," *Brazilian* J. Chem. Eng., vol. 28, no. 3, pp. 353-362, 2011, doi: 10.1590/S0104-66322011000300001.

[93] N. B. V N and H. Ksyadav, "DIFFERENT TECHNIQUES FOR PREPARATION OF POLYMERIC NANOPARTICLES-A REVIEW," *Asian* J. Pharm. Clin. Res., vol. 5, no. 3, pp. 16-23, 2012.

[94] D. Bennet and S. Kim, "Polymer Nanoparticles for Smart Drug Delivery," in *Application of Nanotechnology in Drug Delivery*, InTech, 2014, pp. 257-310.

[95] G. Poovi, U. M. Dhana leks, N. Narayanan, and P. Neelakanta, "Preparation and Characterization of Repaglinide Loaded Chitosan Polymeric Nanoparticles," *Res.* J. Nanosci. Nanotechnol., vol. 1, no. 1, pp. 12-24, Jan. 2011, doi: 10.3923/rjnn.2011.12.24.

[96] M. Lee *et al.*, "Size control of self-assembled nanoparticles by an emulsion/solvent evaporation method," Colloid Polym. Sci., vol. 284, no. 5, pp. 506-512, Feb. 2006, doi: 10.1007/s00396-005-1413-3.

[97] A. Mitra and B. Dey, "Chitosan microspheres in novel drug delivery systems," *Indian* J. Pharm. Sci., vol. 73, no. 4, pp. 355-366, Jul. 2011, doi: 10.4103/0250-474X.95607.

[98] S. Naskar, K. Koutsu, and S. Sharma, "Chitosan-based nanoparticles as drug delivery systems: a review on two decades of research," J. Drug Target., vol. 27, no. 4, pp. 379-393, Apr. 2018, doi: 10.1080/1061186X.2018.1512112.

[99] M. A. Mohammed, J. T. M. Syeda, K. M. Wasan, and E. K. Wasan, "An Overview of Chitosan Nanoparticles and Its Application in Non-Parenteral Drug Delivery," Pharmaceutics, vol. 9, no. 4, pp. 1-26, 2017, doi: 10.3390/pharmaceutics9040053.

[100] V. T. Chivere, P. P. D. Kondiah, Y. E. Choonara, and V. Pillay, "Nanotechnology-Based Biopolymeric Oral Delivery Platforms for Advanced Cancer Treatment," Cancers (Basel)., vol. 12, no. 522, pp. 1-22, 2020, doi: 10.3390/cancers12020522.

[101] A. R. Dudhani and S. L. Kosaraju, "Bioadhesive chitosan nanoparticles: Preparation and characterization," Carbohydr. Polym., vol. 81, no. 2, pp. 243-251, Jun. 2010, doi: 10.1016/j.carbpol.2010.02.026.

[102] B. K. Patel, R. H. Parikh, and P. S. Aboti, "Development of Oral Sustained Release Rifampicin Loaded Chitosan Nanoparticles by Design of Experiment," J. Drug Deliv., vol. 2013, pp. 1-10, 2013, doi: 10.1155/2013/370938.

[103] M. R. Avadi *et al.*, "Ex Vivo Evaluation of Insulin Nanoparticles Using Chitosan and Arabic Gum," *Int. Sch. Res. Netw. Pharm.*, vol. 2011, pp. 1-6, 2011, doi: 10.5402/2011/860109.

[104] H. M. Ibrahim and E. M. R. El-Zairy, "Chitosan as a Biomaterial — Structure, Properties, and Electrospun Nanofibers," in *Concepts, Compounds and the Alternatives of Antibacterials*, InTech, 2015, pp. 81-101.

[105] A. E. Caprifico, E. Polycarpou, P. J. S. Foot, and G. Calabrese, "Fluorescein Isothiocyanate Chitosan Nanoparticles in Oral Drug Delivery Studies," Trends Pharmacol. Sci., vol. 41, no. 10, pp.

686-689, Oct. 2020, doi: 10.1016/j.tips.2020.07.005.

[106] C. Feng *et al.*, "Chitosan/o-carboxymethyl chitosan nanoparticles for efficient and safe oral anticancer drug delivery: In vitro and in vivo evaluation," Int. J. Pharm., vol. 457, no. 1, pp. 158-167, Nov. 2013, doi: 10.1016/j.ijpharm.2013.07.079.

[107] J. Ciccolini, C. Serdjebi, · Godefridus, J. Peters, and E. Giovannetti, "Pharmacokinetics and pharmacogenetics of Gemcitabine as a mainstay in adult and pediatric oncology: an EORTC-PAMM perspective," Cancer Chemother. Pharmacol., vol. 78, pp. 1-12, 2016, doi: 10.1007/s00280-016-3003-0.

[108] H. Hosseinzadeh, F. Atyabi, R. Dinarvand, and S. N. Ostad, "Chitosan–Pluronic nanoparticles as oral delivery of anticancer gemcitabine: preparation and in vitro study," Int. J. Nanomedicine, vol. 7, pp. 1851-1863, Apr. 2012, doi: 10.2147/IJN.S26365.

[109] G.-J. TSAI and W.-H. S. H.-C. C. C.-L. PAN, "Antimicrobial activity of shrimp chitin and chitosan from different treatments and applications of fish preservation," *Fish. Sci.*, vol. 68, pp. 170-177, 2002, doi: https://doi.org/10.1046/j.1444-2906.2002.00404.x.

[110] A. I. Barbosa, A. J. Coutinho, S. A. C. Lima, and S. Reis, "Marine Polysaccharides in Pharmaceutical Applications: Fucoidan and Chitosan as Key Players in the Drug Delivery Match Field," Mar. Drugs, vol. 17, no. 654, pp. 1-21, 2019, doi: 10.3390/md17120654.

[111] F. Y. Alqahtani *et al.*, "Preparation, characterization, and antibacterial activity of diclofenac-loaded chitosan nanoparticles," *Saudi* Pharm. J., vol. 27, no. 1, pp. 82-87, Jan. 2019, doi: 10.1016/j.jsps.2018.08.001.

[112] D. D. Milincic *et al.*, "Application of Polyphenol-Loaded Nanoparticles in Food Industry," Nanomaterials, vol. 9, no. 1629, pp. 1-21, 2019, doi: 10.3390/nano9111629.

[113] R. Sabra, C. J. Roberts, and N. Billa, "Courier properties of modified citrus pectinate-chitosan nanoparticles in colon delivery of curcumin," Colloid Interface Sci. Commun., vol. 32, no. 100192, pp. 1-9, 2019, doi: 10.1016/j.colcom.2019.100192.

[114] R. Sabra, N. Billa, and C. J. Roberts, "Cetuximab-conjugated chitosan-pectinate (modified) composite nanoparticles for targeting colon cancer," Int. J. Pharm., vol. 572, no. 118775, pp. 1-11, 2019, doi: 10.1016/j.ijpharm.2019.118775.

[115] P. Anand, A. B. Kunnumakkara, R. A. Newman, and B. B. Aggarwal, "Bioavailability of curcumin: Problems and promises," Mol. Pharm., vol. 4, no. 6, pp. 807-818, Nov. 2007, doi: 10.1021/mp700113r.

[116] E. Alkhader *et al.*, "Pharmacokinetic and anti-colon cancer properties of curcumin-containing chitosan-pectinate composite nanoparticles," J. Biomater. Sci. Polym. Ed., vol. 29, no. 18, pp. 2281-2298, Dec. 2018, doi: 10.1080/09205063.2018.1541500.

[117] C. Y. Wong, H. Al-Salami, and C. R. Dass, "The role of chitosan on oral delivery of peptide-loaded nanoparticle formulation," J. Drug Target., vol. 26, no. 7, pp. 551-562, Aug. 2017, doi: 10.1080/1061186X.2017.1400552.

[118] Z. He *et al.*, "Scalable fabrication of size-controlled chitosan nanoparticles for oral delivery of insulin," Biomaterials, vol. 130, pp. 28-41, Jun. 2017, doi: 10.1016/j.biomaterials.2017.03.028.

[119] J. S. Ling Tan, C. J. Roberts, and N. Billa, "Mucoadhesive

chitosan-coated nanostructured lipid carriers for oral delivery of amphotericin B," Pharm. Dev. Technol., vol. 24, no. 4, pp. 504-512, 2019, doi: 10.1080/10837450.2018.1515225.

[120] J. S. L. Tan, C. Roberts, and N. Billa, "Pharmacokinetics and tissue distribution of an orally administered mucoadhesive chitosan-coated amphotericin B-Loaded nanostructured lipid carrier (NLC) in rats," J. Biomater. Sci. Polym. Ed., vol. 31, no. 2, pp. 141-154, 2020, doi: 10.1080/09205063.2019.1680926.

Chapter 3

Chitosan-Based Oral Drug Delivery System for Peptide, Protein and Vaccine Delivery

Siti Zuhairah Zainuddin and Khuriah Abdul Hamid

Abstract

Oral delivery is the most common and preferred route of drug administration due to its convenience and ease of administration. However, various factors such as poor solubility, low dissolution rate, stability, and bioavailability of many drugs remain an ongoing challenge in achieving desired therapeutic levels. The delivery of drugs must overcome various obstacles, including the acidic gastric environment, the presence of the intestinal efflux and influx transporters and the continuous secretion of mucus that protects the gastrointestinal tract (GIT). As the number and chemical diversity of drugs has increased, various strategies are required to develop orally active therapeutics. One of the approaches is to use chitosan as a carrier for oral delivery of peptides, proteins as well as vaccines delivery. Chitosan, a non-toxic N-deacetylated derivative of chitin appears to be under intensive progress during the last years towards the development of safe and efficient chitosan-based drug delivery systems. This polymer has been recognised as a versatile biomaterial because of its biodegradability, biocompatibility, and non-toxicity. This chapter reviews the physicochemical characteristics of chitosan and the strategies that have been successfully applied to improve oral proteins, peptides, and vaccines bioavailability, primarily through various formulation strategies.

Keywords: chitin, chitosan, bioavailability, oral delivery, peptides, vaccines

1. Introduction

The extent of drug bioavailability has been shown to be influenced by the route of drug administration. Oral drug route needs travelling through the continuous passageway of the GIT, which makes them susceptible to the harsh environment of GIT. Drugs intended for administration via this route can be formulated in a variety of dosage forms, such as tablets, capsules, solutions, and powders.

Due to its high patient compliance and ease of administration, the oral route of administration is preferred among other routes. Self-administration is possible with great compliance and reduced risk in developing systemic side effects, which is the major concern in the parenteral route [1]. Despite that, the oral delivery system approaches for certain drugs are challenging, especially the delivery of peptide drugs and vaccines [2].

The normal physiological functions of GIT are to digest food and to interfere with pathogen entry. These functions need to be considered as peptide drugs and vaccines tend to be digested together with food in the presence of digestive enzymes. The highly acidic pH in the stomach and the presence of proteolytic enzymes such as protease and pepsin can cause protein degradation [1].

Furthermore, they will have difficulties in permeating the physical barrier of the mucus lining, which prevents pathogenic substances from penetrating the cell [3]. Owing to these challenges, protein and peptide drugs are suitable to be administered via parenteral routes such as intravenous or subcutaneous injection [4]. However, these routes require frequent administration with long-term use which will develop patient incompliance to medication [4]. In such a manner, the approach to improve the oral delivery of peptide drugs and vaccines by using suitable polymers are needed to enhance drug effectiveness and patient compliance.

Chitin is the second most abundant polysaccharide present in nature. However, it has more applications when converted to chitosan by partial deacetylation under alkaline conditions [5]. Chitosan is a positively charged polymer that can improve the bioavailability of the oral drug delivery system. It has been used to improve the formulation of peptide drugs, resulting in enhanced cell permeability, which allows an adequate therapeutic concentration of drugs into the systemic circulation [6].

For protein and peptide therapeutics, factors such as poor permeability, luminal, brush border, cytosolic metabolism, and hepatic clearance mechanisms result in their poor bioavailability from oral and non-oral mucosal routes [7]. Oral vaccination is prone to reduce the adequacy of vaccine to be recognised by the immune system due to the presence of gut microbiota and intestinal barrier. Peptide drugs and vaccines can be protected from the degradative barrier of the GIT by encapsulating the drugs into the polymeric chitosan as potential carrier material. The development of nanotechnology, such as nanoparticle systems to transport peptide drugs through the epithelial membrane has been established [6, 8]. Besides, the modification of chitosan is needed to exert its function as a polymer and to protect the drug from enzymatic degradation [9].

This work reviews the physicochemical properties and numerous applications of chitosan, describes its release mechanisms, challenges in oral peptides and vaccines delivery, and strategies to overcome these barriers to improve oral peptides and vaccines bioavailability.

1.1 Chitosan

Chitosan is a strong base with linear polysaccharides consisting of D-glucosamine, which contains amino groups [10]. The hydrolysis of chitin will produce chitosan through alkaline hydrolysis or N-deacetylation (**Figure 1**). Due to protonable amino groups presence in chitosan, this polymer can be easily

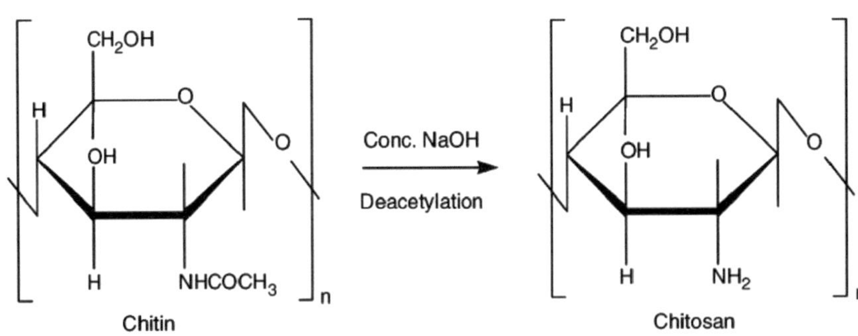

Figure 1.
The N-deacetylation of chitin into chitosan.

dissolved in pH below 6.3. However, both chitosan and chitin are insoluble in an aqueous medium.

Chitin or chitosan is highly available from different species of shrimps, prawns and crabs. These seafood shells release chitosan, which shows properties of antimicrobial and antioxidant activity.

1.2 Characterisation of chitin and chitosan

One of the differences between chitosan and chitin is the presence of amino groups. Amino group in chitosan exhibits high solubility in acidic medium and able to form complexes with metal ions. These positive charges interact with drugs and physiological barriers in the GIT, which is useful in the formulation design of the drug delivery system [9].

Some factors affect chitosan properties, including the degree of deacetylation, degree of substitution, and molecular weight [9, 11]. These factors should be considered before using chitosan as a polymer in a drug delivery system. Most of the chitosan applications are affected by these factors through intermolecular or intramolecular hydrogen bonds [12].

1.2.1 Degree of deacetylation and molecular weight of chitosan

The degree of chitosan deacetylation will affect its biological activity, including swelling rate, molecular weight, crystallinity and polydispersity. The deacetylation process leads to the protonation of the amino groups [13]. A highly positive charge will improve the activity of chitosan as mucoadhesive permeation enhancing [14] and haemostatic agent [15]. Sometimes, the degree of deacetylation can be used to estimate the water solubility of chitosan [11] as shown in **Table 1**.

The degree of deacetylation can influence the particle size and molecular weight of chitosan [13]. The removal of the acetyl group in the structure of chitosan or chitin from deacetylation reduces the interaction between molecules. A low number of acetyl groups minimises the chain length, thus reducing the molecular weight of the polymer [16].

The molecular weight of the polymer will influence the degree of swelling [17]. High molecular weight chitosan (HMWC) tends to have a higher cross-linking ability. Therefore, the drug-coated with HMWC tends to release more slowly [18]. This characteristic is favourable in sustained-release oral drug delivery.

Generally, the lower the molecular weight, the higher solubility of chitosan is obtained [13, 19]. HMWC appears in α-chitin crystalline or antiparallel structure. The structure forms after the release of water, which leads to the loss of entropy during aggregation of the polymeric chain [13]. This phenomenon results in the loss of Gibbs free energy. Gibbs free energy (G) is a way to predict the amount of

Degree of deacetylation	Level	Water solubility
55–70%	Low	Completely insoluble
70–85%	Middle	Partly dissolved
85–95%	High	Good solubility
95–100%	Ultrahigh	Completely soluble

Table 1.
Relationship between degree of deacetylation of chitosan and their water solubility [11].

usable energy in the system. Loss of energy means the reaction in the system tends to be spontaneous.

The α-chitin crystalline form exhibit lower water solubility as compared to β-chitin. The shorter polymeric chain of low molecular weight chitosan (LMWC) is unlikely to aggregate [11]. Interaction between molecules declines due to the formation of the hydrogen bonds is limited. A short chitosan chain contains a low number of amino groups [20].

1.2.2 Crystalline structure

Chitin exists in three different polymorphic forms, which are α-chitin, β-chitin, and γ-chitin (**Figure 2**). The interaction between C=O···NH and C=O···OH maintaining the strength of the polymeric network chain [13]. α-chitin appears in its antiparallel structure and the chain is interacting through both inter-and intramolecular hydrogen bond. β-chitin has a parallel structure, which leads to the formation of the intramolecular hydrogen bond. γ-chitin consists of both antiparallel and parallel structure, as it is the combination of α-chitin and β-chitin [21].

β-Chitin exhibits better water solubility but less common as compared to α-chitin [14]. It has been shown that α-chitosan has a higher crystallinity index as compared to β-chitin. However, the crystallinity index for both forms is lower than the raw chitosan [21]. The crystallinity index will increase when the degree of deacetylation of chitosan increases [16].

The β-chitin has a high affinity to the organic solvent due to its structural flexibility [14]. It exhibits higher reactivity than α-chitosan due to a lack of hydrogen bond. This form has a high capability to swell between crystalline structures while losing its crystalline fraction [21]. The swelling of β-chitin sometimes disrupts the polymeric chain and crystalline structure.

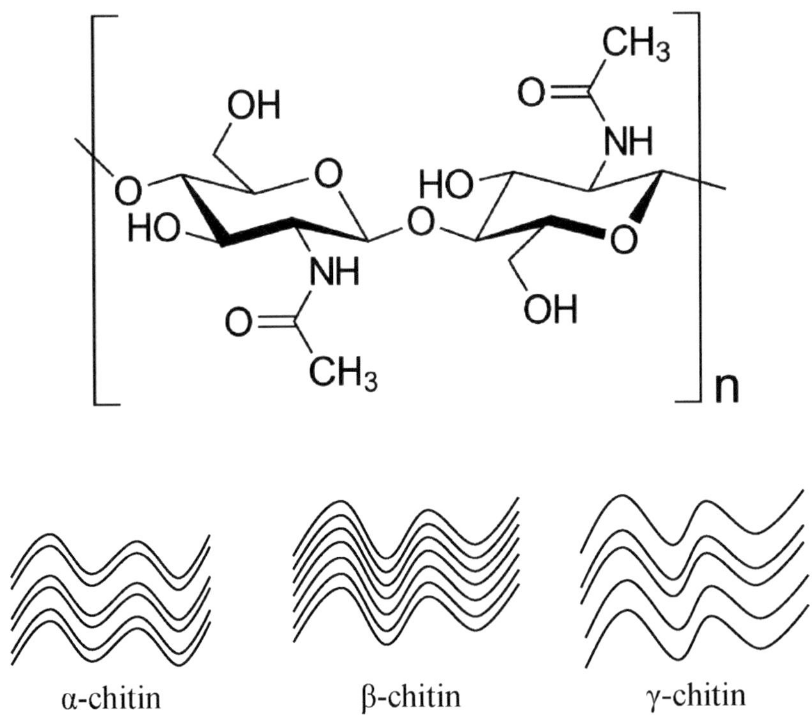

Figure 2.
The different conformational structure between α-chitin, β-chitin and γ-chitin (adapted from [14]).

1.2.3 Polydispersity and particle size of chitosan

The particle size of chitosan plays a major role in developing an efficient carrier for peptide drugs [22]. Monodisperse preparation of nanoparticles is desirable to provide better bioavailability and low toxicity [23]. Polydispersity leads to a larger size distribution, which interferes with the tendency for the nanocarrier to accumulate in the target tissue [23].

Polydispersity describes the degree of non-uniformity of size distribution between molecules due to the aggregation or agglomeration of the polymeric network. It can be estimated using the polydispersity index (PDI), where the ideal index for chitosan nanoparticles is below 0.3 [22, 23]. The degree of deacetylation and molecular weight of chitosan have been proven to influence the polydispersity of the system [24].

PDI increases with an increase in molecular weight. However, it decreases as the degree of deacetylation increases [23]. The increase of amino group protonation and removal of the acetyl group from chitosan structure lead to the enhancement of repulsive forces between molecules and stretch the chitosan to become larger in size [11, 24]. Therefore, the development of chitosan with the optimum degree of deacetylation is needed to minimise the risk of polydispersity. This can be achieved by modifying the time and temperature of the de-*N*-acetylation process [18].

The degree of entanglement for HMWC nanoparticles is higher than LMWC. Therefore, HMWC nanoparticle has a high tendency to aggregate with each other and disrupt the uniformity of the system [25]. However, LMWC cannot be loaded into nanoparticles with smaller size due to its limitation to entangle to the structures of the system [26]. Therefore, maintaining the particle size of the chitosan is crucial in the development of a chitosan nanoparticle.

1.3 Modification of chitosan as biomaterial

Modified chitosan shows greater advantages as compared to unmodified chitosan. The modification of chitosan either chemically or physically may improve its solubility, properties of gelling, and biocompatibility. This modification can be done through cross-linking or substitution [27]. The presence of the various reactive functional groups in the chitosan structure makes it available in many derivatives with different stability properties.

1.3.1 Quaternisation

A quaternary ammonium salt is a hydrophilic group with a permanent positive charge. Therefore, quaternary chitosan does not need an acidic condition to undergo protonation [12, 28]. It allows chitosan quaternary ammonium salts to be soluble in both acidic and basic pH. This is a good approach to increase chitosan solubility in water [28].

The high strength of the positive charge will weaken the hydrogen bond. However, this activity depends on the degree of substitution. The higher the degree of substitution, the higher the water solubility of chitosan. This will improve the quality of chitosan to act as a mucoadhesive agent that aids the penetration into mucus [29].

Trimethyl chitosan (TMC) is an example of a chitosan derivative from quaternisation. This modification is effective in enhancing the bioavailability of antibacterial drugs with antibacterial properties. Moreover, quaternised chitosan also exhibits antibacterial properties by the interaction of its positive charge with the negative charge of Gram-negative bacteria [28–30].

1.3.2 Sulfonated chitosan

Sulfonated chitosan is water-soluble anionic chitosan, which was derived with N-benzyl disulfonated derivative [31]. This modification of chitosan has been shown to be effective, not only as antiviral and antibacterial but also as anticoagulant properties. Sulfonated chitosan interferes with the interaction between the envelope glycoprotein (gp120) and its receptor on the CD4 cells' surface. Therefore, It inhibits the replication of HIV [32].

Sulfonated chitosan has been developed to carry anticoagulant drugs such as reviparin and enhance the anticoagulant activity. Sulfonated chitosan nanoparticles interact with factor Xa and inhibit their function in the blood clotting mechanism [31, 33]. Low molecular weight sulfoethyl chitosan acts as capping of nanoparticles [33]. A capping agent is needed in the nanoparticulate system to prevent agglomeration.

Amphotericin B is used to treat fungal infection by binding to ergosterol on the cell membrane of fungal. It depolarises the membrane and alters its permeability [34]. Sulfonated chitosan has been used in the formulation of amphotericin B to reduce the side effects of the drug by making sure the drug specifically targets the ergosterol of fungal [35].

1.3.3 Thiolation

The structure of mucin that coats the intestinal epithelial cell contains the cysteine-rich domain. This domain easily forms a disulfide bond with a thiolated derivative of chitosan. The bond formations increase the residence time for the chitosan to the mucus and increase the mucoadhesive property of the chitosan [30, 36]. When chitosan covalently bonded with any thiolated moiety, water-soluble carbodiimide is required as a cross-linker [30]. Carbodiimide increases the number of thiol groups. This enhances the immobilisation phenomena, which is the formation of disulfide bonds due to the activation of carboxylic groups.

The thiolation of TMC with the conjugation of cysteine residue increases the strength of covalent bonding between mucin. The covalent bond formation of chitosan with thioglycolic acid (TGA) is an effective carrier in delivering trimethoprim for urinary tract infection [37]. The preparation of chitosan-TGA nanoparticles should be stabilised by covalent cross-linking with polyanion, such as tripolyphosphate [38]. The cross-linking minimise the risk of particle aggregation, increases the disulfide bond, and strengthens the mechanical force between networks, which allows trimethoprim to be released slowly [37].

The conjugation of chitosan with glutathione will protect peptide drugs from aminopeptidase in the GIT [39]. Glutathione has thiol groups which exhibit strong electron-donating properties. It forms an α-peptide bond with cysteine moiety of aminopeptidase [30]. Glutathione also acts as an antioxidant which reduces oxidative stress and increases the adhesion of formulation to the cell [30].

1.3.4 Carboxy alkyl chitosan

The poor water solubility of chitosan makes them less effective as a permeation enhancer. The addition of the carboxyalkyl group will transform the molecule into amphoteric in nature and allow them to react in both basic and acidic conditions [40]. The interaction between the carboxyl group and the primary amino group of chitosan exhibits a promising approach in developing controlled drug release.

The Schiff base reactive gives rise to the formation of the N-carboxymethyl derivative of chitosan [41]. This modification of chitosan has been shown to improve the absorption of low molecular weight heparin (LMWH). After coating into polydopamine, the conjugation of LMWH with carboxymethyl chitosan into polyurethane substrate shows excellent hemocompatibility of heparin. This modification enhances the bioavailability and improves the anticoagulant effect of heparin [42].

Propranolol hydrochloride has a short half-life and requires every 6 to 8 hours in a divided daily dose. The use of carboxymethyl chitosan will coat the drug with a polymer matrix. It controls the release of drug with zero-order kinetics, allowing the constant amount of drug will be eliminated per unit time. The hydration of carboxymethyl chitosan will form a gel layered around the drug, which essential in drug release [43, 44].

1.4 Release mechanisms of chitosan nanoparticle

The release of drugs from the chitosan nanoparticle is influenced by the hydrophilicity of chitosan and pH of the swelling solution. Chitosan release mechanism involves swelling, diffusion of drugs through the polymeric matrix and polymer erosion [45]. Due to the hydrophilicity of chitosan, chitosan nanoparticles exhibit pH-dependent drug and controlled drug release system [6].

Acid and base act as catalysts in the degradation of polymers [46]. Therefore, the behaviour of swelling and the amount of drug released is highly dependent on the pH of the swelling solution. Hence, a modified drug release can be achieved [46]. When polymers get into contact with an aqueous medium, the water will diffuse into the polymer until the polymer swells (**Figure 3**).

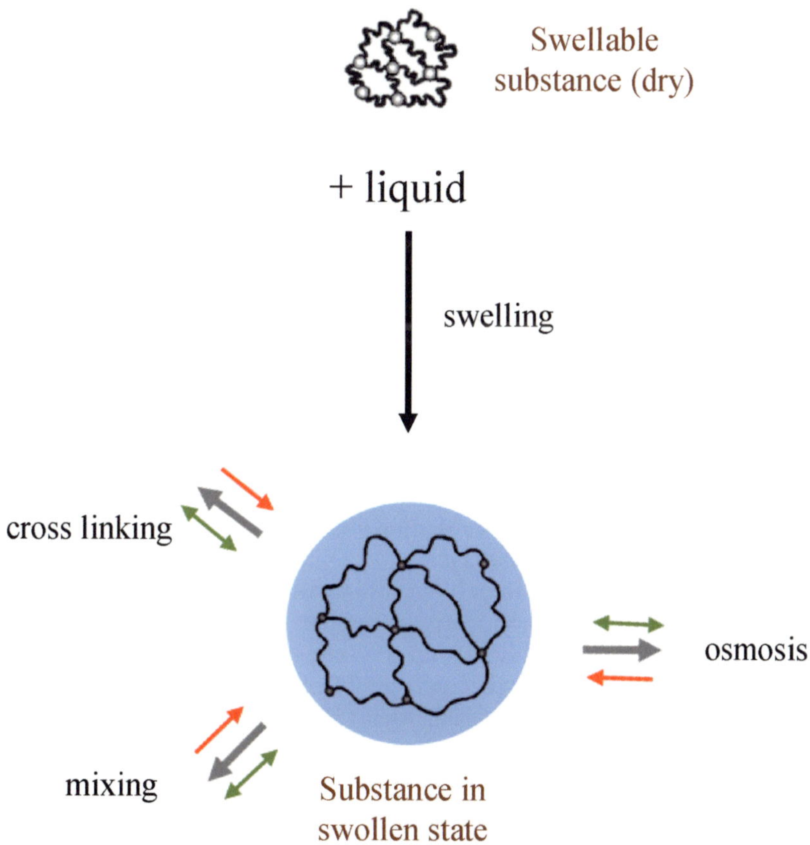

Figure 3.
The swelling of swellablepolymers in aqueous medium (adapted from [49]).

The polymeric chitosan chain will start to detangle. The swelling polymer will form pores which allow drugs to diffuse out of the nanoparticulate system [6, 43, 47]. Therefore, the water solubility of chitosan is crucial in the mechanism of drug release from the nanoparticulate system.

1.5 The use of chitosan to improve drug delivery system

Oral drug administration is the most convenient route, especially among the elderly and children. Unfortunately, some drugs and vaccines cannot withstand the physiological barrier of GIT. In the presence of mucus, proteolytic enzymes, and first-pass metabolism by the liver, drugs tend to be degraded or converted into inactive metabolites [48]. Some drugs will be excreted in the urine lead to low bioavailability.

Due to the challenges aforesaid, chitosan and its derivatives have been used in the development of nanotechnology to improve oral drug delivery [25, 30]. It encapsulates drugs to protect them from degradation in the GIT environment. As a consequence of its excellent biodegradable, biocompatibility, and non-toxic properties, chitosan promotes a stimuli-responsive release of drugs. It allows active ingredients to be released from the formulation in a controlled manner, specifically in enteric-coated drugs [38, 43].

Due to its antimicrobial properties, it was used in the delivery of oral antibiotics to eradicate Gram-negative bacteria such as *E. coli* [49]. This approach not only improves the bioavailability of antibiotic in the body but also indirectly enhance the effectiveness of the drugs in eradicating the infection [15].

2. The properties of protein and peptide

A peptide is made up of short polymers of α-amino acid, which is around 20 to 50 amino acids. The function of small peptides depends on the functional group of various amino acids. Examples of active peptides are glutathione, bradykinin, angiotensin, vasopressin and oxytocin [4].

Protein is a macromolecular and high molecular weight polypeptide, which is made up of long-chain amino acids (more than 50) arranged in a linear chain through peptide bonds [50]. It can exist in four different structural conformations such as primary, secondary, tertiary, and quaternary. The formation of these structures is dependent on the intermolecular interaction between functional groups of amino acids [51], through covalent bonds or non-covalent bonds.

The covalent bonds are strong bonds which include peptide bonds and disulfide bonds [51]. Peptide bonds are interactions between two consecutive amino acids through amino and carboxyl groups. Meanwhile, disulfide bonds link two cysteine residues through sulphhydryl groups [52].

On the other hand, non-covalent bonds are weak bonds that include hydrogen, electrostatic and hydrophobic bonds. Hydrogen bonds link two different peptides with the hydrogen atom of the N-H group and oxygen of the carboxylic group. Hydrophobic bonds will occur if the hydrophobic nature between non-polar side chains of amino acid interacts with each other [51].

3. Oral peptide delivery

The physiological barriers in the GIT responsible for protecting the body from the entry of pathogens. These barriers may reduce the bioavailability of the protein. The barriers aforesaid include biochemical, cellular, and mucus barriers (**Figure 4**) [53].

The entire GIT has been coated with mucus. Mucus also promotes a physical barrier between the lining of epithelial and lumen [54]. It contains mucin protein which secretes proteolytic enzymes and traps peptide drugs through electrostatic interaction [55].

The epithelium of the GIT consists of an intestinal epithelial stem and microfold cells (M-cell) [48]. These cells are responsible for controlling protein uptake from the gut lumen into the bloodstream. Since protein drug is a macromolecule, the presence of protein complexes between adjacent epithelial cells prevents paracellular transport of drug [56]. Meanwhile, transcellular transport is limited only to highly lipophilic molecules, unless the transportation is mediated by P-gp [57].

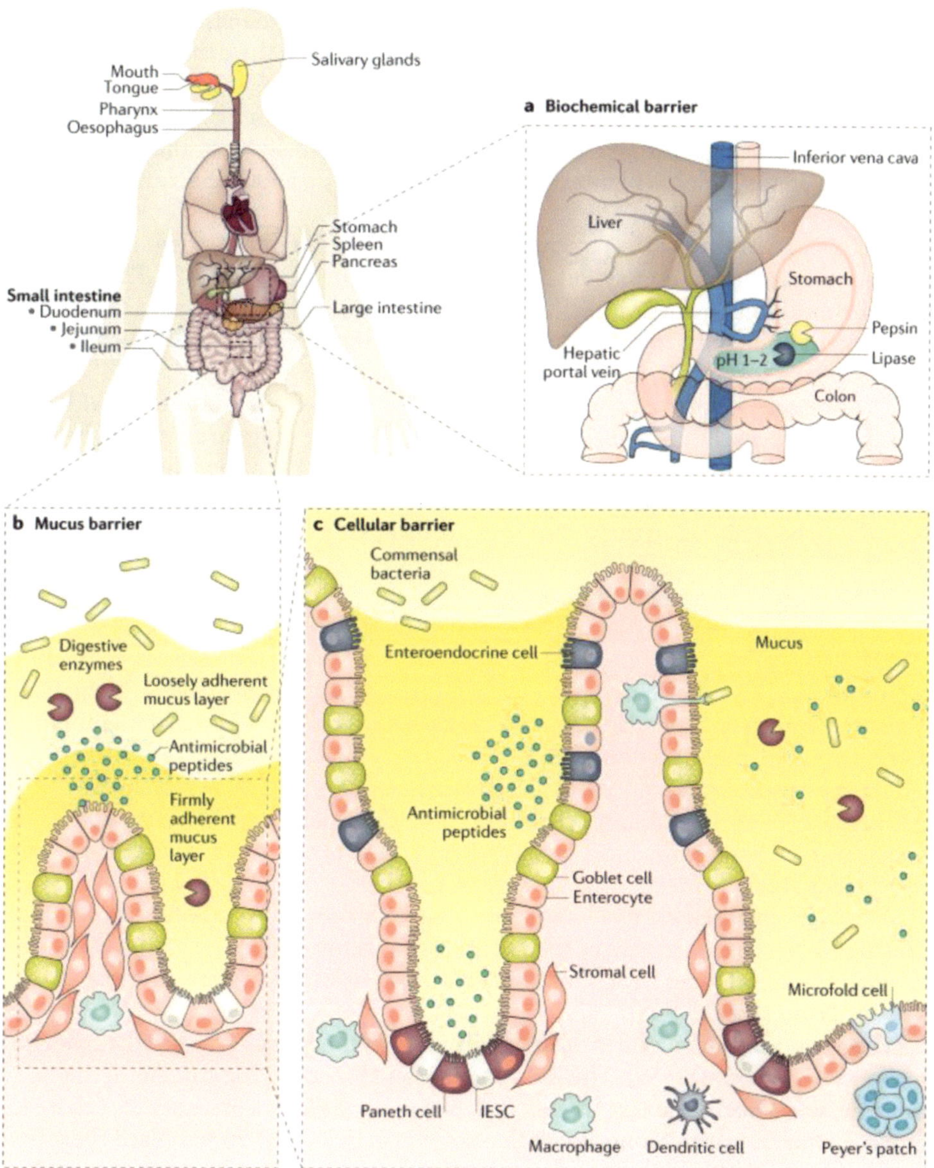

Figure 4.
Physiological barriers to oral protein and peptide delivery (adapted from [53]).

Due to the physical and chemical instability of protein in the GIT, they would not achieve an acceptable therapeutic bioavailability. The nature of the GIT as great physiology to digest food will be a barrier for protein drugs to penetrate through the membrane. The challenges and strategies to improve the protein drug delivery through oral administration need to be considered to ensure the drug achieves adequate therapeutic concentration in the body [58].

3.1 Challenges in oral peptide delivery

3.1.1 The presence of proteolytic enzyme and pH of GIT

GIT is the hollow organs include the oral cavity, stomach, small intestine, large intestine, and colon. Each part of the GIT is varying in pH. However, most proteins are stable at neutral pH and tend to undergo protein denaturation at the extreme changes in pH [48]. A building block of protein is sensitive to pH. In the presence of hydrochloric acid in the stomach, hydrolysis will occur and the disulphide bonds of peptide will be reduced.

Acidic pH environment activates the conversion of pepsinogen into pepsin [59], by transferring hydrogen ion (H+). Pepsin is responsible for the breakdown of peptide bonds which will interfere with the structure and stability of the peptide [59].

In the small intestine, the pancreatic juice is secreted in the duodenum. This juice contains pancreatic enzymes and bicarbonate ions (HCO_3^-) [48]. The pancreatic enzymes consist of amylase, lipase, and protease, which are responsible for the digestion of lipid and peptide. Protease catalyses the proteolysis rate, which cleaves peptide bonds through hydrolysis [4, 58].

The presence of chymotrypsin and peptidase in the jejunum interferes with peptide absorption in the epithelial membrane [48]. Peptide drugs are digested before it reaches the membrane and the fraction of the undigested peptide will reduce. This physiological function will lower the possibility for the therapeutic concentration of peptides to be achieved in the systemic circulation [58].

3.1.2 The intestinal barrier to drug absorption

The layers of the epithelial cell of the intestine are covered by mucus or mucin glycoproteins [55]. These glycoproteins will form a gel layer that covers the surface of the intestinal cell. The diffusion rate for the peptide to the epithelial membrane is restricted in the presence of mucus [54].

One of the intestinal mucosal epithelial cells is the goblet cell. Goblet cells are responsible for secreting mucin 2 in the intestine and Mucin-5b in the colon. Lubricate the passage for chime is the main function of the mucus layer, protecting the epithelium from mechanical damage of GIT [54].

The overexpression of mucin will interfere with the pharmacokinetics of drugs. The higher the concentration of the mucin, the lower the ability of a drug to diffuse through the membrane. Drugs and mucin interact through hydrophobic and Van der Waals interaction.

Large amounts of enzymes present within the mucus layer increase the tendency to digest peptide drugs [58]. The ionic strength, pH, and chyme content in the intestine will affect the charge density in the mucin [54]. The presence of a charged group on mucin interact ionically with charged particles and immobilised them in the mucus. The immobilisation of peptides leads to the clearing from the tract when the layer of mucus is shed [60].

3.1.3 Tight junctions between adjacent epithelial cells

Tight junctions are protein complexes that existed within the adjacent epithelial cells [48]. It prevents leakage and restricts the flux of substances through the paracellular pathway [61]. They consist of transmembrane protein with extracellular domains called Claudin 4 (CLDN4). CLDN4 protects the paracellular physiological function of GIT [62].

The linkage of CLDN4 domains with zonula occludens 1 will result in the connection of cytoskeleton components through linker protein [62]. The components include actin, myosin and microtubules which involve in the contraction of muscle, upon phosphorylation. The contraction leads to cellular tightness, hence, reduce the permeability of substance into the cell.

3.2 Strategies to improve oral peptide delivery system using chitosan

The effective delivery of oral peptide drugs can be achieved by altering the formulation for maximum solubility, avoid enzymatic degradation and enhance the absorption of drugs through the intestinal epithelial cell [63]. For the sake of preventing enzymatic degradation or inactivation, the addition of enzyme or protease inhibitor is a great approach (**Figure 5**). Proteolytic enzymes are responsible for cleaving protein molecules into an inactive amino acid chain. Protease inhibitors such as aprotinin and chromostatin can be used to prevent the inactivation of protein drugs [7].

As discussed earlier, protein molecules show poor permeability through various mucosal surfaces and biological membranes. The improvement of membrane permeability can be achieved by the inclusion of a permeation enhancer into the formulation. Permeation enhancers are either tight junction selective or membrane perturbing [61].

Chitosan and its derivatives have been used as an enzyme inhibitor, permeation enhancer and mucoadhesive agent [30]. With the different mechanism, modification, and preparation technique, this polymer also involves in the encapsulation of peptide drugs into the nanoparticulate system [60], which protect them from harsh GIT environment.

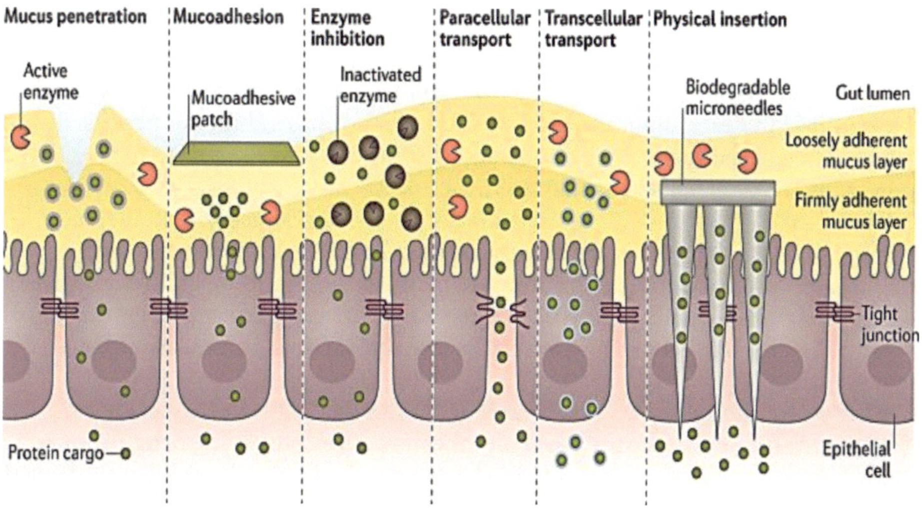

Figure 5.
Strategies to improve oral peptide delivery system (adapted from [53]).

3.2.1 Enzyme inhibition

There are two types of protease, including serine protease and Zn^{2+}-dependent protease. Serine proteases, such as trypsin, chymotrypsin and elastase are pancreatic digestive enzymes. Meanwhile, Zn^{2+}-dependent protease such as matrix metalloproteinase is an insulin-degrading enzyme [64]. Some enzyme inhibitors need chitosan to enhance their anti-protease activity and minimise peptide drug degradation [64]. For example, chymostatin is a protease inhibitor selectively to chymotrypsin-like serine proteases. This inhibitor will covalently be linked to the amino group of chitosan.

The active site of matrix metalloproteinase, such as carboxypeptidase, contains Zn^{2+} binding motif. It requires Zn^{2+} to promote nucleophilic attack by water. This protease cleaves membrane-bound pre-proteins of the cell to release cytokine. The ethylenediaminetetraacetic acid (EDTA) is a complexing agent that is capable of forming a complex with Zn^{2+} and retard the nucleophilic attack of water on carboxypeptidase. To inhibit the Zn^{2+}-dependent protease, the EDTA is covalently bound to the primary amino groups of the chitosan-inhibitor conjugate [27, 65].

Moreover, the study showed that the effect of trypsin inhibitors would be disrupted after the gastric phase. Therefore, the encapsulation of the peptide drug and trypsin inhibitor with chitosan-EDTA conjugates improve the controlled release of the molecules.

3.2.2 Chitosan as a mucoadhesive agent

Chitosan derivatives improve the permeation of water-soluble drug molecules due to their ability to adhere to the mucus [30]. Thiolated chitosan shows a greater effect in improving drug permeation through the cell membrane. With the formation of disulfide bridges, the thiol group of chitosan interacts with the cysteine-rich subdomains of mucus and allows greater mucoadhesion. Thus, the absorption of the peptide drug molecule increases with residence time [30].

It is worth noting that chitosan with a low degree of acetylation and high molecular weight leads to high charge density. The higher positive charge density of chitosan will bind to negatively charged tight junction channels. Ion displacement occurs, leading to intracellular spaces loosening [16].

The integrity and permeability of tight junctions can be illustrated with transepithelial electrical resistance (TER). The ability of TMC [66] and carboxymethyl chitosan [67] in decreasing the TER will increase the permeability of peptide drugs. TMC has been used to formulate buserelin, a synthetic peptide analogue for LHRH agonist, by the oral delivery system [68].

Thiolated chitosan, such as glutathione, cysteine and N-acetylcysteine, have strong mucoadhesive properties due to covalent bonding with cysteine-rich subdomains of the mucus glycoprotein. For chitosan-glutathione, this derivative improves chitosan stability, enhanced mucoadhesion and permeation enhancing effect. This system has been applied to the oral delivery of immunostimulant drug, thymopentin [30].

Chitosan-cysteine shows similar mucoadhesive but improved cohesion as compared to unmodified chitosan. The cohesiveness of polymeric drug formulation is crucial to ensure the stability of the drug and will be released in a controlled manner. Furthermore, chitosan-N-acetylcysteine produces a longer retention time than unmodified chitosan. However, no drugs have been tested yet for these two chitosan derivatives [30].

3.2.3 Encapsulation of peptide into the nanocarrier

By encapsulating peptides into a nanoparticulate system, enzymolysis and peptide aggregation can be avoided. This approach enhances the absorption of peptide drugs through the transmembrane of the small intestinal epithelium [69]. Nanoparticle will provide controlled-release properties in the presence of chitosan as a polymer. This condition will reduce repetitive dose administration and improve drug bioavailability [43]. In the presence of chitosan as a mucoadhesive agent, the retention time between formulation and absorption site will be maximised.

Cyclosporine is a cyclic peptide drug used to suppress the immune system, after organ transplantation. Cyclosporine with high molecular weight (1.32 kDa) shows poor bioavailability with low permeability through the biological barrier. Conventional oral cyclosporine has been shown to have an unpredictable low therapeutic concentration in the bloodstream.

Therefore, a nanoparticle drug delivery system is a promising strategy to improve the oral bioavailability of cyclosporine. Chitosan nanoparticles in the presence of tripolyphosphate, as a cross-linker, make it more convenient as compared to conventional ones. The bioavailability of the nanoparticulate cyclosporine increases by 73% [69]. Exendin-4 is a glucagon-like peptide-1 receptor agonist that has been approved to control type 2 diabetes mellitus. This peptide drug has high susceptible to enzymatic degradation [70]. Chitosan-tripolyphosphate conjugated nanoparticle was used to design oral suspension and enteric-coated capsules of exendin-4 to increase the bioavailability of exendin-4 slightly.

3.2.4 Efflux pump inhibition

Efflux pump is a membrane protein located within the cytoplasmic membrane of a cell. It translocates a variety of substrates across extra- and intra-cellular membranes. Multidrug efflux pump can be one of the drug resistance mechanisms, as it pumps foreign substances (or drugs) out of cells. This active process is an ATP-dependent [71].

P-glycoprotein (P-gp) is a transmembrane glycoprotein and the best example of a multidrug efflux pump. It is expressed and located in the intestinal epithelium, liver cells and proximal tubule cells of the kidney. P-gp is also located within the blood–brain barrier (BBB), which provides an obstacle for drugs to enter the region. Therefore, it must be difficult for antipsychotic drugs to bypass BBB and exert their effect [57, 71, 72].

Chitosan may enhance drug permeation by opening of tight junctions which is highly related to CLDN4 [73]. Chitosan will modulate CLDN4 protein redistribution to the cytosol and disrupt tight junctions [62]. This phenomenon will enhance paracellular permeability and reduce TER. Thus, declining barrier function of epithelial cells to allow drugs to enter the cell [74].

Furthermore, the use of thiolated chitosan (thiomer) has shown to be useful in bypassing the P-gp. The thiol-moiety of thiolated chitosan may allow the formation of disulfide bonds between the cysteine groups of the P-gp [9]. The thiomers then enter the channels of the P-gp pump together with the therapeutic agent, which obstruct the function of the multidrug efflux pump [57].

For infectious disease by Gram-negative bacteria, chitosan plays an important role in facilitating effective delivery of antimicrobials to the infection site [15, 49]. Chitosan will encapsulate drugs and carry them into bacterial cells by attraction forces between polycationic chitosan and negatively charged bacteria [49]. This action avoids the efflux pump at the cell membrane of the bacterial cell.

4. Oral vaccine delivery system

Vaccination is one of the most cost-effective approaches to prevent infectious diseases such as hepatitis B, tetanus, polio, and rabies. Vaccines contain pathogens, either live-attenuated, inactive or killed antigen [75]. These pathogens will be administered in the body and recognised by the immune system.

The oral delivery of vaccines is quite challenging as the pathogen is introduced into the body. It is mandatory to ensure mucosal immune response works effectively to protect the body against the pathogen and their toxin [54].

As the vaccine enters the intestine, its presence will trigger the inductive site, the Peyer's patches. The Peyer's patches consist of M-cell which will allow the entry of the antigen through endocytosis. The antigen then will be transported into intraepithelial dendritic cells or macrophages and be taken up by the cell through phagocytosis [76].

The antigen-loaded dendritic cell will present the antigen fragment on its surface and triggers the activation of naive $CD4^+$ T-cells. The activated $CD4^+$ T-cells will bind to the antigen fragment, MHC class II. This binding releases chemical mediators, interleukin-2 (IL-2), that function to regulate the activity of lymphocytes for immunity. IL-2 stimulates the cell division of CD4+ T-cells, activates B-cells and cytotoxic T-cells. B-cell is responsible for mediating humoral immunity by differentiating into plasma cells. Plasma cells will generate antibodies to fight against pathogens [77].

5. Challenges in oral vaccine delivery system

Viral protein requires the right structural conformation to attach to the host cell and replicate. Highly acidic in the stomach and extreme temperature changes will cause protein denaturation. The denaturation of the virus will alter the conformation of its structure [58]. The high temperature will break the phosphodiester bond. However, at low temperature, the degradation of the nucleic acid will also lead to viral inactivation [78].

Furthermore, to transport vaccines orally, it should be able to overcome the biological barrier of the intestinal epithelial cell such as tight junction and mucus. The hydrophilic antigen cannot cross the phospholipid bilayer to enter systemic circulation due to the function of tight junction in controlling the permeability of the membranes. Therefore, the uptake of the antigen to mucosal tissue is limited with a short time of exposure [78].

The GIT contains normal flora or microbiota which help in maintaining the structure of the gut mucosal barrier [55]. Those microbiotas not only aid nutrient metabolism, but they also possess an action to protect against invading pathogens [79, 80]. Well-balance microbiota is needed to induce the effectiveness of vaccines through oral administration. The delivery of the vaccine will be interrupted in patients with microbiota dysbiosis, leading to blunted vaccine response [80].

The induction of danger signals appropriately by the vaccine is essential to trigger an immune response [81]. Due to these limitations, there is a problem in inducing an adequate immune response against administered pathogens [3]. Consequently, a higher and repetitive dose is required. Nevertheless, the administration of high antigen doses repetitively may develop systemic oral tolerance towards vaccines [3, 78].

6. Strategies to improve oral vaccine delivery system using chitosan

The uptake of antigen by immune cells depends on the particle size of the antigen. The smaller size of pathogens is readily taken up by dendritic cells [75]. The same goes for peptide, encapsulation of pathogens in nanoparticles is a good approach to improve the effectiveness of vaccines in stimulating the immune system [75].

Small particle size is required to penetrate the mucus. The formulation will be excluded out from the layer of mucus if particle size greater than normal mucus pore size (100–500 nm). This leads to the interruption in the bioavailability of antigen to targeted antigen-presenting cell (APC). Therefore, the development of nanoparticulate systems is required to provide a smaller size (20 – 40 nm) of formulation [82].

Medium molecular weight chitosan (MMWC) with the degree of deacetylation of 85% has been shown to improve the delivery of ovalbumin antigen with the presence of alginate and calcium phosphate (CaP). CaP has adjuvant properties by activating the surface expression of B-cell. CaP can be coated with mucosal penetrating polymers, such as chitosan and alginate to avoid biodegradation by enzymes present in the GIT [82]. In the stomach, the alginate-chitosan-coated CaP nanoparticle delays the release of ovalbumin antigen. The antigen then will be released in the intestine and colon with a sustained-release mechanism. This nanocarrier has successfully encapsulated ovalbumin antigen with small size (< 50 nm) [82].

The antigen should be transported to the intestine and directly to the M-cell of the Peyer's patches [6]. Chitosan develops well-protected mucoadhesion by prolonging the residence time at mucosal surfaces. The uptake of antigen by epithelial cells of the intestine will be improved by chitosan. An increase in the activity of macrophages will improve the secretion of mucosal IgA and IgG [76].

7. Conclusions

Chitosan-based drug formulation has gained attention for their ability to serve as a carrier and an enhancer for oral delivery of peptides and vaccines. Although oral delivery is the most convenient and preferred route of administration, however, it has limitations due to the presence of the proteolytic enzyme, pH of GIT and the intestinal barrier to drug absorption. In recent years, there has been considerable research interest in the application of chitosan as an enzyme inhibitor, mucoadhesive agent and efflux pump inhibitor. Interaction of positively-charged amino groups of chitosan with negatively-charged sialic acid groups that exist in mucin prolongs the residence time between drugs and membranes, therefore enhancing the bioavailability of the drugs. Other formulation strategies include encapsulation of proteins, peptides and vaccines into a nanoparticulate delivery system. By encapsulating peptide into a nanocarrier system, the enzymolysis and peptides aggregation can be avoided thus enhances the absorption of peptide drugs in the intestinal epithelium. Similarly, encapsulation of pathogens in nanoparticles is a good approach to improve the effectiveness of vaccines in stimulating the immune system.

Acknowledgements

The authors acknowledge Faculty of Pharmacy, Universiti Teknologi MARA and internal grant (DUCS) 2.0, 600-UiTMSEL (PI. 5/4) (016/2020).

Conflict of interest

The authors declare no conflict of interest.

Appendices and nomenclature

kDa	Kilodalton
nm	Nano-metre
GIT	Gastrointestinal tract
G	Gibbs free energy
PDI	Polydispersity index
LMWC	Low molecular weight chitosan
HMWC	High molecular weight chitosan
MMWC	Medium molecular weight chitosan
TMC	Trimethyl chitosan
TER	Transcellular electrical resistance
TGA	Thioglycolic acid
LMWH	Low molecular weight heparin
P-gp	P-glycoprotein
CLDN4	Claudin 4
EDTA	Ethylenediaminetetraacetic acid
IL-2	Interleukin-2
APC	Antigen-presenting cell
CaP	Calcium phosphate
PEG	Polyethylene glycol
PGC	Pegylated chitosan
HMPCP	Hydroxypropyl methylcellulose phthalate

Author details

Siti Zuhairah Zainuddin and Khuriah Abdul Hamid*
Department of Pharmaceutics, Faculty of Pharmacy, Universiti Teknologi Mara, Cawangan Selangor, Puncak Alam, Malaysia

*Address all correspondence to: khuriah@uitm.edu.my

References

[1] Jin, J. F., Zhu, L. L., Chen, M., Xu, H. M., Wang, H. F., Feng, X. Q., Zhu, X. P., & Zhou Q. The optimal choice of medication administration route regarding intravenous, intramuscular, and subcutaneous injection. Patient Prefer adherence. 2015;9:923-942.

[2] Wang, B., Xie, N., & Li B. Influence of peptide characteristics on their stability, intestinal transport, and in vitro bioavailability: A review. J Food Biochem. 2018;e12571.

[3] Vela Ramirez, J. E., Sharpe, L. A., & Peppas NA. Current state and challenges in developing oral vaccines. Adv Drug Deliv Rev. 2017;114:116-131.

[4] Cao, S., Xu, S., Wang, H., Ling, Y., Dong, J., Xia, R., & Sun X. Nanoparticles: Oral Delivery for Protein and Peptide Drugs. AAPS PharmSciTech. 2019;20(5).

[5] David L, María CM-O. Chitosan: Strategies to Increase and Modulate Drug Release Rate. Biological Activities and Application of Marine Polysaccharides. 2017.

[6] Mohammed, M. A., Syeda, J., Wasan, K. M., & Wasan EK. An Overview of Chitosan Nanoparticles and Its Application in Non-Parenteral Drug Delivery. Pharmaceutics. 2017;9(4):53.

[7] Muheem, A., Shakeel, F., Jahangir, M. A., Anwar, M., Mallick, N., Jain, G. K., Ahmad FJ. A review on the strategies for oral delivery of proteins and peptides and their clinical perspectives. Saudi Pharm J. 2016;24(4):413-428.

[8] Bajracharya, R., Song, J. G., Back, S. Y., & Han HK. Recent Advancements in Non-Invasive Formulations for Protein Drug Delivery. Comput Struct Biotechnol J. 2019;17:1290-1308.

[9] Brasselet, C., Pierre, G., Dubessay, P., Dols-Lafargue, M., Coulon, J., Maupeu J. Modification of Chitosan for the Generation of Functional Derivatives. Appl Sci. 2019;9(7):1321.

[10] Elieh-Ali-Komi, D., & Hamblin MR. Chitin and Chitosan: Production and Application of Versatile Biomedical Nanomaterials. Int J Adv Res. 2016;4(3):411-427.

[11] Cheung, R. C., Ng, T. B., Wong, J. H., & Chan WY. Chitosan: An Update on Potential Biomedical and Pharmaceutical Applications. Mar Drugs. 2015;13(8):5156-5186.

[12] Wu, Y., Rashidpour, A., Almajano, M. P., & Metón I. Chitosan-Based Drug Delivery System: Applications in Fish Biotechnology. Polymers (Basel). 2020;12(5):1177.

[13] Roy, J. C., Salaün, F., Giraud, S., Ferri, A., Chen, G., & Guan J. Solubility of Chitin: Solvents, Solution Behaviors and Their Related Mechanisms. In: Solubility of Polysaccharides. 2017. p. 110-22.

[14] Sabu Thomas, Anitha Pius SG. Volume 1: Preparation and Properties Elsevier. In: Handbook of Chitin and Chitosan. 2020. p. 37.

[15] Radwan-Pragłowska, J., Piątkowski, M., Deineka, V., Janus, Ł., Korniienko, V., Husak, E., Holubnycha, V., Liubchak, I., Zhurba, V., Sierakowska, A., Pogorielov, M., & Bogdał D. Chitosan-Based Bioactive Hemostatic Agents with Antibacterial Properties-Synthesis and Characterization. Molecules. 2019;24(14):2629.

[16] Foster, L. J., Ho, S., Hook, J., Basuki, M., & Marçal H. Chitosan as a Biomaterial: Influence of Degree of Deacetylation on Its Physiochemical, Material and Biological Properties. PLoS One. 2015;10(8):e0135153.

[17] Sukhadeorao Dongre R. Chitosan Formulations: Chemistry, Characteristics and Contextual Adsorption in Unambiguous Modernization of S&T. Hysteresis of Composites. 2019.

[18] Szymańska, E., & Winnicka K. Stability of chitosan-a challenge for pharmaceutical and biomedical applications. Mar Drugs. 2015;13(4):1819-1846.

[19] Nguyen NT, Hoang DQ, Nguyen ND, Nguyen QH, Nguyen DH. Preparation, characterization, and antioxidant activity of water-soluble oligochitosan. Green Process Synth. 2017;6(5):461-468.

[20] Younes, I., & Rinaudo M. Chitin and chitosan preparation from marine sources. Structure, properties and applications. Mar Drugs. 2015;13(3):1133-1174.

[21] Jampafuang, Y., Tongta, A., & Waiprib Y. Impact of Crystalline Structural Differences Between α- and β-Chitosan on Their Nanoparticle Formation Via Ionic Gelation and Superoxide Radical Scavenging Activities. Polymers (Basel). 2019;11(12):2010.

[22] Ataide, J. A., Gérios, E. F., Cefali, L. C., Fernandes, A. R., Teixeira, M., Ferreira, N. R., Tambourgi, E. B., Jozala, A. F., Chaud, M. V., Oliveira-Nascimento, L., Mazzola, P. G., & Souto EB. Effect of Polysaccharide Sources on the Physicochemical Properties of Bromelain-Chitosan Nanoparticles. Polymers (Basel). 2019;11(10):1681.

[23] Masarudin, M. J., Cutts, S. M., Evison, B. J., Phillips, D. R., & Pigram PJ. Factors determining the stability, size distribution, and cellular accumulation of small, monodisperse chitosan nanoparticles as candidate vectors for anticancer drug delivery: application to the passive encapsulation of [(14)C]-doxorubicin. Nanotechnol Sci Appl. 2015;8:67-80.

[24] Qinna, N. A., Karwi, Q. G., Al-Jbour, N., Al-Remawi, M. A., Alhussainy, T. M., Al-So'ud, K. A., Al Omari, M. M., & Badwan AA. Influence of molecular weight and degree of deacetylation of low molecular weight chitosan on the bioactivity of oral insulin preparations. Mar Drugs. 2015;13(4):1710-1725.

[25] Garg, U., Chauhan, S., Nagaich, U., & Jain N. Current Advances in Chitosan Nanoparticles Based Drug Delivery and Targeting. Adv Pharm Bull. 2019;9(2):195-204.

[26] Alhajj, N., Zakaria, Z., Naharudin, I., Ahsan, F., Li, W., & Wong TW. Critical physicochemical attributes of chitosan nanoparticles admixed lactose-PEG 3000 microparticles in pulmonary inhalation. Asian J Pharm Sci. 2019;16(55):6.

[27] Kyzas, G., & Bikiaris D. Recent Modifications of Chitosan for Adsorption Applications: A Critical and Systematic Review. Mar Drugs. 2015;13(1).

[28] Liu, P., Meng, W., Wang, S., Sun, Y., & Aqeel Ashraf M. Quaternary ammonium salt of chitosan: preparation and antimicrobial property for paper. Open Med. 2015;10(1):473-478.

[29] Martins, A., Facchi, S., Follmann, H., Pereira, A., Rubira, A., & Muniz E. Antimicrobial Activity of Chitosan Derivatives Containing N-Quaternized Moieties in Its Backbone: A Review. Int J Mol Sci. 2014;15(11):20800-20832.

[30] M Ways, T. M., Lau, W. M., & Khutoryanskiy V V. Chitosan and Its Derivatives for Application in Mucoadhesive Drug Delivery Systems. Polymers (Basel). 2018;10(3):267.

[31] Sabar, S., Aziz, H. A., Yusof, N. H., Subramaniam, S., Foo, K. Y., Wilson, L. D., & Lee HK. Preparation of sulfonated chitosan for enhanced adsorption of methylene blue from aqueous solution. React Funct Polym. 2020;104584.

[32] PYRC K, MILEWSKA A, NOWAKOWSKA M, SZCZUBIALKA K, KAMINSKI K. The Use of Chitosan Polymer in the Treatment and Prevention Of Infections Caused by Coronaviruses. 2013. p. 2.

[33] Heise, K., Hobisch, M., Sacarescu, L., Maver, U., Hobisch, J., Reichelt, T., ... Spirk S. Low-molecular-weight sulfonated chitosan as template for anticoagulant nanoparticles. Int J Nanomedicine. 2018;13:4881-4894.

[34] Stone, N. R., Bicanic, T., Salim, R., & Hope W. Liposomal Amphotericin B (AmBisome(®)): A Review of the Pharmacokinetics, Pharmacodynamics, Clinical Experience and Future Directions. Drugs. 2016;76(4):485-500.

[35] Karimi, Soroush, Moradipour, Pouran, Azandaryani, Abbas Hemati, & Arkan E. Amphotericin-B and vancomycin-loaded chitosan nanofiber for antifungal and antibacterial application. Brazilian J Pharm Sci. 2019;55:e17115.

[36] Díez-Pascual, A. M., & Díez-Vicente AL. Wound Healing Bionanocomposites Based on Castor Oil Polymeric Films Reinforced with Chitosan-Modified ZnO Nanoparticles. Biomacromolecules. 2015;16(9):2631-2644.

[37] Şenyiğit, Z. A., Karavana, S. Y., İlem-Özdemir, D., Çalışkan, Ç., Waldner, C., Şen, S., Bernkop-Schnürch, A., & Baloğlu E. Design and evaluation of an intravesical delivery system for superficial bladder cancer: preparation of gemcitabine HCl-loaded chitosan-thioglycolic acid nanoparticles and comparison of chitosan/poloxamer gels as carriers. Int J Nanomedicine. 2015;10:6493-6507.

[38] Pant, Anjali; Negi JS. Novel controlled ionic gelation strategy for chitosan nanoparticles preparation using TPP-β-CD inclusion complex. Eur J Pharm Sci. 2018;112:180-185.

[39] Zainab A. Enhancing Oral Delivery of Glutathione Using Chitosan Nanoparticles. 2013.

[40] Li, J., Cai, C., Li, J., Li, J., Li, J., Sun, T., Wang, L., Wu, H., & Yu G. Chitosan-Based Nanomaterials for Drug Delivery. Molecules. 2018;23(10):2661.

[41] Wang, Y., Huang, X., He, C., Li, Y., Zhao, W., & Zhao C. Design of carboxymethyl chitosan-based heparin-mimicking cross-linked beads for safe and efficient blood purification. Int J Biol Macromol. 2018;117:392-400.

[42] Yingying Jin, Zhongqiang Zhu, Lin Liang, Kaiyue Lan QZ, Yuqin Wang, Yishun Guo, Kangning Zhu, Rashid Mehmood BW. A Facile Heparin/Carboxymethyl Chitosan Coating Mediated by Polydopamine on Implants for Hemocompatibility and Antibacterial Properties. Appl Surf Sci. 2020;146539.

[43] Li, J., & Mooney DJ. Designing hydrogels for controlled drug delivery. Nat Rev Mater. 2016;1(12):16071.

[44] Hernawan, Hayati, S. N., Nisa, K., Indrianingsih, A. W., Darsih, C., & Kismurtono M. Development and in vitro evaluation of carboxymethyl chitosan based drug delivery system for the controlled release of propranolol hydrochloride. IOP Conf Ser Earth Environ Sci. 2017;101:012038.

[45] Younis, M. K., Tareq, A. Z., & Kamal IM. Optimization Of Swelling, Drug Loading And Release From Natural Polymer Hydrogels. IOP Conf Ser Mater Sci Eng. 2018;454:012-017.

[46] Suarato G, Li W, Meng Y. Role of pH-responsiveness in the design of chitosan-based cancer nanotherapeutics: A review. Biointerphases. 2016;11(4):04B201.

[47] Benghanem, S., Chetouani, A., Elkolli, M., Bounekhel, M., & Benachour D. Effects of physical and chemical modification on biological activities of chitosan/carboxymethylcelluse based hydrogels. J Chil Chem Soc. 2017;62(1):3376-3380.

[48] M.Camilleri, M.I.Vazquez Roque. Gastrointestinal Functions. Encycl Neurol Sci (Second Ed). 2014;411-416.

[49] Goy, R. C., Morais, S. T. B., & Assis OBG. Evaluation of the antimicrobial activity of chitosan and its quaternized derivative on E. coli and S. aureus growth. Rev Bras Farmacogn. 2016;26(1):122-127.

[50] Are P, Formed M, Acids BYA. Proteins are macromolecules formed by amino acids. 2017;21-71.

[51] Zhou, H. X., & Pang X. Electrostatic Interactions in Protein Structure, Folding, Binding, and Condensation. Chem Rev. 2018;118(4):1691-1741.

[52] Kuhlman, B., & Bradley P. Advances in protein structure prediction and design. Nat Rev Mol Cell Biol. 2019;20(11):681-697.

[53] Brown, T.D., Whitehead, K.A. & Mitragotri S. Materials for oral delivery of proteins and peptides. Nat Rev Mater 5. 2020;127-148.

[54] Pelaseyed, T., Bergström, J. H., Gustafsson, J. K., Ermund, A., Birchenough, G. M., Schütte, A., van der Post, S., Svensson, F., Rodríguez-Piñeiro, A. M., Nyström, E. E., Wising, C., Johansson, M. E., & Hansson GC. The mucus and mucins of the goblet cells and enterocytes provide the first defense line of the gastrointestinal tract and interact with the immune system. Immunol Rev. 2014;260(1):8-20.

[55] Paone, P., & Cani PD. Mucus barrier, mucins and gut microbiota: the expected slimy partners? Gut Published Online First. 2020;322260.

[56] Joël Richard. Challenges in oral peptide delivery: lessons learnt from the clinic and future prospects. Ther Deliv. 2017;8(8):663-684.

[57] Hoosain, F. G., Choonara, Y. E., Tomar, L. K., Kumar, P., Tyagi, C., du Toit, L. C., & Pillay V. Bypassing P-Glycoprote in Drug Efflux Mechanisms: Possible Applications in Pharmacoresistant Schizophrenia Therapy. Biomed Res Int. 2015;2015:484963.

[58] Richard J. Challenges in oral peptide delivery: lessons learnt from the clinic and future prospects. Ther Deliv. 2017;8(8):663-684.

[59] Bardhan, K. D., Strugala, V., & Dettmar PW. Reflux revisited: advancing the role of pepsin. Int J Otolaryngol. 2014;646901.

[60] P. Lundquist PA. Oral absorption of peptides and nanoparticles across the humanintestine: Opportunities, limitations and studies in human tissues. Adv Drug Deliv Rev. 2016;106:256-276.

[61] Zhang J, Zhu X JY. Mechanism study of cellular uptake and tight junction opening mediated by goblet cell-specific trimethyl chitosan nanoparticles. Mol Pharm. 2014;11:1520-1532.

[62] Xin Cong, Yan Zhang, Jing Li, Mei Mei, Chong Ding, Ruo-Lan Xiang, Li-Wei Zhang, Yun Wang, Li-Ling Wu G-YY. Claudin-4 is required for modulation of paracellular permeability by muscarinic acetylcholine receptor in epithelial cells. J Cell Sci. 2015;128:2271-2286.

[63] Homayun, B., Lin, X., & Choi HJ. Challenges and Recent Progress in Oral Drug Delivery Systems for Biopharmaceuticals. Pharmaceutics. 2019;11(3):129.

[64] DeLano, F. A., Hoyt, D. B., & Schmid-Schönbein GW. Pancreatic digestive enzyme blockade in the intestine increases survival after experimental shock. Sci Transl Med. 2014;5(169):11.

[65] Rajasree R and Rahate KP. An overview on various modifications of Chitosan and it's applications. Int J Pharm Sci Res. 2013;4(11):4175-4193.

[66] Sovány YH-EYIGRJEIHT. Review of recently used techniques and materials to improve the efficiency of orally administered proteins/peptides. DARU J Pharm Sci. 2020;28:403-416.

[67] Onakpoya, I., Hung, S. K., Perry, R., Wider, B., & Ernst E. The Use ofGarciniaExtract (Hydroxycitric Acid) as a Weight loss Supplement: A Systematic Review and Meta-Analysis of Randomised Clinical Trials. J Obes. 2011;1-9.

[68] Wang, W., Meng, Q., Li, Q., Liu, J., Zhou, M., Jin, Z., & Zhao K. Chitosan Derivatives and Their Application in Biomedicine. Int J Mol Sci. 2020;21(2):487.

[69] Mohammed, M., Syeda, J., Wasan, K., & Wasan E. An Overview of Chitosan Nanoparticles and Its Application in Non-Parenteral Drug Delivery. Pharmaceutics. 2017;9(4):53.

[70] Lee, S., & Lee DY. Glucagon-like peptide-1 and glucagon-like peptide-1 receptor agonists in the treatment of type 2 diabetes. Ann Pediatr Endocrinol Metab. 2017;22(1):15-26.

[71] Spengler, G., Kincses, A., Gajdács, M., & Amaral L. New Roads Leading to Old Destinations: Efflux Pumps as Targets to Reverse Multidrug Resistance in Bacteria. Molecules. 2017;22(3):468.

[72] Sanchez-Covarrubias, L., Slosky, L. M., Thompson, B. J., Davis, T. P., & Ronaldson PT. Transporters at CNS barrier sites: obstacles or opportunities for drug delivery? Curr Pharm Des. 2014;20(10):1422-1449.

[73] Jaafar MHM, Hamid KA. Chitosan-Coated Alginate Nanoparticles Enhanced Absorption Profile of Insulin via Oral Administration. Curr Drug Deliv. 2019;16:672-686.

[74] Benediktsdóttir, B. E., Gudjónsson, T., Baldursson, Ó., & Másson M. N-alkylation of highly quaternized chitosan derivatives affects the paracellular permeation enhancement in bronchial epithelia in vitro. . Eur J Pharm Biopharm. 2014;86(1):55-63.

[75] Kim, Mi-Gyeong; Park, Joo Yeon; Shon, Yuna; Kim, Gunwoo; Shim, Gayong; Oh Y-K. Nanotechnology and vaccine development. Asian J Pharm Sci. 2014;9(5):227-235.

[76] Turula H WC. The Role of the Polymeric Immunoglobulin Receptor and Secretory Immunoglobulins during Mucosal Infection and Immunity. Viruses. 2018;10(5):237.

[77] Kato T, Fahrmann JF, Hanash SM VJ. Extracellular Vesicles Mediate B Cell Immune Response and Are a Potential Target for Cancer Therapy. Cells. 2020;9(6):1518.

[78] Chen, X. Y., Butt, A. M., & Amin MCIM. Enhanced paracellular delivery of vaccine by hydrogel microparticles-mediated reversible tight junction opening for effective oral immunization. J Control Release. 2019;311-312:50-64.

[79] Ogunrinola, G. A., Oyewale, J. O., Oshamika, O. O., & Olasehinde GI. The

Human Microbiome and Its Impacts on Health. Int J Microbiol. 2020;2020:1-7.

[80] Jandhyala, S. M., Talukdar, R., Subramanyam, C., Vuyyuru, H., Sasikala, M., & Nageshwar Reddy D. Role of the normal gut microbiota. World J Gastroenterol. 2015;21(29):8787-8803.

[81] Pulendran, B., & Ahmed R. Immunological mechanisms of vaccination. Nat Immunol. 2011;12(6):509-517.

[82] Cao, P., Han, F. Y., Grøndahl, L., Xu, Z. P., & Li L. Enhanced Oral Vaccine Efficacy of Polysaccharide-Coated Calcium Phosphate Nanoparticles. ACS omega. 2020;5(29):18185-18197.

Section 2

Tissues Regeneration

Chapter 4

Chitosan Based Biocomposites for Hard Tissue Engineering

Fouad Dabbarh, Noureddin Elbakali-Kassimi and Mohammed Berrada

Abstract

Bone is the second most transplanted organ, just after blood. It provides structural support, protection for organs and soft tissues. It holds some critical biological processes such as the bone marrow blood forming system. It is responsible for storing and supplying minerals such calcium and phosphate. Bone is a connective tissue formed by two predominant phases: an inorganic phase containing mainly apatitic calcium and phosphate and an organic phase made of fibrous type I collagen. This natural biocomposite has many biological features such osteoconductivity, osteoinductivity, osteogenicity and is subject to a continuous remodeling process through osteoclastic and osteoblastic activities. In biomedical engineering, the restoration of damaged hard tissue with autologous bone is not always possible or even the best option. The development of some safe and low-cost alternatives such as biocomposites that mimic organic and calcified bone materials have shown very good results and offer an alternative to autologous bone implants. However, the mechanical properties of biocomposites still present a big challenge as a hard tissue substitute. This chapter reviews the properties of bone substitute materials chitosan and calcium phosphates, discusses strategies used in the treatment of calcified hard tissues as well as new approaches developed in this field.

Keywords: Bone, Chitosan, Calcium phosphate, Bioceramics, Biocomposites

1. Introduction

Bone is the second most transplanted tissue in the world, second only to blood [1]. Hundreds of millions of people worldwide are affected by musculoskeletal conditions which are on the increase with aging population and lifestyle. Bone is a critical tissue within the vertebrates. It is a dynamic organ with many functions. It provides load bearing, body structural support onto which musculature is attached, protection for vital organs and soft tissues (brain, heart, lung, etc), and enables locomotion and motor functions. It is the host of important biological processes critical cells such as postnatal stem cell populations that support hematopoiesis, myelopoiesis, and skeletogenesis. Bone is also responsible for storing and supplying of minerals such as calcium and phosphate [2]. Native bone is a connective tissue made of two predominant components: a mineralized and an unmineralized phase. The mineralized inorganic phase contains mainly crystalline apatitic calcium phosphate (70%), water (20%), and the non-mineralized organic phase (10%) is made of fibrous type I collagen, proteins, polysaccharides and lipids (**Figure 1**) [3].

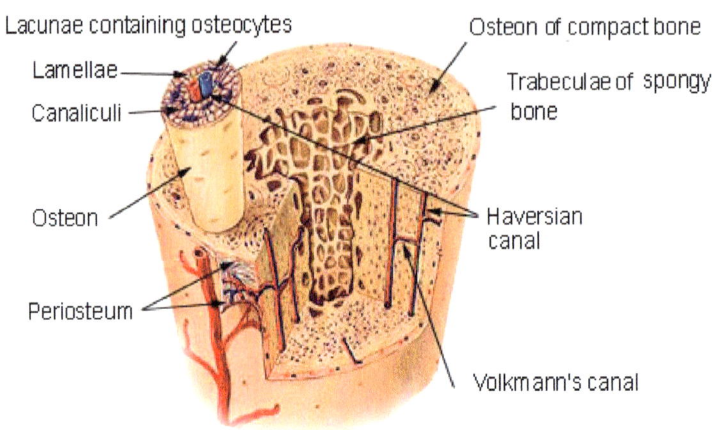

Figure 1.
Cancellous and cortical bone microstructure.

This natural biological "composite" [4] has many biological features such osteoconductivity, osteoinductivity, osteogenicity and is subject to continuous remodeling and regeneration process through osteoclastic and osteoblastic activities. The hard tissues in vertebral is not uniform tissue it could be dense and hard like dental enamel and cortical bone or spongy and highly porous like a foam as the cancellous bone.

Hard tissue repair is a multifaceted, coordinated physiological process that requires new tissue formation and resorption, eventually returning the fractured bone, for example, to its original state. Bone has the capacity of regenerating itself, especially in noncritical size defects. However, large bone defects caused by trauma, injuries, tumor resections, infections, would not heal spontaneously, and would require a bone substitute grafting material to fill the bony void for proper regeneration to take place [5].

The first documented bone transplant was performed in 1668 by a Dutch surgeon, Jacob van Meekeren, when he used dog cranium (xenograft) to repair a soldier's skull defect. The success of the grafting technique was discovered later when the soldier came back asking for removal of the "dog bone," because it cost him excommunication from the church. Meekeren discovered then that the bone healed so well it was impossible to remove the graft. The first human to human bone graft performed was in 1880 by Scottish surgeon William Macewan. He replaced the infected humerus of a 4- year-old boy with a tibia graft taken from a child with rickets [6]. The use of synthetic bone grafts could be traced back to as early as 1892 when Dreesmann reported on the results of filling osseous defects with calcium sulfate [7]. Since then, hundreds of thousands of bone grafting surgeries have been performed on humans and animals.

In 1980, major health issues related to safety of bone donors (Aids and Hepatitis) has brought the associated problems of contamination and spread of dangerous diseases to the spotlight. Some years later, (1986) the discovery of the contagious pandemic bovine spongiform encephalopathy (BSE) and the porcine endogenous retrovirus (PERVs) [8], made more obvious the necessity of alternative safe bone substitute materials in bone transplant procedures. This provided a considerable boost to research and development in the usage of synthetic bone substitutes as a safe and an affordable alternative to natural bone materials. Since then, many technologies have been adopted and used to produce bone-like products with tailored biological, physical, and chemical properties, including plasma projection [9], sol-gel [10], composites, foaming, nanotechnology,

3D-printing techniques, additive manufacturing [8, 11] and some biological therapies that involve usage of growth factors, proteins, peptides, stem cells or gene therapies [12, 13].

Nowadays, many options are available to regenerate or replace bone in clinical conditions. The main clinical approach is using a natural or a man-made bone or bone induction materials (see **Table 1**). There are three categories of natural bone, a large family of synthetic bone substitutes and biological factors-based approaches. (**Table 1**).

In bone regeneration therapy, the gold standard has been the autograft (patient's own harvested bone) [1, 9, 14]. However, autograft treatment is not always possible or even the best option. It is also limited by the volume of bone that can be harvested from the iliac crest and subsequently transplanted into the defect site. Furthermore, post-operative complications include morbidity at the harvest site, chronic pain, infection, local hematoma and, in some cases, remodeling issues of the implanted bone [8].

The established safety, efficacy, and abundance of supply of advanced synthetic bone-substitute materials made them stand as an attractive and effective alternative to the autogenous bone gold standard. On-going research in the field and a growing body of clinical data points to an even more promising future for these substitutes. Some calcium phosphate bioceramics, for example, display remarkable clinical performances and research and technological developments keep intensifying with the aim of bridging the gap to the ideal bone grafting material which would possess the three principal characteristics of the gold standard: osteogenicity, osteconduction and osteoinduction.

In human and animal medicine, orthopedic and dental surgeries, alloplastic biomaterials for hard tissue repair are divided in two categories that can be classified as per their biological responses:

i. **The bio-inert materials category:** They can be permanent or implanted for short-term and removed or replaced, like metallic dental or orthopedic implants. They are generally made of titanium, stainless-steel, nickel, zirconia or made of synthetic polymers, e.g., polymethylmethacrylate (PMMA) or Polyether ether ketone (PEEK).

Category	Advantage	Limitations
Autografts	No biological risk, osteogenic, osteoinductive, contain live cells	Limited supply Donor site, inflammation and chronic pain, site morbidity, Requires a second surgery, No mechanical
Allograft	Greater supply compared to autograft tissue	• Reduced osteogenic, -Immune rejections, Disease transmission (AIDS, Prion), Slow resorption
Xenograft (demineralized)	Unlimited supply could be osteoconductive	• Reduced osteoinductive, osteogenic properties, Immune rejections, Disease transmission (Mad cow)
Synthetic (Alloplastic)	Pure, Unlimited supply, Longer shelf life, tuneable properties	Do not have any biological factor,
Metal and polymeric based implants	Biocompatible, Load bearing applications	Not biodegradable, Bioinert, some toxicity (monomer, metal debris)
Cells, growth factors, BMP, PePgen, Ifactors	Natural	Requires biomaterial carrier, Limited applications, side effects (BMP) No mechanical

Table 1.
Available therapies used to regenerate/support bone tissue.

ii. **The bioactive biomaterials**: They are mostly resorbable at different levels. It is a large family of bone substitutes that vary in type and composition such as bioglass, calcium sulfate, calcium phosphate bioceramics (CaP), biopolymers and biocements. They could also be in tunable forms such as powder, granules, blocs, paste or injectables. They offer a dynamic choice of material and applications.

In this chapter we will review some interesting development and advancement made in biomaterial sciences in regeneration of natural hard tissues through man made products. We will focus on the polysaccharide polymer, chitosan, similar to the organic phase of natural bone and cartilage and calcium phosphate based bioceramics, similar to the inorganic phase of natural bone. We will present some tested biocomposites formulations made out the combination of the two biomaterials to mimic the composition and structure of natural bone and discuss the success and limitation of the technology.

2. Chitosan biomedical and regenerative features

Many biomaterials are available in the market for medical, cosmetic, and pharmaceutical applications. In tissue engineering, synthetic or natural biopolymers make one of the fastest growing niche segments of biomaterials. The growth is probably driven by the wide range of possibilities offered by their chemistry for different applications and the increasing demand of the biotechnology industry market.

After cellulose, chitin is the most abundant and beneficial structural non soluble organic biopolymers found on Earth. Chitin, a long polymer of N-acetylglucosamine, is the primary compound naturally found in the exoskeleton of arthropods such as crabs and shrimps, and in the cell membranes of fungi, yeasts, and other microorganisms. Deacetylation of some of acetylglucosamine units of chitin has brought a very interesting polysaccharide biopolymer to the biotechnology field, especially the biomedical area; chitosan (CS). CS is a polysaccharide composed of successive acetylglucosamine and N-glucosamine units, where the number of N-glucosamine units is called the degree of deacetylation (DDA) [15] (usually $55\% < DDA < 99\%$). CS macromolecules gained increasing attraction during the last three decades in research and industrial fields, especially in water treatment processes, pharmaceutical and biomedical engineering. Its chemistry with three reactive functional groups of amin/acetamido groups and primary and secondary hydroxyl groups allows for a large spectrum of possible chemical modifications and substitutions of its functional groups (ex: OH, and NH_2 by $-COCH_3$, $-CH_3$, $-CH_2COOH$, SO_3H, $-PO(OH)_2$, etc) [16, 17]. This improves and creates additional functional properties and features and facilitates its adaptability to different applications such as antimicrobial agency in food processing and its packaging industries, as a fungicide, as a blood sugar and pressure reducing agent, as a dietary supplement. Other applications are also found in veterinary medicine, microbiology, immunology, and agriculture, and most importantly, in highly innovative areas such as pharmaceuticals (e.g., drug delivery systems) and tissues/organs regeneration medicine in the biomedical field [18].

The global chitosan market size was valued at USD 6.8 billion in 2019 and is expected to expand and reach USD 28.93 billion by 2027. After water treatment, the second largest market is the pharmaceutical and biomedical market [19].

In the chitosan manufacturing process, the degree of deacetylation (DDA) and the molecular weight (MW) are critical parameters, as the final properties and applications of the CS biomaterial will depend on it.

Many parameters could affect the degree of deacetylation (DDA) and, consequently, the physical, chemical, and biological properties of chitosan. Parameters include the source raw material (animal, insect, fungi, mollusca, cephalopod, etc) [20] and processing conditions (pH, temperature, processing time). After manufacturing, batch parameters such as the degree of deacetylation (DDA), molecular weight (MW), molecular mass (MM), viscosity, solubility, pH, purity, protein content, endotoxin, ash content, contaminants should be carefully evaluated to ensure a safe and an adequate utilization.

Chitosan and chitosan derivative biopolymers were found to be non-toxic, biocompatible, osteogenic [21, 22] antibacterial, biodegradable, bioresorbable, antioxidant, immunoenhancing and anticancer [23]. In addition, they were found to promote cell adhesion, proliferation, and differentiation, which are important processes in tissue repair. It is, then, no coincidence that chitosan is one of the most extensively investigated polymers in tissue engineering to replace or restore the structure and function of damaged organs or tissues [24–34].

2.1 Chitosan and hard tissues

In the biomedical field, particularly in the tissue engineering domain, the main goal is to replace or substitute, repair maintain or improve tissue function through the use of isolated living cells, cells substitute tissue inducers on/or in a matrix to repair and regenerate tissue by combining engineering principles and life sciences [24, 25]. To reach that goal, there are critical properties that candidates biomaterials need to have. They are summarized in **Table 2**:

These imply that the biomaterial should allow the proliferation, adhesion and differentiation of the cells, the basic elements of any living tissue. Chitosan biomaterial can be processed in different forms such as film, mesh and fibers, freeze dried beads or scaffolds, as composite, as thermal, light, or chemical sensitive injectable gel solution or crosslinked polymer. Alone or grafted with other biopolymers (e.g., alginate, polyvinyl alcohol, polyacrylic acid, etc) [27]. Among all the possibilities, researchers and physicians have to select the formulations that are most compatible with the targeted tissue environment and function.

Characteristics	Description of the characteristic
Biocompatibility	They must be accepted by the receptor and must not lead to rejection mechanisms because of its presence.
Absorbability and degradability	Absorbable, with controllable degradation and resorption rate to be the same as the in vitro and in vivo cell/tissue growth
Not to be toxic or carcinogenic	Its degradation products cannot cause local or systemic adverse effect on a biological system
Chemically stable	Chemical modifications not being present in a biological system implant or biodegradable in nontoxic products, at least during the scheduled time to regenerate tissue
Chemically adequate surface	To have a chemically adequate surface for cell access, proliferation and cell differentiation
Adequate resistance and mechanical properties	Resistance and mechanical properties, superficial characteristics, fatigue time, and weight, according to the receptor tissue needs, as well
The proper design, size, and shape of the scaffolding	Which allows having a structure with properties according to the needs of the receiving tissue to regenerate or repair.

Table 2.
Main characteristics that biomaterial should have.

Hard tissues like bone and cartilage require some specific formulations, with specific chemical and physical properties to withstand the regeneration of the native hard tissues process.

2.1.1 Chitosan in cartilage tissue therapy

Osteoarthritis affects 7% of the global population. That is more than 500 million people worldwide. It is considered one of the critical causes of disability over the world population (28.) Cartilage, a connective tissue forming the skeleton, is a complex tissue, not vascularized and is made of chondral cells that produce extracellular matrix proteins [29]. It is composed of a dense network of collagen fibers embedded in a firm, gelatinous ground substance that has the consistency of plastic. This structure gives the tissue tensile strength, enabling it to bear weight while retaining greater flexibility than bone [30].

Cartilaginous connective tissues are highly involved into biomechanical function. They are subject to high load bearing stress. Critical size defects cannot heal on their own, so there is a need for tissular therapy to regenerate the cartilage tissue [28]. Different therapies are available, such as autograft (the gold standard) allograft (cardioviral tissue), mosaicplasty (autograft), autologous chondrocytes or tough tissue engineering procedures such as use biopolymer templates that are chitosan based.

chitosan has shown good success in regeneration of cartilage lesion, because it has structural similarity with various glycosaminoglycans found in articular cartilage [31].

A clinical study with 80 patients over a period of 1 and 5 years of a marketed thermosensitive hydrogel formulation BST-CarGel® (Smith & Nephew) has been reported. BST-CarGel® act as a scaffold and matrix that stabilize the blood clot in the cartilage lesion by dispersing a soluble and adhesive polymer scaffold containing chitosan throughout uncoagulated whole blood [32]. The gel is recommended for all synovial joints (knee, hip, and ankle) and on size defects ranging from 0.3cm^2 to 7cm^2. The Product has two components: a soluble chitosan powder, and a solution of glycerophosphate salt. It is used arthroscopically using a microfracture techniques (bone marrow simulation). Patients were divided in two groups; one for the baseline where no product was used after the microfracture and the second was treated with the product mixed with autologous blood. The red viscous mixture was injected in the cartilaginous defect area to set. Following treatment periods, regeneration of cartilaginous tissue of 92.37% compared to 85.54% for baseline was observed after 12 months (**Figure 2**) and 93.79% vs. 86.96% respectively after 5 years. The difference was statically significant [33].

In another study, layered highly porous nano structured 3D scaffold using chitosan and chondroitin sulphate was developed. It was loaded in vitro with bovine chondrocytes (BCH) and bone marrow derived stroma cells (hMSCs). The experiment was conducted for 21 days. It has shown that cells attached, proliferated and were metabolically active over the entire scaffold. Cartilaginous extracellular matrix (ECM) formation was further assessed, and results showed that glycosaminoglycan secretion occurred indicating the maintenance of the chondrogenic phenotype and the chondrogenic differentiation of bone marrow derived stromal cells. The mechanical properties were poor and not comparable to natural cartilage. The authors mentioned the need of improving the mechanical shortfall by adding growth factors, nanotubes, or crosslinked template polymers that would reduce the degradation rate [35].

With low mechanical performance and lack of clinical data for long periods (>5 years), it is difficult to fully assess the efficiency of the products.

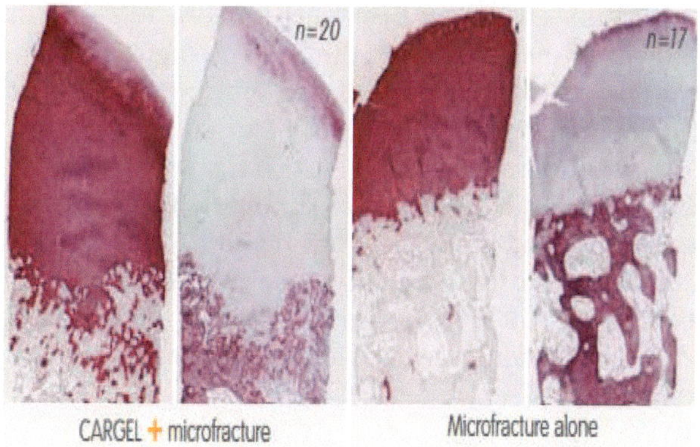

Figure 2.
Biopsy histology of the best repairs of the BST-CarGel and microfracture (MFx) groups at 13 months post treatment, the BST-CarGel biopsies show superior tissue quality and organization compared with the MFx biopsies [34].

2.1.2 Chitosan in bone tissues therapy

Chitosan formulations were also used in bone tissue regeneration as a delivery system for bone morphogenic proteins, peptides, or growth factors for cells. The chitosan is tailored in general in the form of a 3D structure (e.g., freeze dried scaffold and injectable gel), which is loaded with biological elements. In **Table 3**, we

Tested formulation	Form	Animal Model	Results	Ref
Chitosan Scaffold	zzfreeze dried scaffold	Rat calvarialosteoblasts	Increased biomineralization and osteogenesis	[36]
Chitosan–poly(lactide-co-glycolide) modified with heparin	Microsphere scaffold	Rabbit ulnar critical-sized-defect model	The in vivo section of study: promotion early bone formation	[37]
Chitosan-polylactic acid	Composite scaffold	Preosteoblast (MC3T3-E1) cells	Improvement of the interface of tissue engineering scaffold	[38]
Chitosan Scaffold	Freeze dried scaffold	omental adipose-derived stromal cells implanted in mandibular	Significantly earlier regeneration of bone than the use of the scaffold alone	[39]
Chitosan nanoparticle	Nano particles	Rats model femur defect	In-vitro chitosan induces osteogenic differentiation in MSCs in vitro, increases osteoblast viability in vitro, reduces osteoclast numbers in vitro, assists bone fracture healing,	[40]
Chitosan-Collagen type I	Electrospun / casted barrier membranes in guided bone regeneration	Calvaria defect in New Zealand rabbits	Found to be biocompatible osteoconductive, osteoinductive, and has osteogenesis properties	[41]

Table 3.
Example of studies that have used chitosan-based formulation to treat bone defect.

have reported some studies that have been performed with chitosan polymer or its derivatives to treat bone defects. Despite the good biological properties of chitosan formulations developed till now for hard tissues, the poor mechanical properties and lack of certain bioactivity proper to bone tissue such as osteoconduction, chitosan and derivatives are so far not the best clinical choice to treat bone defects.

Researchers have tried and are still trying to overwhelm the shortfalls. The most hopeful ones are those that combine chitosan-based formulations and synthetic inorganic biomaterial similar to the calcified phase of natural bone [42].

In the next section, we will review some of interesting options related to bone substitutes' candidates that could be used along with chitosan to achieve a biomimicry of natural bone tissue.

3. Bone-like calcium phosphates

Bone and teeth are the hardest human tissues. Bone provides support and protection to organs. When skeletal system is damaged, an immediate fix is required to avoid any complications, physiological function and mobility impair and even death.

An ideal bone substitute should be biomechanically stable, able to resorb as natural bone within an appropriate time frame while new bone regenerate, exhibit osteoconductive (interconnected porous scaffold onto which bone cells can attach, migrate, differentiate, and grow new bone tissue). osteogenic and osteoinductive properties (ability to stimulate differentiation of a progenitor cells toward an osteoblast lineage) and provide a favorable environment for invading blood vessels and bone forming cell [43].

When it comes to bone substitution, autogenous bone is typically considered as the gold standard for bone defect regeneration since it is living tissue and contains osteogenic cells, still involves harvesting bone from one part of the patient's body and putting it in a damaged bone area. It was and still the method of choice in reconstructing bone either for dental or orthopedic applications. It provides perfect biocompatibility along with the body's own growth factors and structural proteins. Because of limited supply, the need of a second surgery associated with site morbidity and infection risks, negative effect on the mechanics, autograft is not always possible or the best option.

Calcium phosphate (CaP) is the main constituent of inorganic phase of natural bone and teeth and it play essential roles in our daily lives. Damaged calcified natural tissues would be best repaired with something similar. CaP biomaterials are the most legitimate candidates when it comes to regenerate bone. They have been extensively used for decade with great success in orthopedic and dental fields [44]. CaP bone substitutes materials are safe and efficient. They are biocompatible with bone tissues. When implanted they have the particularity to go over the same biological osteoclastic resorption and new bone regeneration processes as the natural bone. They are highly bio-similar to the inorganic phase of autologous bone tissue. Their resorbability and solubility depend in general in their ratio Ca/P (**Table 4**) empirical formulations were proposed to describe the mineral composition of natural bone [45]. The chemical formula of Calcium Phosphate materials eq. (1) is shown below:

$$Ca_{8,31,7}\left(PO_4\right)_{4,3}\left(HPO_4 \text{ or } CO_3\right)_{1.7}\left(OH \text{ or } {}^{1/2}CO_3\right)_{0,31.7} \quad (1)$$

Actually, mineral bone composition is more versatile, it has many other minor chemical elements such: Mg, Sr., Si, F, Na, and others (**Table 4**).

% Element	Enamel	Dentine	Bone	HA
Ca	37,6	40,3	36,6	39
P	18,3	18,6	17,1	18,5
CO_2	3,0	4,8	4,8	/
Na	0,7	0,1	1,0	/
K	0,05	0,07	0,07	/
Mg	0,2	1,1	0,6	/
Sr	0,03	0,04	0,05	/
Cl	0,4	0,27	0,1	/
F	0,01	0,07	0,1	/
Ratio Ca/P	1,59	1,67	1,65	1,67
Crystallinity	good	low	low	good

Table 4.
Chemical composition of calcified hard tissues vs. stochiometric synthetic Hfydroxyapatite (HA).

The composition and crystallinity of bone tissues depends on many parameters (location: cortical, cancellous, dental enamel, dentine, the age, biological metabolism, etc).

Many studies have reported development of CaP products that would be used as potential bone substitutes. In the **Table 5**, a list of the main most popular products used in the development or formulation of CaP biomaterials.

One of the furthermost interesting CaP biomaterials are the osteoconductive biphasic calcium phosphate (BCP) containing Hydroxyapatite (HA) and Beta TCP. These two phases have different resorption rates. HA (less soluble) will provide short- and long-term physical stability to the bone defect and scaffold for bone ingrowth, whereas Beta TCP (more resorbable) will provide locally Ca and phosphate ions to regenerate new bone and activate the osteogenesis process [46]. To enhance the physical, physiological and/or therapeutical properties, CaP biomaterials could be easily assorted with polymers, drug, proteins, Growth Factors, cells, blood cells, bone marrow and even autologous bone tissue.

CaP biomaterials are relatively easy to make osteoconducteurs by different methods, to mimic the trabecular structure of natural bone (**Figures 3** and **4**).

CaP Biomaterial	Formula; Abbreviation	Ca/P	Solubility at 25C mg/L)
Dicalcium phosphate dihydrate	$CaHPO_4{\cdot}2H_2O$; DCPD	1.00	88
Octocalcium phosphate	$Ca_8H_2(PO_4)_6{\cdot}5H_2O$;OCP	1.33	8.1
Hydroxyapatites	$Ca_{10}(PO_4)(OH)_2$; *HA*	1.67	9.4
α-Tricalcium phosphate	α-$Ca_3(PO_4)_2$; α-TCP	1.50	2.5
β-Tricalcium phosphate	β-$Ca_3(PO_4)_2$; β-TCP	1.50	0.5
Biphasic Calcium phosphate	xβ-$Ca_3(PO_4)_2$ + y$Ca_{10}(PO_4)_6(OH)_2$; BCP	1.50–1.67	0.3–0.5
Tetracalcium phosphate	$Ca_4(PO_4)_2O$; TTCP	2.00	0.7

Table 5.
Short list of calcium phosphates with biological interest.

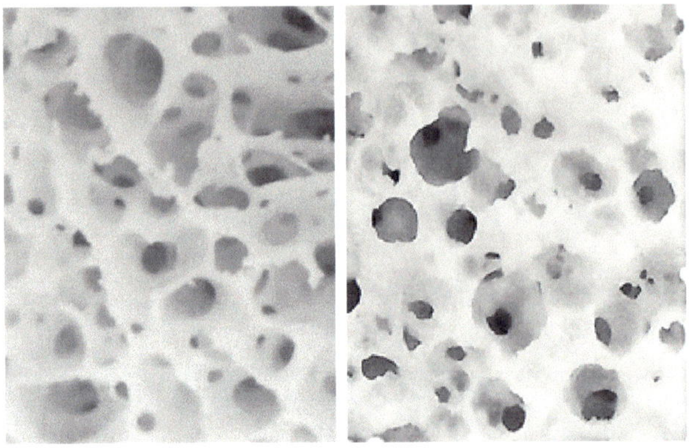

Figure 3.
Optical microscopic pictures (x8) of natural cancellous bone (left) and synthetic osteoconductive bioceramics (BCP 50–50%) (right, porosity >70%, Biomatcan).

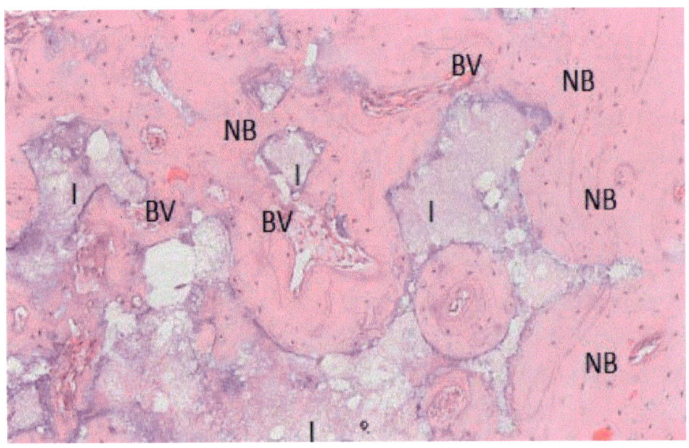

Figure 4.
Histological picture of Osteoconductive bone graft implanted in rabbit tibia bone after 12 weeks. Pores are filled with new bone (Osteoconduction). BV: Blood vessels, NB: New bone, I: Implant (x50) (Biomatcan).

New developed formulations were found to have some outstanding properties similar to biological growth factor in autologous bone such bone morphogenic proteins (BMPs). They are osteoinductive. The osteoinduction is trigged either by the addition of chemical elements such silicates ions (Actifuse bone graft, by Baxter) or by tailored sub-micron surface topography and porosity [47] that has the capability to induce bone formation in ectopic or heterotopic location such in muscle or under skin. (**Figures 5** and **6**). The mechanism through which a Ca-P graft mediates an osteoinduction in the host bed is still an active subject of research.

This approach has more benefit. It is less expensive and safer than the BMPs therapy that has limitations (e.g.: not recommended in bone joints or small bones, serious complication, and side effects (cancer, unpredictable ectopic bone growth, neurological impairment, fertility problem ...) [48].

A study conducted by Van Dijk et al., showed that in spine fusing in ovine model, formulation of osteoinductive submicron surface topography of BCP bone graft (Magnetos) outperforms Bioglass and monophasic Tricalcium phosphate CaP bioceramics (Vitoss) mixed with Bioglass. The induced bone growth was found similar when using autologous bone (**Figure 5**) [48]. Unlike the other natural substitutes, there is no risks of incompatibility, allergy, or transmission of diseases.

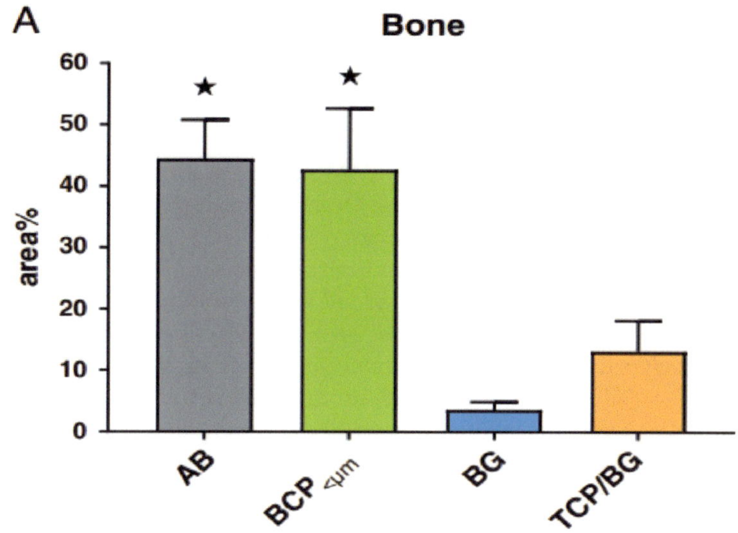

Figure 5.
Posterolateral fusion on ovine model: Histomorphometry diagrams of bone performed on low-magnification micrographs of histologic sections. Data are presented as area%, in mean and SD. ★, significantly different from BG and TCP/BG ($P < 0.001$). ($P < 0.005$) and TCP/BG. AB: Autograft bone; BCP < µm, biphasic calcium phosphate with submicron topography; BG, bioglass; TCP, tricalcium phosphate. [47].

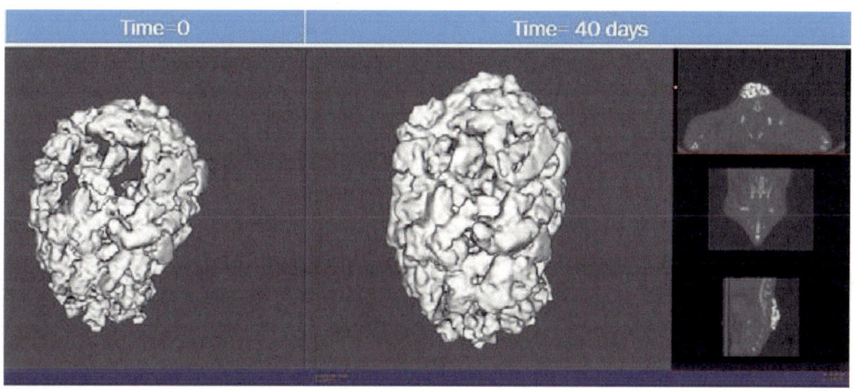

Figure 6.
Ct-scan of heterotopic implantation of Osteoinductive BCP (50–50%) in mice model, noticeable increase (10.6%) of implant size after 40 days. (Biomatcan).

We can confirm that synthetic CaP biomaterials are safe and a reliable alternative for autograft or allograft. With a history of safety and effectiveness in clinical both human and animal health, they are gaining more attention and started to be considered the new gold standard in bone regeneration therapy.

4. Biocomposites: Chitosan-CaP bioceramics

Many researchers have worked on development of biocomposites containing CS [49–55] and CaP biomaterials. If the biological properties were improved in some cases, the mechanical properties still not comparable to natural bone. In this section we are going to report some testing and results on the developed biocomposites:

An injectable bone graft formulation and hardening injectable bone cements. The mechanical properties were evaluated in both of cases.

4.1 Bone graft biocomposites

The bone graft was prepared as follow: A solution of chitosan (1,7%) (DDA 83% ± 3%, supplied by Biomolecules and Organic Synthesis Laboratory, Ben M'Sick University, Casablanca) was prepared in diluted chloric acid solution (0.2 N). The chitosan was dissolved under ultrasonic agitation. Disodium glycerophosphate solution (0.5 N) was added slowly under agitation at low temperature. The pH was maintained between (6.5–7). The chitosan solutions were then autoclaved. Porous Biphasic calcium phosphate bioceramics (BCP) (50%Beta TCP-50%HA, porosity = 76%, Biomatcan) with average granules size of 135 microns was added slowly and gently homogenized. It was found during the preliminary tests, that the best formulation that preserve homogeneity and injectability have a ratio of BCP comprising between 35% and 50%. Low concentration led to aggregation of the granules and high concentration affects the injectability and the structural stability of the biocomposites. The obtained products were kept at cold temperature till use. The mechanical properties of the obtained biocomposites were measured at physiological temperature (37°C) with rheometer (Brookfield DV3T). The obtained results are reported in the table and figures bellow (**Table 6**, **Figures 7** and **8**).

In this case we notice that the increases of the BCP mass in the chitosan solution increase the mechanical properties of mixture. This increase is not linear. The maximum is obtained for L/S = 40%. Over this limit the biocomposite is less injectable and less elastic. 0.4% of BCP represent the maximum load for this formulation with optimal mechanical properties.

BCP (%)	**0**	**0.36**	**0.40**	**0.44**	**0.5**
Chitosan solution (%)	1.7	1.7	1.7	1.7	1.7
Elastic modulus (Kpa)	1.8	3.8	14.2	5.2	2.8
Time (min)	27	115	40	63	62

Table 6.
Eleastic moduls of biocomposite bone graft with different BCP.

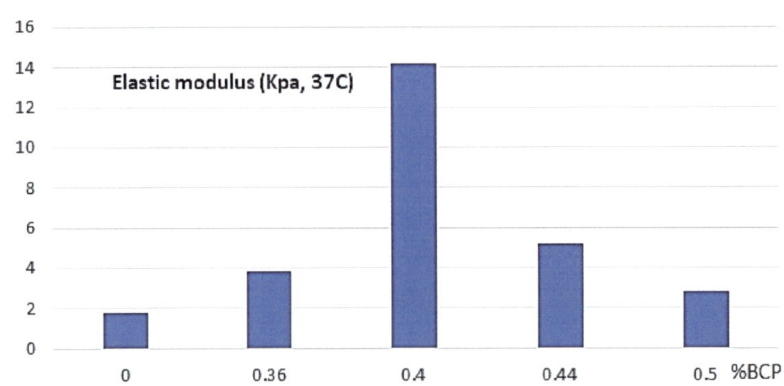

Figure 7.
Elastic modulus of biocomposites formulations (KPa, 37°C).

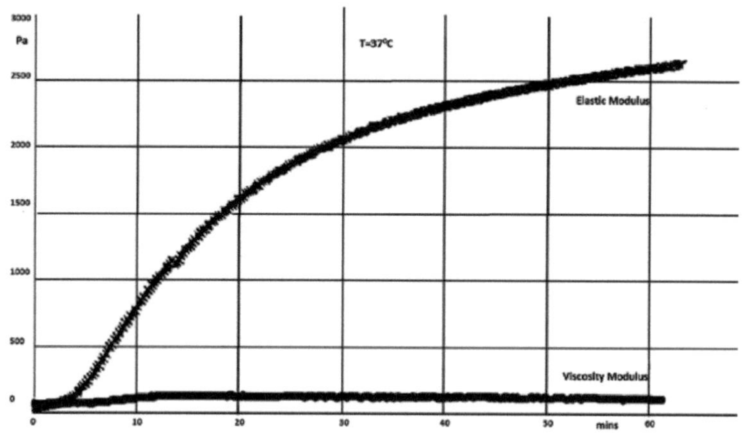

Figure 8.
Representative example of rheological test obtained at 37°C.

4.2 Injectable bone substitute material- biocements

The biocements are made by mixing solid (S) and liquid phases (L) they are known to harden in certain conditions, the mechanical properties depend on the solid and liquid compositions. They are used in bone augmentation situations like joint fixation, maxillofacial surgeries, and others. We have tested biocomposites made by two different chitosan solutions.

4.2.1 Self-hardening *biocomposites*

These materials are made out of a grafted chitosan mixed with Alpha PTC bioceramics fine powder. The biocomposites has the advantage that when it is mixed with the CS solution it forms an injectable paste that turns to rubber-like material. It should provide a good initial mechanical stability for the bone defect and the implant. The hardening of the biocomposites occurs progressively over time. The biocomposites was prepared as follow:

Grafted chitosan solution: a mPEG-grafted-chitosan [49] transparent and homogeneous gel was prepared from a liquid chitosan aqueous solution (chitosan 2.0% w/v, pH < 6) and Monomethoxypolyethyleneglycol-N-hydroxysuccinimidylsuccinate (mPEG-suc- NHS). The obtained polymer solution was mixed with fine powder CaP ceramic powder (PTC alpha, Ca/P = 1.50, D50 = 4microns, Biomatcan). The Liquid/powder ratio (L/S) varies from 0.4, to 0.6. The biocomposites cement pastes were injected in a rubber made cylindrical molds (6 mm in diameter x 12 mm height). The elastic silicone-like articles were demolded and stored at 37°C in humid atmosphere for 24 h to harden. The solid blocs were matured in Simulated Body Fluid (SBF) solution at 37°C for 3, and 7 days. Then washed with cold distilled water and dried at 40oC for 24 h. The obtained biocomposites articles were mechanically tested (Zwick Z010 mechanical testing machine, with a crosshead speed of 1 mm/min). 10 specimens were tested for each test formulation. The measured compressive strength (MPa) for different ratio L/S is reported in **Table 7**.

4.2.2 Self hardening *CaP biocements*

The biocements are made with crosslinked CS formulations and without chitosan solution were prepared and compared side by side. Chitosan (83% ± 3 DDA)

	Ref	0.4 ml/g	0.5 ml/g	0.6 ml/g
3 days	23.22 + 3.58	8.51 + 1.76	7.73 + 1.95	5.51 + 1.30
7 days	29.68 + 4.23	9.82 + 0.26	5.69 + 0.94	4.04 + 1.66

Table 7.
Compressive strength (MPa) obtained for different bone cement with modified chitosan solution after 3 and 7 days of maturation (Ref = PTC alpha with water only, L/S = 0.5).

Formulations	Ca/P	L/S (ml/g)	Compressive strength (Mpa)	Compressive strength (Mpa) (1% chitosan solution)	Variation (%)
αTCP-DCPA	1.33	0.5	17.3 + 3.1	7.9 + 2.2	54%
αTCP-MCPM	1.37	0.72	12.8 + 3.9	11.8 + 1.6	7%
αTCP-HA	1.52	0.5	29.0 + 4.9	11.3 + 4.8	61%
αTCP-HA-MCPM	1.55	0.46	12.7 + 3.9	11.2 + 1.5	11%
TTCP-MCPM	1.66	0.55	8.3 + 1.0	2.9 + 0.4	65%
TTCP-DCPA-MCPM	1.50	0.60	6.8 + 2.5	2.2+ 0.5	67%

Table 8.
Compressive strength comparison of biocement formulations prepared with water vs. 1 of chitosan solution.

was dissolved in 1%HCl). The pH was maintained 6.7 to 7 with Sodium glycerophosphate (Sigma Aldrich). The solid phases were selected from different sources of CaP material. The tricalcium alpha (alpha)TCP and Hydroxyapatite (HA) supplied by Biomatcan, tetra calcium phosphate (TTCP, Cambioceramics, NL), Brushite (DCPD) and monocalcium phosphate (MCPM) from Sigma-Aldrich.

The biocomposites were prepared by mixing powder and solutions with predetermined ratio L/S. The paste was handled as mentioned before. When the cements harden, the cylindrical blocs were put in phosphate buffer saline solution at 37°C, pH 7.4 for 24 hours, then washed with cold water and dried at 40°C for 24 hours. The formulations and obtained results are summarized in **Table 8**.

The results of the mechanical tests on both formulations show that the addition of mPEG- grafted-chitosan solution or crosslinked chitosan solution decreases dramatically the mechanical properties of self-herding biocements. It could be explained by the effect of chitosan on the CaP crystal growth during maturation of the biocements, or by the heterogenous structure of the biocements, where chitosan polymer creates some discontinuity in the physical structure. Moreover, the shrinkage of the chitosan network during the drying process could induce a distortion of the article volume thus reducing its mechanical properties. In-vivo testing would be the best approach to assess the mechanical properties of such formulations.

5. Conclusion

In conclusion we have presented some works done related to the development of chitosan, CaP biomaterials that mimic the composition of natural bone. Despite the proven biological benefits and the huge number of research, publications and patents done on the use of chitosan in medical field and especially in hard tissues replacement, there is a big discrepancy between research, commercial and market reality. Less than handful products are marketed mainly for cartilage repair.

The principal obstacles are proper to the material itself and processing. No validated manufacturing process, variability in the raw material, the formulations developed up to date have low mechanical properties, regulatory burden associated with the endotoxin content that require additional steps and control in the manufacturing process, the sterilization that affect the polymer, the storage, shelf life and stability conditions especially for the liquid and gel formulations. However, some new technologies have been tested to solve some of these problems, such plasma sterilization that delivers free endotoxin chitosan raw material [56]. It is still at early stage and need to be validated technically and economically at large scale. Other improvements still have to come before chitosan and derivative become attractive solution in bone tissue regeneration for the bioindustry players.

Author details

Fouad Dabbarh[1*], Noureddin Elbakali-Kassimi[2] and Mohammed Berrada[3]

1 Biomatcan Ltd, Fredericton, Nouveau-Brunswick, Canada

2 Department of Chemistry, University of New Brunswick, Fredericton, New Brunswick, Canada

3 Faculty of Sciences Ben M'Sick, Laboratory of Biomolecules and Organic Synthesis, (BIOSYNTHO), Department of Chemistry, University Hassan II of Casablanca, Morocco

*Address all correspondence to: fdabbarh@biomatcan.com

References

[1] Williams, Amy; Szabo, Robert M. Bone transplantation. / In: Orthopedics, Vol. 27, No. 5, 05.2004, p. 488-495.

[2] Jay R. Lieberman M.D. AAOS Comprehensive Orthopaedic Review 3, 2019

[3] Fourie, J., Taute, F., du Preez, L. et al. Chitosan Composite Biomaterials for Bone Tissue Engineering—a Review. (2020) Regen. Eng. Transl. Med.

[4] Vaz, M.F. & Canhao, Helena & Fonseca, Joao. (2011). Bone: A Composite Natural Material. 10.5772/17523.

[5] Jahan, K., Manickam, G., Tabrizian, M. et al. In vitro and in vivo investigation of osteogenic properties of self- contained phosphate-releasing injectable purine-crosslinked chitosan-hydroxyapatite constructs. (2020) Sci Rep 10, 11603.

[6] Williams, Amy & Szabo, Robert. (2004). Bone transplantation. Orthopedics. 27. 488-495; quiz 496.

[7] Arnav M. Calcium sulfate: An unconventional bone graft in the management of furcation involvement a case series. J Int Clin Dent Res Organ 2019;11:36-42.

[8] R.Singh et al. International Journal of Oral Implantology and Clinical Research, May-August 2013;4(2): 68-71.

[9] Arjun Dey , Anoop Kumar Mukhopadhyay Microplasma Sprayed Hydroxyapatite Coatings, 1st Edition, 2015 10 Eduardo J.et al, (2011).

[10] Eduardo J. Nassar, Katia J. Ciuffi, Paulo S. Calefi, Lucas A. Rocha, Emerson H. De Faria, Marcio L. A. e Silva, Priscilla P. Luz, Lucimara C. Bandeira, Alexandre Cestari and Cristianine N. Fernandes (September 15th 2011). Biomaterials and sol–gel process: A methodology for the preparation of functional materials, Biomaterials Science and Engineering, Rosario Pignatello, IntechOpen,

[11] B. Sharma, S. Sharma and P. Jain, Leveraging advances in chemistry to design biodegradable polymeric implants using chitosan and other biomaterials, International Journal of Biological Macromolecules (2018).

[12] Ho-Shui-Ling A, Bolander J, Rustom LE, Johnson AW, Luyten FP, Picart C. Bone regeneration strategies: Engineered scaffolds, bioactive molecules and stem cells current stage and future perspectives. Biomaterials. 2018 Oct;180:143-162.

[13] Kobbe P, Laubach M, Hutmacher DW, Alabdulrahman H, Sellei RM, Hildebrand F. Convergence of scaffold-guided bone regeneration and RIA bone grafting for the treatment of a critical-sized bone defect of the femoral shaft. Eur J Med Res. 2020 Dec 21;25(1):70.

[14] Bouler et al Biphasic calcium phosphate ceramics for bone reconstruction: A review of biological response, Acta Biomaterialia 53 (2017) 1-12.

[15] Mohebbi, Shabnam & Nasiri Nezhad, Mojtaba & Zarrintaj, Payam & Jafari, Seyed & Gholizadeh, Saman & Saeb, Mohammad Reza & Mozafari, Masoud. (2018). Chitosan in biomedical engineering: A critical review. Current Stem Cell Research & Therapy. 13.

[16] Ahmadi F, Oveisi Z, Samani SM, Amoozgar Z. Chitosan based hydrogels: Characteristics and pharmaceutical applications. Res Pharm Sci. 2015 Jan-Feb;10(1):1-16.

[17] Li S, Tian X, Fan J, Tong H, Ao Q, Wang X. Chitosans for tissue repair and organ three-dimensional (3D) bioprinting. Micromachines (Basel). 2019 Nov 11;10(11):765.

[18] Sultankulov B, Berillo D, Sultankulova K, Tokay T, Saparov A. Progress in the development of chitosan-based biomaterials for tissue engineering and regenerative medicine. Biomolecules. 2019 Sep 10; 9(9):470.

[19] Chitosan Market Size, Global Industry Analysis Report, 2020-2027, Allied Market Research, June 1, 2020, Report ID: 978-1-68038-798-8.

[20] Roberts G. A. F., Chitin chemistry, Houndmills. MacMillan Press Ltd., 1992, 1-5.

[21] Zhu Y, Wang X, Cui FZ, Feng QL, de Groot K. In vitro Cytocompatibility and Osteoinduction of phosphorylated chitosan with osteoblasts. Journal of Bioactive and Compatible Polymers. 2003;18(5):375-390.

[22] Zhang H, Neau SH. In vitro degradation of chitosan by a commercial enzyme preparation: Effect of molecular weight and degree of deacetylation. Biomaterials. 2001 Jun;22(12):1653-1658.

[23] Berrada, Mohamed & Serreqi, Alex & Dabbarh, F & Owusu, A & Gupta, Ajay & Lehnert, Shirley. (2005). A novel non-toxic camptothecin formulation for cancer chemotherapy. Biomaterials. 26. 2115-2120.

[24] Langer, R., and Vacanti, J. P. (1993). Tissue engineering. Science 260, 920-926.

[25] Y. Fung, "A proposal to the national science Foundation for an Engineering Research Centre at USCD," UCSD 865023, Center for the Engineering of Living Tissues, 2001.

[26] Rodríguez-Vázquez M, Vega-Ruiz B, Ramos-Zúñiga R, Saldaña-Koppel DA, Quiñones-Olvera LF. Chitosan and its potential use as a scaffold for tissue engineering in regenerative medicine. Biomed Research International. 2015 ;2015:821279.

[27] Zhensheng Li, Hassna R. Ramay, Kip D. Hauch, Demin Xiao, Miqin Zhang, Chitosan–alginate hybrid scaffolds for bone tissue engineering, Biomaterials, Volume 26, Issue 18, 2005, Pages 3919-3928.

[28] Global Burden of Disease Collaborative Network Global Burden of Disease Study 2019 (GBD 2019) results.

[29] Slam MM, Shahruzzaman M, Biswas S, Nurus Sakib M, Rashid TU. Chitosan based bioactive materials in tissue engineering applications-a review. Bioact Mater. 2020 Feb 12;5(1):164-183.

[30] https://www.britannica.com/science/cartilage.

[31] Daniela Izzo, Barbara Palazzo, Francesca Scalera, Fabiana Gullotta, Velia lapesa, Stefania Scialla, Alessandro Sannino & Francesca Gervaso Chitosan scaffolds for cartilage regeneration: influence of different ionic crosslinkers on biomaterial properties, International Journal of Polymeric Materials and Polymeric, Biomaterials, (2019), 68:15, 936-945.

[32] Matthew S. Shive; Caroline D. Hoemann; Alberto Restrepo; Mark B. Hurtig; Nicolas Duval; Pierre Ranger; William Stanish; Michael D. Buschmann. BST-CarGel: In Situ ChondroInduction for Cartilage Repair., (2006),16(4), 0-278.

[33] Shive, Matthew & Stanish, William & Mccormack, Robert & Forriol, Francisco & Mohtadi, Nicholas & Pelet, Stéphane & Desnoyers, Jacques & Méthot, Stéphane & Vehik, Kendra & Restrepo, Alberto. (2014).

BST-CarGel(R) treatment maintains cartilage repair superiority over microfracture at 5 years in a multicenter randomized controlled trial. Cartilage. 6. 62-72.

[34] Méthot, Stéphane, Adele Changoor, Nicolas Tran-Khanh, Caroline D. Hoemann, William D. Stanish, Alberto Restrepo, Matthew S. Shive, and Michael D. Buschmann. "Osteochondral biopsy analysis demonstrates that BST-CarGel treatment improves structural and cellular characteristics of cartilage repair tissue compared with microfracture." CARTILAGE 7, no. 1 (January 2016): 16-28.

[35] Silva JM, Georgi N, Costa R, Sher P, Reis RL, et al. Nanostructured 3D constructs based on chitosan and chondroitin Sulphate multilayers for cartilage tissue engineering. (2013) PLoS ONE 8(2): e55451.

[36] Seol, YJ., Lee, JY., Park, YJ. et al. Chitosan sponges as tissue engineering scaffolds for bone formation. Biotechnology Letters (2004) 26, 1037-1041.

[37] Jiang T, Nukavarapu SP, Deng M, Jabbarzadeh E, Kofron MD, Doty SB, Abdel-Fattah WI, Laurencin CT. Chitosan-poly(lactide-co-glycolide) microsphere-based scaffolds for bone tissue engineering: in vitro degradation and in vivo bone regeneration studies. Acta Biomater. 2010 Sep;6(9):3457-3470.

[38] Xu T, Yang H, Yang D, Yu ZZ. Polylactic acid nanofiber scaffold decorated with chitosan Islandlike topography for bone tissue engineering. ACS Appl Mater Interfaces. 2017 Jun 28;9(25):21094-21104.

[39] Mohammadi R, Amini K. Guided bone regeneration of mandibles using chitosan scaffold seeded with characterized uncultured omental adipose-derived stromal vascular fraction: An animal study. Int J Oral Maxillofac Implants. 2015 Jan-Feb;30(1):216-222.

[40] Mei L. Tan, Peng Shao, Anna M. Friedhuber, Mallory van Moorst, Mina Elahy, Sivanjah Indumathy, Dave E. Dunstan, Yongzhong Wei, Crispin R. Dass, The potential role of free chitosan in bone trauma and bone cancer management, Biomaterials, Volume 35, Issue 27, 2014, Pages 7828-7838.

[41] Lotfi G, Shokrgozar MA, Mofid R, Abbas FM, Ghanavati F, Baghban AA, Yavari SK, Pajoumshariati S. Biological evaluation (In vitro and In vivo) of Bilayered collagenous coated (Nano electrospun and Solid Wall) chitosan membrane for periodontal guided bone regeneration. Ann Biomed Eng. 2016 Jul;44(7):2132-2144.

[42] Hu D, Ren Q, Li Z, Zhang L. Chitosan-based biomimetically mineralized composite materials in human hard tissue repair. Molecules. 2020 Oct 19;25(20).

[43] Patricia Janicki, Gerhard Schmidmaier, What should be the characteristics of the ideal bone graft substitute? Combining scaffolds with growth factors and/or stem cells, injury, volume 42, Supplement 2, 2011, Pages S77-S81.

[44] Wouter Habraken, Pamela Habibovic, Matthias Epple, Marc Bohner, Calcium phosphates in biomedical applications: Materials for the future?, Materials Today, Volume 19, Issue 2, 2016, Pages 69-87,

[45] Legros R., Balmain N. and Bonel G.. Structure and composition of the mineral phase of periosteal bone. (1986) J. Chem. Res(S) 8-9.

[46] Zetao Chen, Chengtie Wu, Wenyi Gu, Travis Klein, Ross Crawford, Yin Xiao, Osteogenic differentiation of bone marrow MSCs by β-tricalcium

phosphate stimulating macrophages via BMP2 signalling pathway, Biomaterials, Volume 35, Issue 5, 2014, Pages 1507-1518.

[47] Van Dijk, Lukas & Groot, Florence & Rosenberg, A.j.w.p & Pelletier, Matthew & Christou, Chris & de Bruijn, Joost & Walsh, William. (2020). MagnetOs, Vitoss, and Novabone in a Multi-Endpoint Study of Posterolateral Fusion: A True Fusion or Not?. Clinical Spine Surgery.

[48] Boraiah S, Paul O, Hawkes D, Wickham M, Lorich DG. Complications of recombinant human BMP-2 for treating complex tibial plateau fractures: A preliminary report. Clin Orthop Relat Res. 2009;467(12):3257-3262.

[49] Abdellatif Chenite, Mohammed Berrada, Cyril Chaput, Fouad Dabbarh, Amine Selmani , Composition and method to homogeneously modify or cross-link chitosan under neutral conditions, US Patent 2002 US7098194B2.

[50] F. Damiri, Y. Bachra, A. Grouli, .A. Ouaket, A. Bennamara, N. Knouzi and M. Berrada, A Novel Drug Delivery System Based on Nanoparticles of Magnetite Fe3O4 Embedded in an Auto Cross-Linked Chitosan, In Book: Chitin and Chitosan - Physicochemical Properties and Industrial Applications, 2020 DOI: 10.5772/intechopen.94873.

[51] F. Damiri, Y. Bachra, C. Bounacir, .A. Laaraibi, and M. Berrada,, Synthesis and Characterization of Lyophilized Chitosan-Based Hydrogels Cross-Linked with Benzaldehyde for Controlled Drug Release, Journal of Chemistry, 2020, Chemistry and Applications of Polysaccharide-Based Material, DOI: 10.1155/2020/8747639.

[52] A Laaraibi, S. Hamdouch, A. Abourriche, A. Bennamara, N. Knouzi, M. Berrada, Chitosan-Clay based (CS-NaBNT) biodegradable nanocomposites films for potential utility in food and environment, INTECHOPEN, 2018. ISBN 978-1-78923-406-0.

[53] A. Laaraibi, I. Charhouf, S. Hamdouch, A. Bennamara, A. Abourriche, M. Berrada, Synthesis and characterization of new nanocomposite chitosan /clay for food applications, international Eurasian conference on biological and chemical sciences, EurasianBioChem, p. 174-182, 2018

[54] A. Laaraibi, A. Bennamara, A. Abourriche, A. Chenite, J. Zhu, M. Berrada, Valorization of marine wastes in a preserving film based on chitosan for food applications, journal of materials and environmental science, J. Mater. Environ. Sci., V. 6, No. 12, p.3511-3516, 2015.

[55] I. Charhouf, A. Bennamara, A. Abourriche, A. Chenite, J. Zhu, M. Berrada, International Journal of Sciences: Basic and Applied Research, V. 16, No. 2, p.336-348, 2014.

[56] Andrew R. Crofton, Samuel M. Hudson, Kristy Howard, Tyler Pender, Abdelrahman Abdelgawad, Daniel Wolski, Wolff M. Kirsch, Formulation and characterization of a plasma sterilized, pharmaceutical grade chitosan powder, Carbohydrate Polymers, Volume 146, 2016, Pages 420-426.

Section 3

Chitosan Nanocomposites

Chapter 5

Wound Dressing Application of Ch/CD Nanocomposite Film

Ranju Kandra and Sunil Bajpai

Abstract

In this work, carbon dots (CDs), obtained through microwave assisted synthesis from butane tetra carboxylic acid (BTCA), was introduced into chitosan film via simple solvent casting approach. The CDs had an average diameter of 40 to 60 nm as determined by Transmission Electron Microscopy (TEM) analysis. They possessed a zeta potential of −20.2 mV. The X-ray photon spectroscopy (XPS) confirmed presence of carboxylate groups on the surface of carbon dots. The XRD of both the plain sample Ch/CD (0) and carbon dots loaded sample Ch/CD(2) showed two crystalline sharp peaks at 14.6 and 18.1 degree, along with presence of amorphous region also. The moisture absorption data was well fitted on GAB isotherm and the profiles obtained were sigmoidal. The water vapor permeation rates for the sample Ch/CD(0) and Ch/CD(2) were found to be 1758 and 956 g/m^2 /day respectively. The film samples Ch/CD(0) and Ch/CD(20 expanded 2.8 and 103 times when immersed in 4% gelatin solution for 4 h. The % hemolysis for the samples Ch/CD(0) and Ch/CD(2) was 2.12 and 1.11 respectively, thus indicating biocompatible nature of the films. In the ex-vivo mucoadhesion study, the maximum detachment force ($F_{max)}$ was 88.22 and 46.28 mN for the samples Ch/CD(0) and Ch/CD(2) respectively. Finally, both of the samples, namely Ch/CD (0) and Ch/CD(2) scored "0", suggesting their non-cell cytotoxic nature.

Keywords: chitosan film, carbon dots, biocompatible, percent hemolysis

1. Introduction

In recent past, carbon dots have gained considerable importance as a versatile material formultiple applications [1, 2]. Usually they possess a sp^2 conjugated core with various oxygencontaining functionalities such as carboxyl, hydroxyl, aldehyde groups etc. [3, 4]., ease of functionalization, and production of fluorescence on UV exposure,carbon dots (CDs) find a number of biomedical applications like bio-imaging [5–7], targeteddrug delivery [8, 9], wound dressings [10], cancer theranostics [11], screening the purinemetabolic disorders in human fluid etc. [12]. Carbon dots (luminescent) nanoparticles can be usedto track biological processes inside cells. A thorough literature survey reveals that there has not been even a single study which discussesthe changes in physico-chemical properties of chitosan film due to impregnation of carbon dotsinto film matrix. With this objective, we have previously reported synthesis and characterizationof carbon dots from butane tetra carboxylic acid (BTCA) and preliminary investigation of waterabsorption behavior of chitosan/carbon dots nanocomposite film [13]. In continuation, we hereby report a detailed investigation of physico-chemical properties and biocompatibility of

Ch/CD nanocomposite films, taking plain chitosanfilm as control. As chitosan is a biopolymer with a number of biomedical applications, it may beinteresting to see alteration in its properties upon addition of carbon dots [14].

2. Materials and methods

2.1 Preparation of CDs from BTCA

The CDs were synthesized Via microwave method .

2.1.1 Preparation of CD/Chitosan composite film

For fabrication of the chitosan/CDs nanocomposite, 3 wt% homogenous solution was prepared by dissolving chitosan in 0.3 M acid acetic. The sterile square chitosan patterns were produced using a standard square mold with the size of 30 × 40 mm. Various proportions of CDs, including The films were designated as CD/Ch(0), CD/Ch (1), CD/Ch (2) and CD/Ch (3) and Ch/CD(4) were gently added to the chitosan solution to prepare the desirable antibacterial mixture. Respectively, where the number in Parent hesis denotes the volume of CDs solution present in 20 ml of the film forming chitosan solution. Subse- quently, the mixture was injected into the square mold for the formation of composite films. Then, the molds were lyophi- lized in a freeze dryer to complete the molding process.

2.1.2 Characterization of CDs and CD/Ch composite films

The size of the carbon dots was analyzed by Transmission Electron Microscopy (TEM).The TEM samples were made by placing a drop of nanoparticle ethanol suspension on acarbon-coated copper grid. The X-ray photoelectron spectra (XPS) were performed on aVGESCALAB 220-IXL spectrometer using an Al Kα X-ray source (1486.6 eV). The crystallinenature of the plain and CD loaded chitosan film was investigated by X-ray diffraction analysisusing a Rikagu Diffractometer (Cu radiation = 0.1546 nm) operating at 40 kV and 40 mA. The X-ray Photon Spectroscopy (XPS) was also carried out.

2.1.3 Protein adsorption study

In order to study the adsorption of therapeutic protein Bovine serum albumin(BSA) on the filmsurface, the test films, namely Ch/CD(0) and Ch/CD(4), were cut into 1 x 1 cm^2 pieces andimmersed in BSA solution, prepared in phosphate buffer saline(PBS) at a concentration of 5 mg/ml, for a period of 24 h at 37 °C.

2.1.4 Antioxidant properties of hydrogel wound dressings

In order to evaluate the anti-oxidant property of the carbon dots loaded wound dressing film, sample Ch/CD(2) was taken as a representative. We followed two methods to evaluate antioxidant property of the CDs-loaded chitosan film, namely, 2,2-diphenyl-1- picrylhydrazyl(DPPH) radical scavenging assay, and superoxide radical ($O_2.^{\bullet -}$) scavenging activity assay. Inthe first method, definite quantity of powdered sample Ch/CD(2) was added in to methanolicsolution of DPPH radical (100 μM) and allowed to be kept in dark for a period of 18 h [15]. The DDPH solution, without containing film sample, was taken as control to calculate the percentage scavenging activity [16].

$$\text{DPPH radical scavenging activity}\,(\%) = \frac{A_0 - A_1}{A_0} \times 100$$

Where, A_0 is the absorbance of the control and A_1 is the absorbance of test solution respectively.

The Fenton reagent was employed to test the hydroxyl radicals scavenging capacity of plain chitosan film Ch/CD(0) and CDs-loaded sample Ch/CD(2), as described elsewhere [17]. In a typical experiment, pre-weighed quantity of grinded film was added in to a solution which contained 50 mL of 1.0 mM $FeCl_2$, 100 ml of 1 mM 1,10-phenanthroline, 3.6 ml of 0.2 M phosphate buffer (pH 7.8), 175 mL of 0.18 M H_2O_2. The reaction was initiated by addition of pre-calculated quantity of hydrogen peroxide. The reaction mixture was incubated at room temperature for a period of 15 min under mild stirring so as to keep the grinded film powder constantly exposed to free radicals generated. Finally, the absorbance was recorded at 560 nm. The film free solution was taken as control.

The activity of plain Ch/CD(0) and CDs-loaded sample Ch/CD(2) to scavenge superoxide free radicals was investigated in riboflavin/methionine-light system [18]. Results were expressed as percent inhibition of superoxide radicals. All the experiments were carried out in triplicate and average data were given.

3. Results and discussion

3.1 Preparation of CDs

The microwave assisted synthesis of carbon dots is an effective method to prepare CDs. When a solution of BTCA in aqueous medium, is allowed to get exposure of microwaves for a definite time, there occurs uniform heating and finally the volume of the solution is almost reduced to one tenth of the original volume. Now, the residue is taken out and is diluted by the addition of distilled water. The CDs solution, so obtained is centrifuged under high resolution speed of 10000 rpm and the supernatant is collected. The passage of laser beam through the solution is preliminary indication of the formation of carbon dots. The overall synthetic procedure and passage of laser beam through the solution are shown in **Figure 1(a)** and **(b)** respectively.

3.2 Size determination of carbon dots

As per conventional definition, carbon nanoparticles with a diameter of less than 10 nm are called carbon dots [19, 20]. Such small carbon dots are required for for cell imaging and other related biomedical applications [21]. However, in the present work, CDs with relatively larger size were required so that they could be used as efficient crosslinker to control the permeation properties of the chitosan film. The results of TEM analysis are shown in **Figure 2**.

It can be seen that the particles are almost spherical and bear size in the range of 40 to 60 nm. The relatively bigger size range could be attributable to the fact that we did not use any stabilizer in the preparation of CDs and hence chances of formation of bigger carbon nanoparticles could not be ruled out. There are

Figure 1.
(a) Scheme showing formation of carbon dots, (b) passage of laser beam through the pale yellow solution of CDs.

several reports which describe formation of bigger carbon nanoparticles. For example, Smagulova et al. [22] have synthesized carbon dots from birch soot via hydrothermal approach and reported their size in the range of 10 to 60 nm with a maximum percent of particles with diameter of around 25 nm. Similarly, Hou et al. [23] synthesized carbon dots from human hair for detection of Hg (II) ions and reported their diameter in the range of 29 to 80 nm. Similarly, Runa et al. [24] reported synthesis of carbon dots from spider silk via hydrothermal approach and reported an average diameter of 178 nm. Therefore, it appears that it is possible

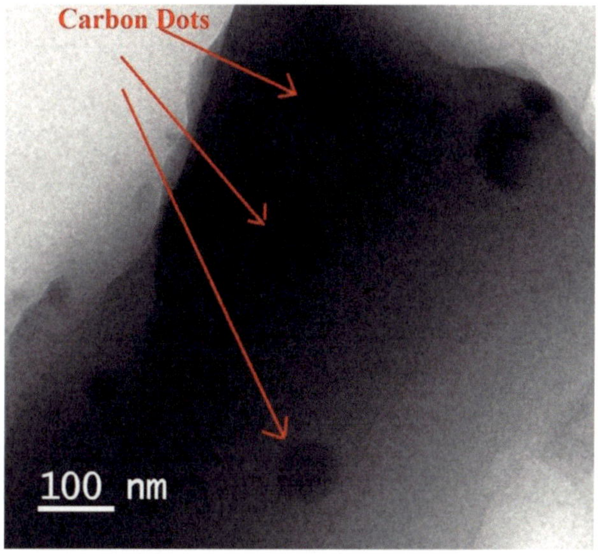

Figure 2.
TEM image of CDs with 100 nm bar length.

to prepare carbon dots of different sizes, depending upon their applicability. For example, CDs mentioned in the aforesaid examples cannot be used for bio imaging and intracellular sensing.

We also measured zeta potential of as –prepared carbon dots which was found to be −20.2 mV, thus indicating the negatively charged surface of carbon dots. This is simply attributable to the presence of carboxylate groups on the surface of carbon dots, thus rendering them negatively charges surface.

3.3 XPS analysis of carbon dots

A survey scan, ranging from 0 to 1200 eV is shown in **Figure 3(a)**. It is evident that two elements, namely C (286 eV) and O (543 eV) are present in the spectrum [25]. The detailed XPS spectra of C 1S and O 1S are shown in **Figure 3(b)** and **(c)** respectively.

The C 1S signal peak can be divided in to three peaks, located at 283.8, 284.83and 287.4 eV. These peaks correspond to C-C, C=C and C=O respectively [26]. Finally, in the XPS spectrum of O1S, the peak at 532.25 eV refers to oxygen singly bound to aliphatic carbon while the smaller or low intensity peak at 535.53 eV refers to COO^- thus confirming the presence of carboxylate groups on the surface of carbon dots.

3.4 Preparation of Ch/CD films

In this work, we prepared Ch/CD films by solvent evaporation method. When the film forming solution is placed in an oven (see experimental section), the solvent is evaporated and semi-transparent film with dark brown appearance is obtained. The crosslinking of chitosan chains by carbon dots may be described as follows: In the film forming solution, chitosan exists in dissolved state with protonated $–NH_3^+$ groups along the macromolecular chains. The final pH of the solution was found to be 6.2. At this pH, -COOH groups present on the surface of the carbon dots, ionize to give negatively charged $–COO^-$ groups, thus rendering negative charges on the surface of CDs. These negative charges bind electrostatically to the protonated amino groups of chitosan chains and a crosslinked network is formed.

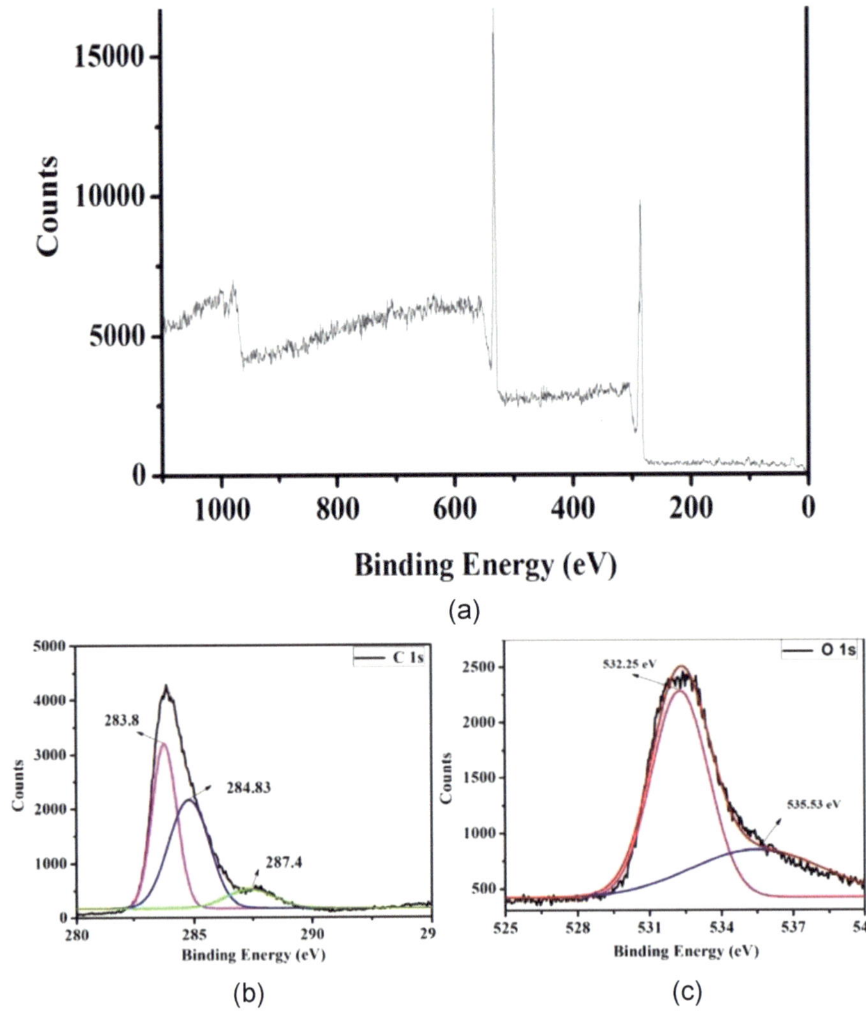

Figure 3.
(a) A survey scan, ranging from 0 to 1200 eV; XPS spectra of (b) C 1S and (c) O 1S.

The plain Ch film appeared to be semi-transparent with pale yellow appearance while the Ch/CD film was dark brownish. An optical photograph of the plain Ch/CD(0) and the Ch/CD(2) films and mode of crosslinking is shown in **Figure 4(a)**.

It is well known that carbon dots emit fluorescence on exposure to UV radiations of suitable Wavelength [27]. This makes them a potential candidate for imaging and other related applications [28]. The optical image of the sample Ch/CD(2), exposed to UV radiations, is shown in **Figure 4(b)**. It can be noticed that the film appears green, due to the fluorescence exhibited by carbon dots present within the film matrix.

3.5 XRD analysis of films

The crystalline nature of the plain and CDs loaded films was investigated by XRD analysis. The XRD patterns of plain film Ch/CD(0) and carbon dots loaded film Ch/CD(2) are shown in **Figure 5(a)** and **(b)** respectively. It can be seen that both of the samples, namely plain sample Ch/CD(0) and CDs-loaded sample Ch/CD(2) exhibit two peaks at 2θ values of 14.6 and 18.1, indicating presence of crystalline region within

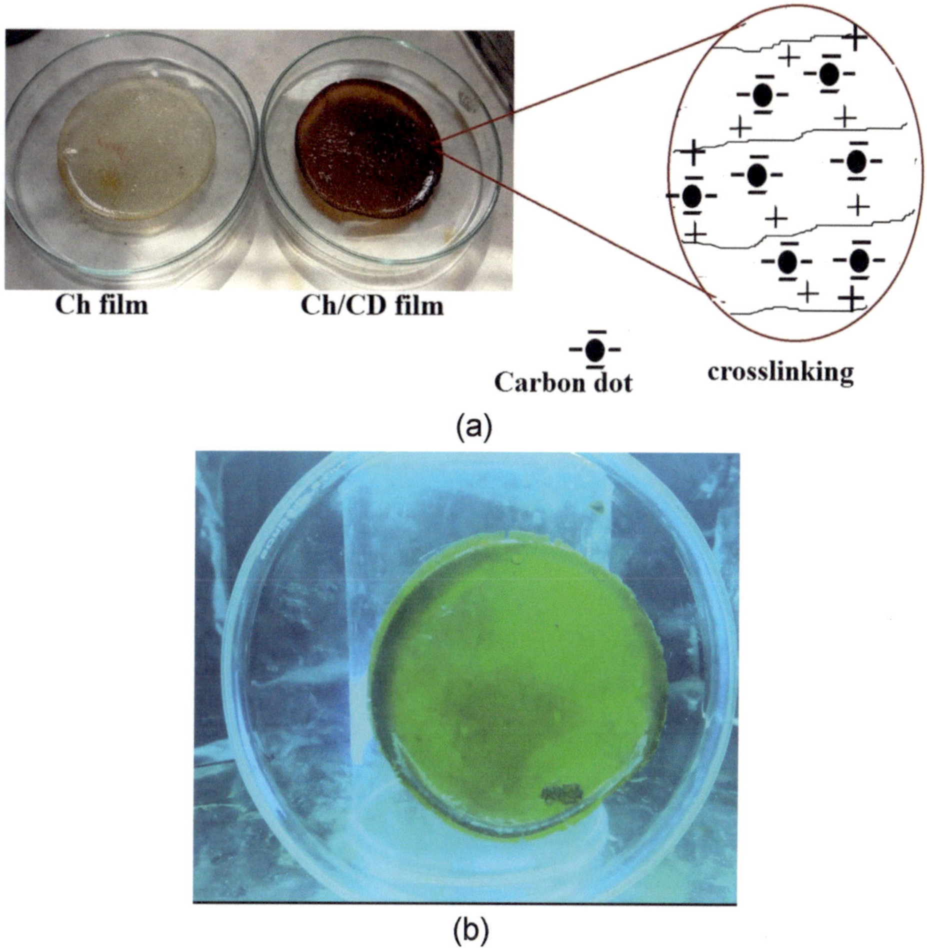

Figure 4.
(a) Optical photographs of the samples Ch/CD(0) and Ch/CreD(2) and mode of crosslinking by carbon dots; (b) optical image of the sample Ch/CD(2) when exposed to UV radiations.

the film matrix. These values are very close to those reported elsewhere [29, 30]. In addition, a scattered broad pattern is also visible, suggesting amorphous region too.

The XRD pattern of the nano-composite film Ch/CD(2), as shown in **Figure 5(b)**, also shows similar pattern with the two peaks, occupying almost the same positions. However, the difference lies in the fact that in the case of CD/Ch(2) film the intensities of the two peakes have decreased remarkably, probably due to presence of amorphous carbon dots within the film matrix. It is also noticable that the amorphous scattered bump is much more pronounced in the XRD pattern of composite film.In this way, it may be concluded that presence of carbon dots within the chitosan film has resulted in increase in the amorphous nature of the composite film.

3.6 Film expansion study

The use of a polymeric film for wound dressing requires fair structural integrity in the presence of exudate coming out from wound. The reason is that when a film is placed over the wound, it comes in contact with the exudate and begins to undergo expansion in its size. If the film expands appreciably, then it may lose its integrity, become soft and sticky, and ultimately may cause inconvenience to the patient. It

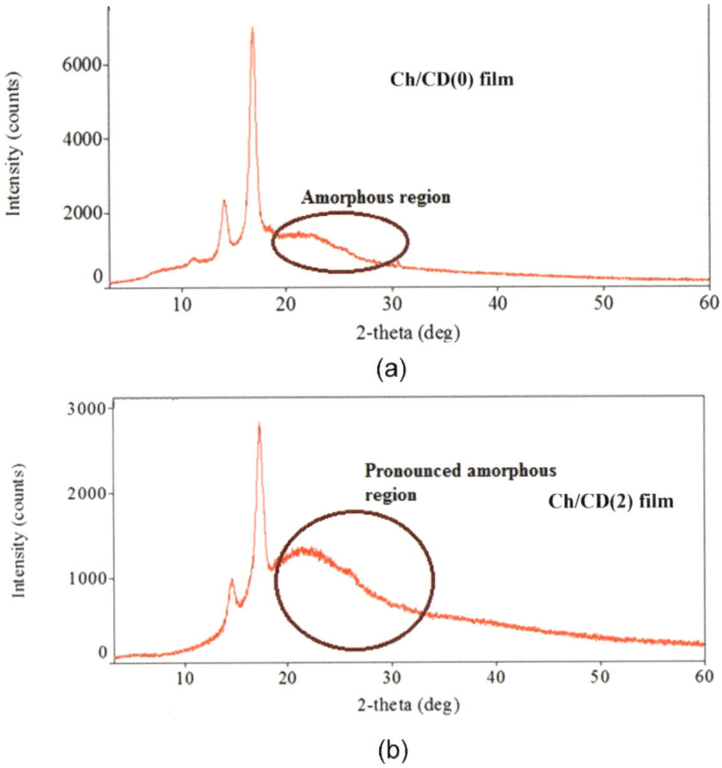

Figure 5.
X-ray diffraction of film samples (a) Ch/CD(0) and (b) Ch/CD(2) .

may also be torn and leave the wound surface. Hence, it becomes essential to test its expansion limit and structural integrity. The results of expansion study are shown in **Figure 6**.

It is noticeable that the plain Ch/CD(0) film undergoes 2.8. Fold expansion in its diameter in duration of 60 min, while in the same time frame, the Ch/CD(2) film expands to only 1.3 times and it maintains its structural integrity throughout. It is also worth mentioning here that the plain film sample Ch/CD(0) gets hydrated, slippery and becomes difficult to handle properly Thus it may be concluded from this study that addition of pre-calculated quantity of carbon dots in to chitosan film can render it enough mechanical strength and control its water absorption capacity as per requirement.

3.7 Biocompatibility tests

Blood compatible nature of the films Ch/CD (0) and Ch/CD(2) was evaluated in the terms of % hemolysis. The % hemolysis of these films was found to be (2.12 ± 0.02) and (1.13 ± 0.18) respectively.

3.8 Protein adsorbed study

The adsorption of protein on a wound dressing film indicates its cell adhesion behavior. Albumin, a multifunctional transporter protein, is the most abundant protein found in the plasma (approx. 50 $mg.ml^{-1}$) [31]. Its adsorption is related to the inhibition of the coagulation cascade and consequently, platelet adsorption. As per reports [32], albumin has high adsorption affinity for hydrophobic surfaces,

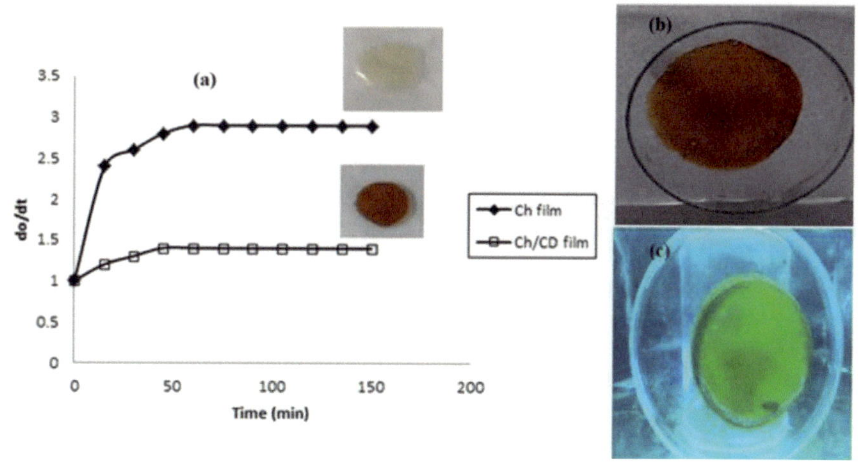

Figure 6.
d/d_o (versus time profiles for the samples) Ch/CD (0) and Ch/CD(2) at 37° C.

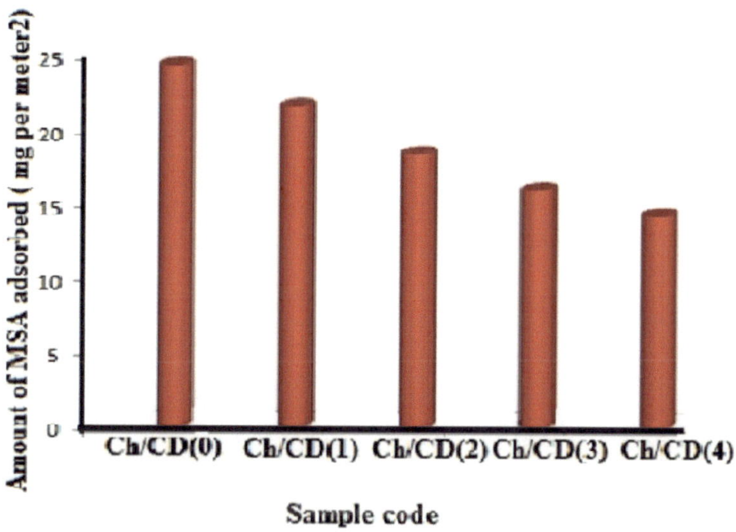

Figure 7.
BSA adsorption on the various samples Ch/CD (0), Ch/CD (1), Ch/CD (2), Ch/CD (3) and Ch/CD (4) at 37° C.

mainly because of hydrophobic interactions between the protein and the surface. Indeed, a wound dressing film with a poor tendency to adsorb albumin is desirable. The results of BSA adsorption study are shown in **Figure 7**.

It can be seen that plain sample Ch/CD(0), and CDs-loaded samples Ch/CD(1), Ch/CD(2), Ch/CD(3) and Ch/CD(4) show BSA adsorption of 24.2, 21.5, 18.3, 15.9 and 14.1 mg/m^2 respectively. The extremely low values of BSA adsorption may be indicative of the hydrophilic nature of the film surfaces for all the samples studied. It has been reported [33] (**Figure 8**).

3.9 Antioxidant properties of hydrogel wound dressings

In this work, % scavenging capacity of the plain film sample Ch/CD(0) and carbon dots loaded samples Ch/CD(1), Ch/CD(2), Ch/CD(3) and Ch/CD(4) for the various free radicals i. e. DPPHR, SOR and HR are shown in **Figure 9**. It can be seen

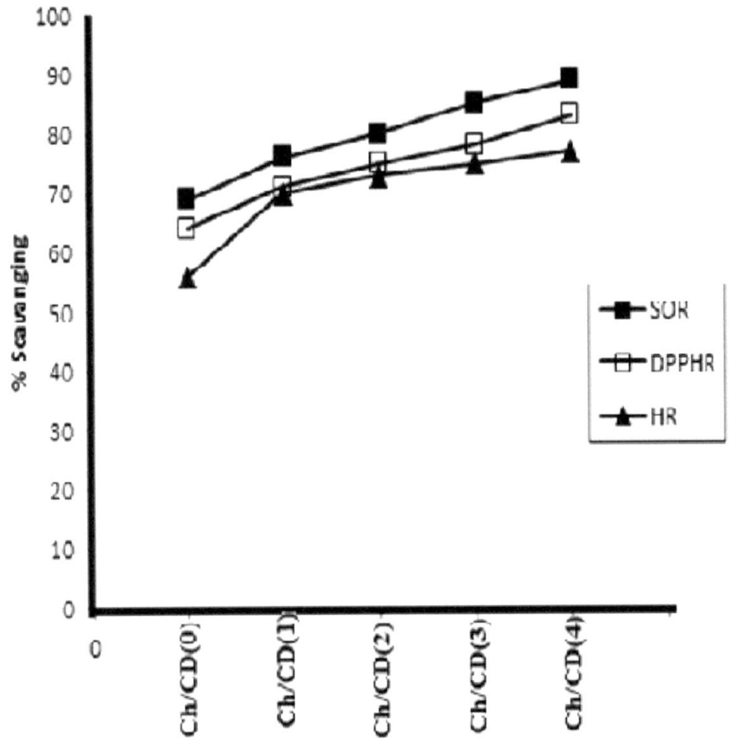

Figure 8.
Percent scavenging for DDPH, superoxide and hydroxyl free radicals by various film samples.

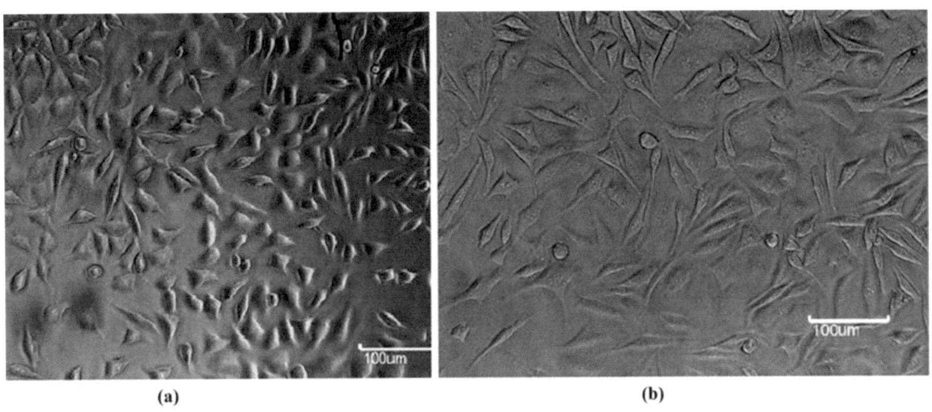

Figure 9.
Images showing completely undamaged L929 cells after 24 h contact with extract of film samples (a) Ch/CD (0) and (b) Ch/CD(2).

that plain chitosan film sample shows fair % Scavenging capacity and it increases slightly due to addition of carbon dots in the film matrix.

Owing to the fair stability, DPPH radical is usually employed to determine the antioxidant or free radical scavenging activity of the wound dressing films. The method involves reduction of methanolic DPPH radical by hydrogen donating antioxidant polymeric film. Topical wound healing formulations with DPPH scavenging ability have been reported to improve skin repair and regeneration [34]. The observed increase may probably be due to the contribution of –COOH groups present on the surface of carbon dots towards reduction of DPPH radicals. It is here also worth mentioning that chitosan has already been reported to have free radicals scavenging activity [35]. Therefore it appears that addition of functionalized carbon dots into chitosan film matrix improves its antioxidant property.

Hydroxyl, an oxygen centered radical, attacks proteins, DNA, polyunsaturated fatty acid in membranes, and most biological molecules with which it comes in contact [36]. It abstracts hydrogen atoms from membrane lipids [37] and brings about peroxidic reaction of lipids. As per reports [38], the scavenging action of chitosan against hydroxyl radicals may probably be due to following reactions (i) The hydroxyl groups, present within the chitosan macromolecular chains, may react with. OH via hydrogen abstraction, and (ii). OH can react with the residual free amino groups NH_2 to form stable macromolecule radicals. Finally, when superoxide anions come in to contact with a biomolecule, they damage it directly or indirectly by forming H_2O_2, ^{-}OH, peroxy nitrite or singlet oxygen during aging and pathological events such as ischemic reperfusion injury. It is also reported that Super oxide radicals initiate lipid peroxidation [39]. Therefore, scavenging of superoxide radicals by chitosan based wound dressing might be helpful for preventing delayed wound healing induced by superoxide radicals in pathological conditions. The scavenging of SOR will also diminish formation of hydroxyl radicals, which is considered main factor for delayed healing of chronic wounds.

3.10 Ex-vivo mucoadhesion studies of films

The adhesion of a wound dressing film on the wounded skin is a significant parameter and requires a perfect balance between the adhesion capacity of the dressing film and comfort level of patient. In case, the film has a very strong adhesion tendency, it may cause discomfort and pain during removal of the dressing. However, its poor adhering property may also be uncomfortable from the point of view of wound healing. In this work, maximum detachment force (F_{max}) required to detach the film from mucosal surface, was determined for all the film samples namely, Ch/CD(0), Ch/CD(1), Ch/CD(2), Ch/CD(3) and Ch/CD(4). The values of F_{max} were found to be 88.22 ± 11.52, 45.15 ± 8.61, 43.29 ± 7, 44.25 ± 6.97, and 46.22 ± 5.94mN.

In a report, F_{max} value for gum acacia-cl-(poly(HEMA-co-carbopol hydrogel film was found to be 70.20 ± 17.57 m N. The higher value was attributed to the fair hydrophilic nature of the film [39]. In the present work, F_{max} values for various CDs-loaded chitosan films were relatively low, probably due to the presence of functionalized carbon dots within the film matrix. The poor water absorption tendency of these films did not induce polymeric chain relaxation and hence there were no new active sites available for exposure to mucus surface. In addition, the physical crosslinks between carboxylate groups present on the carbon dots surface (as discussed earlier) and protonated amino groups on chitosan chains also made the chitosan chains rigid and prevented them from relaxation.

A plausible explanation for mucoadhesion behavior of plain chitosan film Ch/CD(0) and CDs-loaded sample Ch/CD(2) may be given on the basis of the most commonly proposed Diffusion-Interpenetration Theory [40, 41]. In the case of pure chitosan film sample Ch/CD(0), an appreciable water content within the film matrix causes polymeric chains to relax or un-fold,thus resulting in exposure of new active sites. These polar segments come in contact with mucuschains and get entangled with them. This results in stronger bio-adhesion as shown in **Figure 10(a)**.

However, in the case of CDs-loaded sample Ch/CD(2), the film absorbs very small quantity of water due to physical crosslinks and therefore polymeric chains remain folded or non-relaxed. This minimizes the interaction with mucus chains as shown in **Figure 10(b)**. As a result, value of F_{max} is quite low.

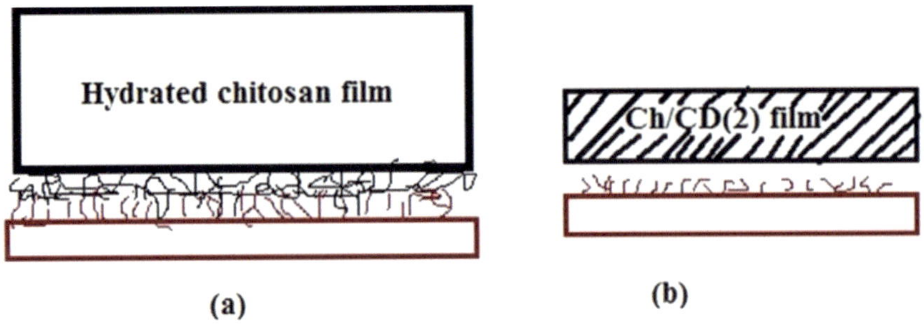

Figure 10.
Bio adhesion in the film samples Ch/CD(0) and Ch/CD(2).

3.11 Cell cytotoxicity studies

The results of cell cytotoxicity for the plain sample Ch/CD (0) and the CDs loaded sample Ch/CD (2) are shown in **Figure 9(a)** and **(b)** respectively.

As per ISO 10993-5, the achievement of numerical grade more than 2 is considered as toxic. The score was obtained as 0 for both of the samples, thus indicating non-cell cytotoxic nature of both of the film samples. It can be seen that there are discrete intra cyto plasmatic granules, no cell lysis and there is no reduction of cell growth.

4. Conclusions

It may be concluded from the above study that BTCA, a tetra carboxylic acid compound, can conveniently be used as precursor to synthesize negatively charged carbon dots. These carbon dots, when loaded into plain chitosan film, cause a drastic fall in the water absorption capacity of the resulting nanocomposite films. A detailed investigation of these CDs-loaded chitosan film is under progress and a detailed report shall be documented soon.

Author details

Ranju Kandra* and Sunil Bajpai
Polymer Research Laboratory, Department of Chemistry, Government Model Science College, Jabalpur, M.P., India

*Address all correspondence to: kandraranju@gmail.com

References

[1] Diao H, Li T, Zhang R, Kang Y, Liu W, Cui Y, Wei S, Wang N, Li L, Wang H, Niu W, Sun T, (2018) Jul 5 Facile and green synthesis of fluorescent carbon dots with tunable emission for sensors and cells imaging 200:226-234, Spectrochim Acta A Mol Biomol Spectrosc.

[2] Shi H, Wei J, Qiang L, Chen X, Meng X, (2014) Fluorescent carbon dots for biolmaging and biosensing applications, 2677-99. J Biomed Nanotechnol

[3] Li Y, Ren J, Sun R, Wang X, (2018) Fluorescent Lignin Carbon Dots for Reversible Responses to High-Valence Metal Ions and Its Bioapplications. J Biomed Nanotechnol. 14(9), 1543-1555

[4] Lin H, Ding L, Zhang B, Huang J. ,R Soc(2018) Detection of nitrite based on fluorescent carbon dots by the hydrothermal method with folic acid. Open Sci, 172149

[5] Sharma V, Kaur N, Tiwari P, Mobin SM, J Full color emitting fluorescent carbon material as reversible pH sensor with multicolor live cell imaging (2018) PhotochemPhotobiol B. (182)137-

[6] Kumar SU, Bhushan B, Gopinath P, Bioactive carbon dots lights up microtubules and destabilises cell cytoskeletal framework (2017)A robust imaging agent with therapeutic activity., Colloids Surf B Biointerfaces159 ,662-672 https://doi.org/10.1016/j.colsurfb.2017.07.054

[7] Park SY, Lee CY, An HR, Kim H, Lee YC, Park EC, Chun HS, Yang HY, Choi SH, Kim HS, Kang KS, Park HG, Kim JP, Choi Y, Lee J, Lee HU, (2017) Advanced carbon dots via plasma-induced surface functionalization for fluorescent and bio-medical applications Nanoscale 9(26) 9210-9217

[8] Sarkar C, Chowdhuri AR, Kumar A, Laha D, Garai S, Chakraborty J, Sahu SK, (2018) One pot synthesis of carbon dots decorated carboxymethyl cellulose-hydroxyapatite nanocomposite for drug delivery, tissue engineering and Fe3+ ion sensingCarbohydrPolym. 181, 710-718

[9] Jha S, Mathur P, Ramteke S, Jain NK, (2017) Pharmaceutical potential of quantum dots Artif Cells NanomedBiotechnol 1-9

[10] Harroun SG, Lai JY, Huang CC, Tsai SK, Lin HJ,(2017) Reborn from the Ashes: Turning Organic Molecules to Antimicrobial Carbon Quantum DotsACS Infect Dis 3(11):777-779

[11] Yao C, Tu Y, Ding L, Li C, Wang J, Fang H, Huang Y, Zhang K, Lu Q, Wu M, Wang Y., (2017)Tumor Cell-Specific Nuclear Targeting of Functionalized Graphene Quantum Dots In Vivo., Bioconjug Chem. Oct 18;28(10):2608-2619

[12] Cui L, Li CC, Tang B, Zhang CY, (2018) Advances in the integration of quantum dots with various nanomaterials for biomedical and environmental applications, Analyst. 29;143(11):2469-2478.

[13] Bajpai SK, Kandra R, (2019) "Synthesis, characterization of carbon dots from BTCA and their application as an effective crosslinker to control the swelling of chitosan films Nano Materials (Accepted)

[14] S. K. Bajpai, Pradeep Daheriya, Sonam Ahuja & K. Gupta, Water absorption and antimicrobial behavior of physically cross linked poly (vinyl alcohol)/ carrageenan films loaded with minocycline, Designed Monomers and Polymers, 2016

[15] Rajani Kanta Sahu ,ManoranjanKar , RasmiraniRoutray, (2013) DPPH Free Radical Scavenging Activity of Some Leafy Vegetables used by Tribals of Odisha, India. Journal of Medicinal Plants Studies, 1(4),21-27

[16] Fischbacher A1, von Sonntag C2, Schmidt TC, Hydroxyl radical yields in the Fenton process under various pH, ligand concentrations and hydrogen peroxide/Fe(II) ratios., Chemosphere. 2017 ;182:738-744.

[17] Sanchez-Moreno C. (2002) Methods used to evaluate the free radical scavenging activity in foods and biological systems. Food Sci Tech Int.;8:121-

[18] C. Lau, M. J. Cooney and P. Atanassov, *Langmuir*, 2008, 24, 7004-7010.

[19] Jia Zhang, Shu-Hong Yu, (2016) Review : Carbon dots: large-scale synthesis, sensing and 600 bio imaging, Materials Today, Volume 19, Issue 7, Pages 382-601 393]

[20] Ghosal K, Ghosh A., Carbon dots: The next generation platform for biomedical 603 applications., Mater Sci Eng C Mater Biol Appl. 2019 Mar;96:887-903.

[21] Smagulova S A, Egorova M N, Tomskaya A E and Kapitonov A N, Synthesis of Carbon 606 Dots with Tunable Luminescence,Journal of Material Sciences & Engineering, J Material Sci 607 Eng. 6: 376.

[22] J. Hou, J. Li, J. Sun, S. Ai and M. Wang, RSC Adv, 2014, 4, 37342-37348.

[23] S. Ruan, B. Zhu, H. Zhang, J. Chen, S. Shen, J. Qian, Q. He and H. Gao, J. Colloid Interface Sci., 2014, 422, 25-29.

[24] Rosemary L. Calabro, Dong-Sheng Yang , Doo Young Kim, (2018) Liquid-phase laser ablation synthesis of graphene quantum dots from carbon nano-onions: Comparison with chemical oxidation, Journal of Colloid and Interface Science 527 132-140

[25] Hongye Huang, Meiying Liu, XunTuo, Junyu Chen, Liucheng Mao, Yuanqing Wen, 617 JianwenTian, Naigen Zhou, Xiaoyong Zhang, Yen Wei, (2018) A novel thiol-ene click reaction 618 for preparation of graphene quantum dots and their potential for fluorescence imaging,Materials 619 Science & Engineering C 91 631-637

[26] Yifan Wang, Yanwu Zhu, Shaoming Yu and Changlong Jiang,(2017) Fluore scent carbon 621 dots: rational synthesis, tunable optical properties and analytical applications, RSC Adv 7, 622 40973-40989

[27] Dan Wang ,Zhiyong Wang , Qiuqiang Zhan , Yuan Pu , Jie-Xin Wang , Neil R. Foster , 624 Liming Dai,((2017) Facile and Scalable Preparation of Fluorescent Carbon Dots for 625 Multifunctional Applications, Engineering 3 402-408

[28] N.F.Mohd Nasir, 2N. Mohd Zain, 2 M.G. Raha, 2N.A. Kadri,(2005) Characterization of 628 Chitosan-poly (Ethylene Oxide) Blends as Haemodialysis Membrane American Journal of 629 Applied Sciences 2 (12): 1578-1583,

[29] Parichat Norranattrakul, Krisana Siralertmukul and Roongkan Nuisin, (2013) Fabrication of 6 chitosan/ titanium dioxide composites film for the photocatalytic degradation of dye Journal of Metals, Materials and Minerals, Vol.23 No.2 pp.9-22,

[30] Blahovec J., Vanniotis S, (2008) GAB generalized equation for sorption phenomena,. Food and Bioprocess Technology 1,82-90.

[31] JiříBlahovec and Stavros Yanniotis, (2010) 'Gab' Generalized Equation

as a Basis for Sorption Spectral AnalysisCzech J. Food Sci, 28 (5): 345-354.

[32] NurulMujahidah Ahmad Khairuddin, Amalina Muhammad afifi, NurAwanisHashim, Shaza 6 Eva Mohamad&KatayoonKalantari, (2018) Immobilization of Bovine Serum Albumin on the Chitosan/PVA Film, SainsMalaysiana 47(6) 1311-1318

[33] Lackner, J. M. and W. Waldhauser, (2010) Inorganic PVD and CVD Coatings 6 inMedicine— A Review of Protein and Cell Adhesion on Coated Surfaces. J AdhesSciTechnol 24, 925-961

[34] Zhao Y, Yu L, Dong X, Sun Y., Protein adsorption to poly(ethylenimine)-modified sepharose FF: VII. Complicated effects of pH., J Chromatogr A. 2018 Dec 14;1580:72-79.

[35] Karina Kubiak-Ossowska , Karolina Tokarczyk, Barbara Jachimska , and Paul A. Mulheran, (2017) Bovine Serum Albumin Adsorption at a Silica Surface Explored by Simulation and Experiment,J. Phys. Chem. B, , 121 (16), pp 3975-3986, DOI:10.1021/acs.jpcb.7b01637

[36] D. Draganescu, C. Ibanescu, B. I. Tamba, C. V. Andritoiu, G. Dodi and M. I. Popa, (2015)Int. J. Biol. Macromol., , 72, 614-623

[37] Ballester-Costa C, Sendra E, Fernández-López J, Viuda-Martos M (2016)Evaluation of the antibacterial and antioxidant activities of chitosan edible films incorporated with organic essential oils obtained from four Thymus species J Food Sci Technol. 53(8):3374-3379..

[38] Sharma SK, Singh AP, (2012) In Vitro Antioxidant and Free Radical Scavenging Activity of Nardostachysjatamansi DC, J Acupunct Meridian Studies,;5(3):112e118.

[39] Ismael Tejero, Angels Gonza'lez-Lafont, Jose´ M. Lluch, and Leif A. Eriksson, (2007) 6 Theoretical Modeling of Hydroxyl-Radical-Induced Lipid Peroxidation Reactions, J. Phys. Chem. B , 111, 5684-5693

[40] Tamer M Tamer, Katarina Valachová, Mohamed Samir Mohyeldin, Ladislav Soltes, (2016) Free radical scavenger activity of chitosan and its aminated derivative, Journal of Applied 6 Pharmaceutical Science Vol. 6 (04), pp. 195-201,

[41] Sharma SK, Gupta VK. (2008) In vitro antioxidant studies of Ficusracemosa Linn Root. Phcog Mag. 4:70e74.

Section 4

Physicochemical Properties

Chapter 6

Ternary Solid Dispersion Strategy for Solubility Enhancement of Poorly Soluble Drugs by Co-Milling Technique

Marouene Bejaoui, Hanen Oueslati and Haykel Galai

Abstract

Amorphous ternary solid dispersion has become one of the strategies commonly used for improving the solubility and bioavailability of poorly water soluble drugs. Such multicomponent solid dispersion can be obtained by different techniques, this chapter provides an overview of ternary solid dispersion by co-milling method from the perspectives of physico-chemical characteristics in vitro and in vivo performance. A considerable improvement of solubility was obtained for many active pharmaceutical ingredients (e.g., Ibuprofen, Probucol, Gliclazid, Fenofibrate, Ibrutinib and Naproxen) and this was correlated to the synergy of multiple factors (hydrophilicity enhancement, particle size reduction, drug-carrier interactions, anti-plasticizing effect and complexation efficiency). This enhanced pharmacokinetic properties and bioavailability of these drug molecules (1.49 to 15-folds increase in plasma drug concentration). A particular focus was accorded to compare the ternary and binary system including Ibuprofen and highlighting the contribution of thermal and spectral characterization techniques. The addition of polyvinylpyrrolidone (PVP K30), a low molecular weight molecule, into the binary solid dispersion (Ibuprofen/β-cyclodextrin), leads to a 1.5–2 folds increase in the drug intrinsic dissolution rate only after 10 min. This resulted from physical stabilization of amorphous Ibuprofen by reducing its molecular mobility and inhibiting its recristallization even under stress conditions (75% RH and T = 40°C for six months).

Keywords: amorphous ternary solid dispersion, co-milling, PVP, physical stability, aqueous solubility, bioavailability, Ibuprofen

1. Introduction

In recent years, solid dispersion technology by milling technique was largely utilized by researchers in order to enhance dissolution rate, bioavailability and thus therapeutic efficiency of several poorly water-soluble drugs (**Table 1**), as it represents a simple, economic and environmental process without using solvents [1–9]. In fact, millings enable particle size reduction and promote the formation of drug nanoparticles, which enhance solubility, flow properties and content uniformities of pharmaceutical dosage forms [10]. However, this process might induced

Drug molecules	Authors
Piroxicam	Anna Penkina et al. [1]
Praziquantel	Alessio Gaggero et al. [2]
Telaprevir	Xinnuo Xiong et al. [3]
Hydrochlorothiazide	Sakib M. Moinuddin et al. [4]
Indomethacin	Georgia Kasten et al. [5]
Nifedipine	Pooja J. Patel et al. [6]
Ibuprofen	Amjad Hussain et al. [7]
Bicalutamide	Joanna Szafraniec et al. [8]
Celecoxib	Zhuang Ding et al. [9]

Table 1.
Some examples of drug molecules exhibiting solubility enhancement by binary solid dispersion using co-milling technique [1–9].

drug transition from crystalline to amorphous state which is more soluble in water but physically unstable in some cases [11]. Physical stabilization of such unstable amorphous material required an optimization strategy using additives (milling time and rate, compatible carriers with optimized proportion) in order to preserve its chemical integrity (absence of degradation) and inhibiting phase transformations or polymorphic conversion towards unstable forms [11]. In some cases, the stabilization and solubilization efficiency of binary solid dispersion is weak by exhibiting limited bioavailability enhancement [12] and required a large amount of carriers. In order to further enhance drug dissolution rate, several researchers have introduced third compound in drug formulations, this led to simultaneous enhancement of drug solubility and physical stability [13–19]. In this chapter, the challenges and strategies in developing robust ternary solid dispersion of high stability and performance are briefly discussed.

2. Ternary solid dispersion: a new alternative to promote drug dissolution rate

Solubility and stability enhancement of drug molecules in ternary solid dispersion resulted from various mechanisms (**Table 2**) including intermolecular interactions (drug/carrier, carrier/carrier) and synergetic effects of excipients. This required the use of appropriate carrier showing compatibility in ternary mixtures (e.g. polymer, surfactant, crosslinked polymer, adsorbent, aminoacids, cyclodextrin molecules) and reinforcing stabilization of amorphous drug by preventing its recristallization or chemical degradation.

Watanabe et al. (2002) have reported the physical stabilization of amorphous Indomethacin (IM) by ternary solid dispersion using Mg $(OH)_2$ and SiO_2 as carriers [13]. DRIFT (Diffuse Reflectance Infrared Fourier Transform Spectroscopy) results have shown the disappearance of the OH group (carboxylic group) band of IM (**Table 3**) in ternary system and this was attributed to mechanical dehydration generated by ball milling in presence of Mg $(OH)_2$ and SiO_2. Such dehydration leads to the increase in the acidity of the carriers and enhances the acid–base interaction which ensures the rapid amorphization of IM (**Table 3**). The presence of these two carriers has increased the glass transition temperature of Indomethacin, the Tg of ternary system was higher than that of binary ground mixtures (IM-SiO_2,

Drug molecules	Carriers	Mechanisms	Pharmacokinetic properties (plasma drug concentration)	Physical stability	Drug release (at 37.0 ± 0.5°C)	Authors
Indomethacin	$Mg(OH)_2/SiO_2$	Acid–base interaction and formation of C - O -Si bridging bond	—	At 30°C and 11% RH for 2 months	—	[13]
Ibrutinib	Oxalic acid (OXA)/ microcrystalline cellulose (MCC)	Ionic Interactions	1.49-fold greater than crystalline Ibrutinib	At 75% RH and T-40°C for 6 months	5.33-folds higher than crystalline Ibrutinib	[14]
Probucol	HPC, pluronic P6g and SDS	Micelle solubilization, increase in hydrophilicity	15-folds higher than that of coarse Probucal suspension	7 days at 4 °C or 25 °C	About 40% at 2 h	[15]
	Kollidon® VA64, Cremophor® RH40	Hydrophilicity, wettability and particle size reduction	6.0-folds higher than commercial tablet	—	30% at 2 h (commercial tablets was only 0.68% at 2 h)	[16]
Naproxen	Hydroxypropyl-β-cyclodextrin, arginine	Hydrogen bond formation, Electrostatic interactions	—	—	—	[17]
Fenofibrate	Hydroxypropyl-β-cyclodextrin, hydroxypropyl methylcellulose (HPMC)	Formation of inclusion complexes	—		90% in 20 min	[18]
Gliclazid	Sodium lauryl sulfate, crosslinked polyvinylpyrrolidone	Prevention of drug agglomeration, wettability and hydrophilicity enhancement	—	—	90% in 2 h	[19]

Table 2.
Specificities of several ternary systems involving poorly water soluble drugs (Indomethacin, Ibrutinib, Probucol, Naproxen, Fenofibrate, and Gliclazid).

		IM + $Mg(OH)_2$	IM + $Si(O)_2$	IM + $Si(O)_2$ + $Mg(OH)_2$
Amorphization time		60 min	60 min	10 min
DRIFT	OH (IM)/at 3377 cm^{-1}	Decreased	Decreased	Disappeared
Bands	Silanol group/at 3747 cm^{-1}	—	Disappeared	Disappeared
	OH (Mg $(OH)_2$)/at 3377 cm^{-1}	Unchanged	—	Unchanged

Table 3.
Observed changes for Indomethacin ternary solid dispersion compared to binary mixtures [13].

IM-$Mg(OH)_2$). After storage at 30°C and 11% of RH, IM-SiO_2 ground mixtures have shown rapid crystallization of amorphous Indomethacin. However, in the ternary co-grinding system, no crystallization was observed and this was attributed to higher acid–base interaction between Indomethacin and SiO_2 -Mg $(OH)_2$ interface that promotes the formation of bridging bonds (Si-O-C or Mg-O-C) which inhibited the molecular mobility of amorphous drug.

Mura et al. (2003), have shown that aqueous dissolution of Naproxen (a poorly water-soluble anti-inflammatory drug) can be considerably enhanced by combination with hydroxypropyl- β-cyclodextrin and some aminoacids (Arginine) [14]. Such ternary system exhibited higher stability constant and drug solubility (at pH ≈ 7and T = 25°C) than binary system (Naproxen/Arginine). The synergetic effect in Naproxen solubility (a 13.4-fold increase compared to pure drug) can be attributed to the establishment of electrostatic interactions between Arginine and the carboxylic group of Naproxen, as well as hydrogen bond formation between Arginine and the hydroxyl groups of HPβCD [14].

Lauretta et al. (2015) have shown that amorphous ternary solid dispersion of Gliclazid (poorly soluble drug used in the treatment of patients with type 2 diabetes) with crosslinked polyvinylpyrrolidone and SLS (Sodium Lauryl Sulfate) by co-milling method exhibited higher dissolution rate compared to the commercial product "Diabrezide" (Drug release of Gliclazid reached 90% in 2 h) [15]. Such solubility enhancement resulted from prevention of drug agglomeration in solution and improvement in wettability and hydrophilicity of co-milled particles.

In the case of Fenofibrate (FNB), a lipid-lowering drug (Class II) used in the treatment of hypertriglyceridemia and mixed hyperlipidemia, Xizhao Ding et al. (2018) have shown that the addition of Hydroxypropyl methylcellulose (HPMC) to the binary solid dispersion of Fenofibrate and Hydroxypropyl-β-cyclodextrin (molar ratio of 1:1), has shown a considerable enhancement of drug dissolution rate [16]. This ternary system has shown a percentage of 90% of drug release in 20 min (at 37 ± 0.5°C/pH = 7) which is higher than pure drug (24%) or binary system (60%). Such ternary solid dispersion was obtained by ball milling and led to the formation of inclusion complexes (increase in stability constants and complexation efficiency). This was attributed to the strong interactions established between FNB and HP-β-CD in presence of HPMC such as Van Der Waals dispersion forces, hydrophobic and hydrogen bonds, following the release of high-energy water molecules from HP-β-CD cavity [16].

Co-milling Ibrutinib (poorly water soluble antitumor drug) with oxalic acid (OXA) and microcrystalline cellulose (MCC) for six hours (Man Zhang et al. 2019) led to a simultaneous improvement in drug dissolution rate (5.33-fold higher than crystalline Ibrutinib) and physical stability of amorphous drug under stress conditions (75% RH and T = 40°C for six months) [17]. Plasma drug concentration of the ternary system (Ibrutinib, OXA, and MMC) exhibited also an increase of

1.49-fold compared with crystalline Ibrutinib. This was attributed to wettability and hydrophilicity enhancement, as well as the presence of ionic interactions between drug and carriers as suggested by XPS (X-ray photoelectron spectroscopy) analysis showing the appearance of protonated amines (N^+) and high binding energy (+2.75 eV) in the ternary system (Ibrutinib/OXA/MCC) [17].

The combination of polymer (Kollidon® VA64) and surfactant (Cremophor®RH40) was effective for Probucol (poorly water soluble anti-oxidative drug/BCS II) solubility enhancement by ball milling technique (Lijia et al. 2017) [18]. This considerably enhanced in vitro dissolution and in vivo bioavailability in rats. Such enhancement was attributed to the greater hydrophilicity, increased wettability and particle size reduction. The local solubilization effect of surfactant contributed also for preventing the aggregation of drug particles during dissolution. Otherwise, pharmacokinetic study has shown an increase of maximum plasma drug concentration for the ternary co-milled system which was 6.0-folds greater than that of Probucol commercial tablets [18].

On the other hand, Fang Li et al. (2019) have shown that dissolution rate and oral absorption of Probucol can be further improved by the formation of drug nanosuspensions (planetary beads-milling technique) using ternary stabilizers mixtures (Hydroxypropyl cellulose, an anionic surfactant (Sodium dodecylsulfate, SDS) and a nonionic surfactant (Pluronic F69)) [19]. Such Probucol nanosuspensions were physically stable after storage during 7 days at 4°C or 25°C, with highest dissolution rate (more than 60% at 2 h). The in vivo pharmacokinetic study has also shown 15-folds higher value of the plasma Probucol concentration compared to that obtained for coarse Probucol suspension. Probucol dissolution enhancement was attributed to particle size reduction and the characteristics of the polymeric chain which is dependent on the nature of polymeric stabilizer used in the mixture [19].

3. Solubility enhancement of Ibuprofen by formation of physically stable amorphous ternary system (Ibuprofen, PVP, β-cyclodextrin)

In recent years, researchers have used several techniques to improve the dissolution rate and bioavailability of Ibuprofen (IB), a non-steroidal anti-inflammatory drug (NSAID) which is poorly water soluble. The manipulation of the solid state of ibuprofen (**Figure 1**) remains a challenge for researchers because of its lower glass transition ($T_g = -42 \pm 1°$ C) and its tendency to recrystallized at room temperature [20]. The solubility improvement of ibuprofen was achieved by solid dispersion with different excipients (HPMC, Soluplus, PVP, Kaolin) [21–23], which differ in their solubilization abilities and their interactions with drug molecules. Several researchers have used complexation in the presence of βCD to improve the bioavailability of IB [24]. A water-soluble complex IB/βCD was obtained by co-precipitation or granulation (wet process) [25, 26]. Ibuprofen can also form an inclusion complex

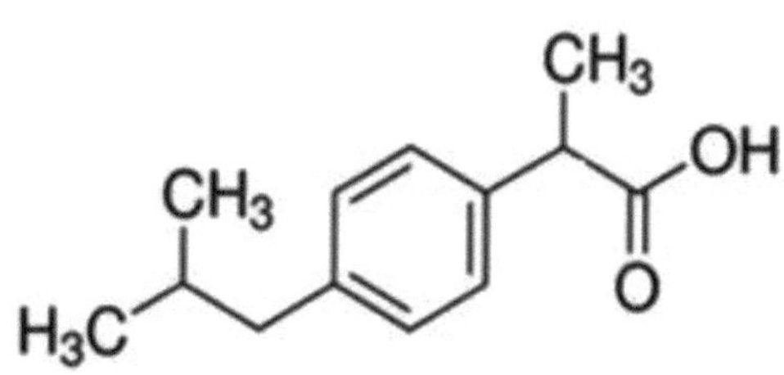

Figure 1.
Ibuprofen molecule.

with βCD mechanically [27]. It has been shown that the complexing efficiency of the IB/βCD system formed in the solid state depends on the techniques applied [28]. In our previous published work [29], binary solid dispersion of IB was achieved by co-milling the drug and βCD molecules, and then PVP was added to the binary mixture (IB/βCD) in order to evaluate physico-chemical changes in solid state.

3.1 Milling method

Ibuprofen was milled in presence of β-Cyclodextrin and then PVP at different weight ratios, physical mixtures (PM) were prepared by homogenization of pure components using ceramic mortar. Milling procedure was performed in a planetary ball mill (Pulverisette 7, Fritsch) using two milling jars (45 cm3)/7 balls (ø = 1 cm) in ZrO_2. The rotation rate was set to 300 rpm and the ball/sample weight ratio was 82.5:1. The milling procedure was optimized for 10 h at room temperature (≈298 K) constituted by 20 min milling periods with pause periods (10 min) in order to minimize the overheating of the sample.

3.2 Results

In the case of the binary system (IB/βCD), the X-ray diffraction results showed the complete amorphization of the βCD (**Figure 2**), while the crystallinity of the IB has been slightly modified [29]. Amorphization of βCD by milling at room temperature is predictable, as milling is carried out at a temperature sufficiently below the Tg of βCD (Tg ≈ 292°C) [30]. The enlargement of the Bragg peaks is explained by the reduction in the size of the crystallites and slight distortions of the crystal lattice generated by ball milling. As shown by SEM micrographs (**Figure 2**), the birefringence and the crystallinity of drug particles were moderately affected in the binary system (IB/βCD). By adding 20% of PVP to this binary mixture, the formation of amorphous agglomerations can be observed [29]. This aspect is similar to that obtained in the case of inclusion complexes between IB and βCD by different methods (co-precipitation, lyophilization) [31]. As shown by X-ray diffraction results (**Figure 2**), the ternary mixture (IB/βCD/PVP) was totally amorphous and showed physical stability after storage under stress conditions at RH: 75% and T = 40° C for 6 months, while the quenched IB (**Figure 2**) recrystallized at room temperature after few minutes [29].

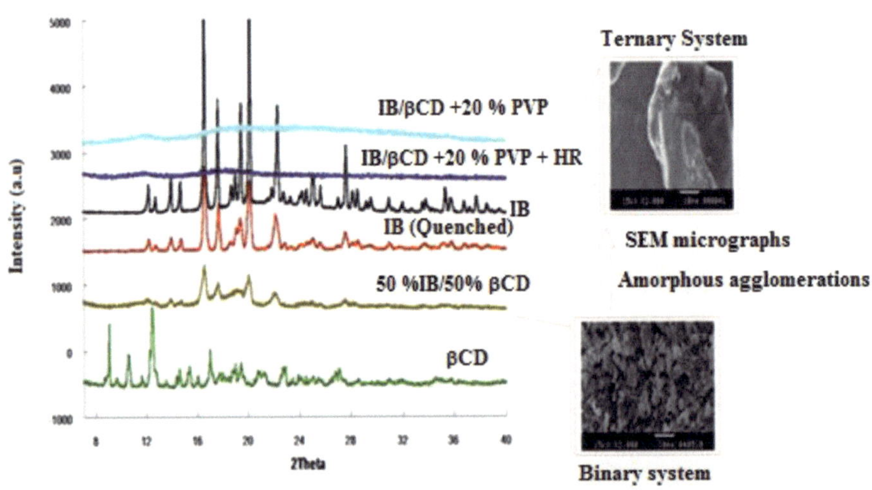

Figure 2.
X-ray diffraction results and SEM micrographs of binary and ternary system [29].

Different mechanisms were involved in the physical stabilization of amorphous IB in the ternary system, this required the use of several techniques (XRD, FTIR, SEM, DSC, and NMR) according to ICH recommendations [32]. Noticeable changes were resumed in **Tables 4** and **5**. In the case of the binary mixture (IB/βCD), the frequency shift of the carbonyl group of IB (**Table 4**) [29], could be attributed to the breaking of certain intermolecular bonds (hydrogen bonds) associated with IB dimmers, largely described in the literature [33]. The shift of the CO (primary alcohol) band and the OH (associated) band of βCD compared to physical mixtures, as well as the disappearance of the NMR proton of the carboxylic group of IB and the hydroxyls peaks of βCD (2, 3, 6) suggested the presence of hydrogen bonds between the carboxylic group of IB and hydroxyl groups of βCD [29]. DSC results (**Table 5**) have shown that the ternary mixture (IB/βCD/PVP) exhibited higher glass transition temperature (Tg˃250°C). This effect contributed to the reduction of molecular mobility of amorphous drug molecules and prevented its recristallization [29]. The addition of PVP to the binary system (IB/βCD) generated higher shifts for several infrared bands of IB, PVP and βCD (**Table 4**), this was accompanied also by an upfield shift of proton peaks of IB and βCD (located inside and outside the cavity), and a downfield shift of carbonyl peaks of IB and PVP [29]. All these changes can be attributed to the formation of multiple hydrogen bonds between the carboxylic group of IB and the carbonyl group and nitrogen of the pyrrolidone ring of PVP [34], as well as intermolecular hydrogen bonds between βCD (**Figure 3**) and PVP that were frequently observed in ternary systems [29, 35].

Mixtures/ infrared bands	Shifts					
	C=O(IB)	C=O(PVP)	OH/associated (βCD)	C-O/ secondary alcohol (βCD)	C-O/ primary alcohol (βCD)	C-N (PVP)
Binary system	8 cm^{-1}	—	11 cm^{-1}	2 cm^{-1}	3 cm^{-1}	—
Ternary system	12 cm^{-1}	5 cm^{-1}	22 cm^{-1}	5 cm^{-1}	6 cm^{-1}	12 cm^{-1}

Table 4.
Shifts of Infrared Bands compared to physical mixtures [29].

	DSC	RMN 1H	RMN 13 C
Binary system	• A decrease of melting temperature Tf (IB)	• Disappearance of the proton peaks of IB carboxylic group and hydroxyl groups of βCD (2, 3, 6)	• Upfield shirt of IB carbon peaks (Δδ [0.04:0.08] ppm)
Ternary system	• Disappearance of IB melting event	• Upfield shift of all proton peaks of βCD (located inside and outside the cavity) except hydroxyl peak (6) which downfielded	• Downfield shift of carbonyl peaks of IB and PVP
	• Tg (glass transition temperature) >250°C	• Upfield shift of IB protons (higher than binary system)	

Table 5.
NMR and DSC characterization of co-milled mixtures (IB/βCD, IB/βCD/PVP) [29].

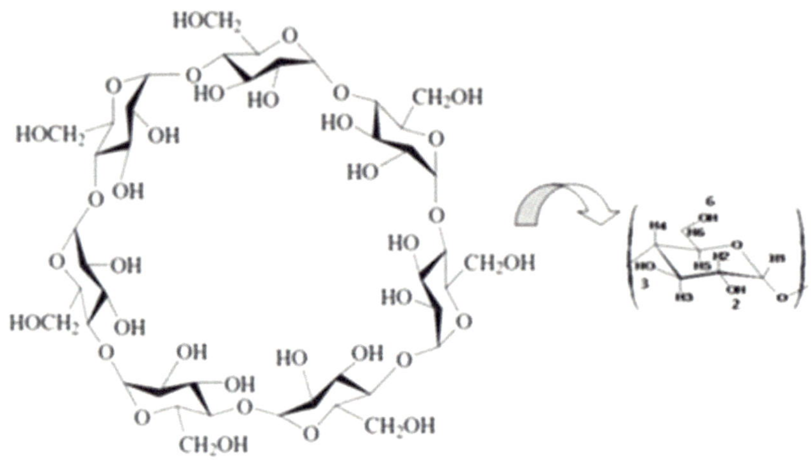

Figure 3.
βCD molecule [29].

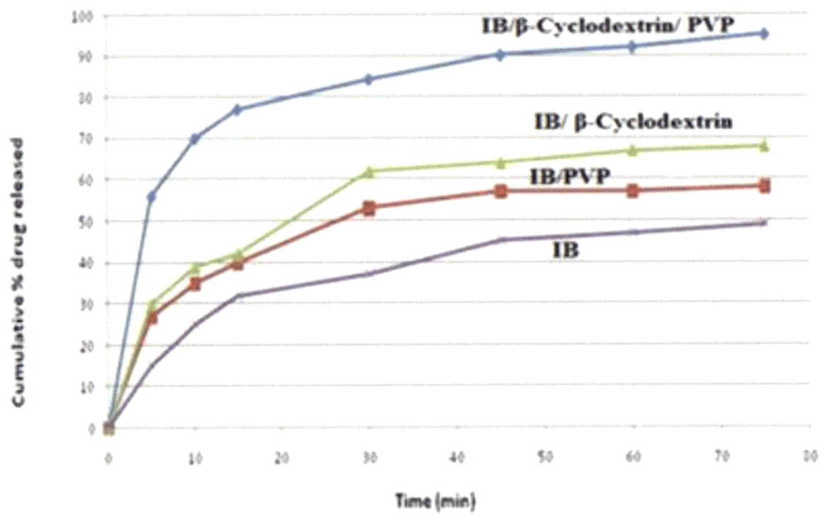

Figure 4.
Dissolution test results of co-milled mixtures (each point represents mean ± S.D., n = 3 [29]).

As a result of physical stabilization of amorphous Ibuprofen via different factors described previously, the IB dissolution rate (**Figure 4**) in the ternary mixture (IB/βCD/PVP) in 1:1:0.5 w/w ratio was greater than that obtained in the case of the binary mixture (IB/βCD) or (IB/PVP) in 1:1 w/w ratio [29]. The formation of such a water-soluble system can be explained not only by the synergistic effect of PVP and βCD via their mutual intermolecular interactions, but also by the ability of PVP to solubilize and promote formation of βCD complexes in the solid state [29]. Loftsson et al., have shown the role of PVP as a solubilizer in the case of several βCD complexes [36]. A ternary system (salt formation) was also obtained for IB and showed a considerable improvement in drug solubility and stability compared to the IB/βCD binary system [37]. Thus, the combination of PVP and β-Cyclodextrin molecules represents a scalable alternative for dissolution enhancement of IB which weakly interacted with β-Cyclodextrin by ball milling at ambient temperature [29].

4. Conclusion

In summary, the formation of physically stable ternary amorphous system by solid dispersion method using optimized ball milling technique, represents a promising alternative for drug solubility and stability enhancement. This succeeded for improving the dissolution rate of several active pharmaceutical ingredients (e.g., Probucol, Gliclazid, Fenofibrate, Ibrutinib and Naproxen). Their pharmacokinetic properties and in vivo bioavailability were considerably improved in comparison to pure drug molecules (up to 15-folds increase in plasma drug concentration). Ibuprofen dissolution rate was considerably enhanced in presence of PVP and βCD (release of 90% in 1 h), such ternary system (IB/βCD/PVP) in 1:1:0.5 w/w ratio exhibited higher drug release than binary systems (IB/PVP, IB/βCD) in 1:1 w/w ratio. This was resulted from various mechanisms (intermolecular interactions, synergetic effects of carriers, anti-plasticizing effect, hydrophilicity enhancement, particle size reduction, inclusion of IB molecules in βCD cavity) promoting stabilization of amorphous Ibuprofen even under stress conditions (75% RH and T = 40°C for six months). However, such scalable strategy requires the association of several analytical techniques in order to fully understand the solubilization and stabilization processes involved.

5. Perspectives

A development of stability assay method should be performed to evaluate the absence of drug impurities in the ternary system. Moreover, it is necessary to further investigate the nature of interactions between drug molecules and carriers in such complex system.

Acknowledgements

The authors thank Pr. Abdessalem Ben Haj Amara (Faculty of Sciences of Bizerte, Tunisia) for his considerable inputs and helpful discussions. On the other hand, the authors confirmed that this research Work did not receive any specific funding.

Conflict of interest

The authors declare that they have no conflict of interest.

Author details

Marouene Bejaoui[1,2]*, Hanen Oueslati[3] and Haykel Galai[2]

1 Département de Recherche, Faculté des Sciences de Bizerte, Université de Carthage, Jarzouna, Tunisia

2 Institut National de Recherche et d'Analyses Physico-Chimiques de Sidi-Thabet, Sidi Thabet, Tunisia

3 Laboratoire National de Contrôle des Médicaments, Bab Saadoun, Tunis, Tunisia

*Address all correspondence to: marouene2010@gmail.com

References

[1] Penkina A, Semjonov K, Hakola M, Vuorinen S, Repo T, Yliruusi J, Aruväli J, Kogermann K, Veski P, Heinämäki J. Towards improved solubility of poorly water-soluble drugs: cryogenic co-grinding of piroxicam with carrier polymers. Drug development and industrial pharmacy. 2016; 42:378-388. DOI: 10.3109/03639045.2015.1054400

[2] Gaggero A, Dukovski BJ, Radić I, Šagud I, Škorić I, Cinčić D, Jug M. Co-grinding with surfactants as a new approach to enhance in vitro dissolution of praziquantel. Journal of Pharmaceutical and Biomedical Analysis. 2020;189:113494. DOI: 10.1016/j.jpba.2020.113494

[3] Xiong X, Zhang M, Hou Q, Tang P, Suo Z, Zhu Y, Li H. Solid dispersions of telaprevir with improved solubility prepared by co-milling: formulation, physicochemical characterization, and cytotoxicity evaluation. Materials Science and Engineering: C. 2019;105:110012. DOI: 10.1016/j.msec.2019.110012

[4] Moinuddin SM, Ruan S, Huang Y, Gao Q, Shi Q, Cai B, Cai T. Facile formation of co-amorphous atenolol and hydrochlorothiazide mixtures via cryogenic-milling: enhanced physical stability, dissolution and pharmacokinetic profile. International Journal of Pharmaceutics. 2017;532:393-400. DOI: 10.1016/j.ijpharm.2017.09.020

[5] Kasten G, Nouri K, Grohganz H, Rades T, Löbmann K. Performance comparison between crystalline and co-amorphous salts of indomethacin-lysine. International Journal of Pharmaceutics. 2017;533:138-144. DOI: 10.1016/j.ijpharm.2017.09.063

[6] Patel PJ, Gajera BY, Dave RH. A quality-by-design study to develop Nifedipine nanosuspension: examining the relative impact of formulation variables, wet media milling process parameters and excipient variability on drug product quality attributes. Drug Development and Industrial Pharmacy. 2018;44:1942-1952. DOI: 10.1080/03639045.2018.1503296

[7] Hussain A, Smith G, Khan KA, Bukhari NI, Pedge NI, Ermolina I. Solubility and dissolution rate enhancement of ibuprofen by co-milling with polymeric excipients. European Journal of Pharmaceutical Sciences. 2018 ;123:395-403. DOI: 10.1016/j.ejps.2018.08.001

[8] Szafraniec J, Antosik A, Knapik-Kowalczuk J, Kurek M, Syrek K, Chmiel K, Paluch M, Jachowicz R. Planetary ball milling and supercritical fluid technology as a way to enhance dissolution of bicalutamide. International Journal of Pharmaceutics. 2017;533:470-479. DOI: 10.1016/j.ijpharm.2017.03.078

[9] Ding Z, Wang L, Xing Y, Zhao Y, Wang Z, Han J. Enhanced oral bioavailability of celecoxib nanocrystalline solid dispersion based on wet media milling technique: formulation, optimization and in vitro/in vivo evaluation. Pharmaceutics. 2019;11:328. DOI: 10.3390/pharmaceutics11070328

[10] Loh ZH, Samanta AK, Heng PW. Overview of milling techniques for improving the solubility of poorly water-soluble drugs. Asian journal of pharmaceutical sciences. 2015 ;10:255-274. DOI: 10.1016/j.ajps.2014.12.006

[11] Karagianni A, Kachrimanis K, Nikolakakis I. Co-amorphous solid dispersions for solubility and absorption improvement of drugs: Composition, preparation, characterization and formulations for oral delivery. Pharmaceutics. 2018;10:98. DOI: 10.3390/pharmaceutics10030098

[12] Baghel S, Cathcart H, O'Reilly NJ. Polymeric amorphous solid dispersions: a review of amorphization, crystallization, stabilization, solid-state characterization, and aqueous solubilization of biopharmaceutical classification system class II drugs. Journal of pharmaceutical sciences. 2016;105:2527-2544. DOI: 10.1016/j.xphs.2015.10.008

[13] Watanabe T, Ohno I, Wakiyama N, Kusai A, Senna M. Stabilization of amorphous indomethacin by co-grinding in a ternary mixture. International journal of pharmaceutics. 2002 ;241:103-111. DOI: 10.1016/S0378-5173(02)00196-5

[14] Zhang M, Suo Z, Peng X, Gan N, Zhao L, Tang P, Wei X, Li H. Microcrystalline cellulose as an effective crystal growth inhibitor for the ternary Ibrutinib formulation. Carbohydrate polymers. 2020;229:115476. DOI: 10.1016/j.carbpol.2019.115476

[15] Li F, Li L, Wang S, Yang Y, Li J, Liu D, Zhang S, Wang S, Xu H. Improved dissolution and oral absorption by co-grinding active drug probucol and ternary stabilizers mixtures with planetary beads-milling method. Asian Journal of Pharmaceutical Sciences. 2019 ;14:649-657.DOI: 10.1016/j.ajps.2018.12.001

[16] Li J, Yang Y, Zhao M, Xu H, Ma J, Wang S. Improved oral bioavailability of probucol by dry media-milling. Materials Science and Engineering: C. 2017;78:780-786. DOI: 10.1016/j.msec.2017.04.141

[17] Mura P, Maestrelli F, Cirri M. Ternary systems of naproxen with hydroxypropyl-β-cyclodextrin and aminoacids. International journal of pharmaceutics. 2003;260:293-302. DOI: 10.1016/S0378-5173(03)00265-5

[18] Ding X, Zheng M, Lu J, Zhu X. Preparation and evaluation of binary and ternary inclusion complexes of fenofibrate/hydroxypropyl-β-cyclodextrin. Journal of Inclusion Phenomena and Macrocyclic Chemistry. 2018;91:17-24. DOI: 10.1007/s10847-018-0793-1

[19] Maggi L, Canobbio A, Bruni G, Musitelli G, Conte U. Improvement of the dissolution behavior of gliclazide, a slightly soluble drug, using solid dispersions. Journal of Drug Delivery Science and Technology. 2015;26:17-23. DOI: 10.1016/j.jddst.2015.01.002

[20] Dudognon E, Danède F, Descamps M, Correia NT. Evidence for a new crystalline phase of racemic ibuprofen. Pharmaceutical research. 2008;25:2853-2858. DOI: 10.1007/s11095-008-9655-7

[21] Hussain A, Smith G, Khan KA, Bukhari NI, Pedge NI, Ermolina I. Solubility and dissolution rate enhancement of ibuprofen by co-milling with polymeric excipients. European Journal of Pharmaceutical Sciences. 2018;123:395-403. DOI: doi.org/10.1016/j.ejps.2018.08.001

[22] Najib NM, Suleiman M, Malakh A. Characteristics of the in vitro release of ibuprofen from polyvinylpyrrolidone solid dispersions. International journal of pharmaceutics. 1986;32:229-236.DOI: 10.1016/0378-5173(86)90183-3

[23] Mallick S, Pattnaik S, Swain K, De PK, Saha A, Ghoshal G, Mondal A. Formation of physically stable amorphous phase of ibuprofen by solid state milling with kaolin. European Journal of Pharmaceutics and Biopharmaceutics. 2008;68:346-351. DOI: 10.1016/j.ejpb.2007.06.003

[24] Loftsson T, Brewster ME. Pharmaceutical applications of cyclodextrins. 1. Drug solubilization and stabilization. Journal of pharmaceutical sciences. 1996;85:1017-1025. DOI: 10.1021/js950534b

[25] Hładoń T, Pawlaczyk JA, Szafran BA. Stability of Ibuprofen in its Inclusion Complex with β-cyclodextrin. Journal of inclusion phenomena and macrocyclic chemistry. 2000;36:1-8. DOI: 10.1023/A:1008046724527

[26] Ghorab MK, Adeyeye MC. Enhancement of ibuprofen dissolution via wet granulation with β-cyclodextrin. Pharmaceutical development and technology. 2001;6:305-314. DOI: 10.1081/PDT-100002611

[27] Pereva S, Sarafska T, Bogdanova S, Spassov T. Efficiency of "cyclodextrin-ibuprofen" inclusion complex formation. Journal of Drug Delivery Science and Technology. 2016;35:34-39. DOI: 10.1016/j.jddst.2016.04.006

[28] Yang H, Bohne C. Effect of amino acid coinclusion on the complexation of pyrene with β-cyclodextrin. The Journal of Physical Chemistry. 1996;100:14533-14539. DOI: 10.1021/jp9607531

[29] Bejaoui M, Galai H, Amara AB, Rhaiem HB. Formation of Water Soluble and Stable Amorphous Ternary System: Ibuprofen/β-Cyclodextrin/PVP. Glass Physics and Chemistry. 2019;45:580-588. DOI: 10.1134/S1087659619060130

[30] Tabary N, Garcia-Fernandez MJ, Danede F, Descamps M, Martel B, Willart JF. Determination of the glass transition temperature of cyclodextrin polymers. Carbohydrate polymers. 2016;148:172-180. DOI: 10.1016/j.carbpol.2016.04.032

[31] Hussein K, Türk M, Wahl MA. Comparative evaluation of ibuprofen/β-cyclodextrin complexes obtained by supercritical carbon dioxide and other conventional methods. Pharmaceutical research. 2007;24:585-592. DOI: 10.1007/s11095-006-9177-0

[32] Pannala R. Impurity Profiling of Solid Oral Drug Products to Sail through GDUFA-II Requirements. American Journal of Analytical Chemistry. 2018;9:187. DOI: 10.4236/ajac.2018.94016

[33] Bogdanova S, Pajeva I, Nikolova P, Tsakovska I, Müller B. Interactions of poly (vinylpyrrolidone) with ibuprofen and naproxen: experimental and modeling studies. Pharmaceutical research. 2005;22:806-815. DOI: 10.1007/s11095-005-2598-3

[34] Najib NM, El-Hinnawi MA, Suleiman MS. Physicochemical characterization of ibuprofen-polyvinylpyrrolidone dispersions. International journal of pharmaceutics. 1988;45:139-144. DOI: 10.1016/0378-5173(88)90042-7

[35] Valero M, Pérez-Revuelta BI, Rodríguez LJ. Effect of PVP K-25 on the formation of the naproxen: β-ciclodextrin complex. International journal of pharmaceutics. 2003;253:97-110. DOI: 10.1016/S0378-5173(02)00664-6

[36] Loftsson T, Friðriksdóttir H. The effect of water-soluble polymers on the aqueous solubility and complexing abilities of β-cyclodextrin. International journal of pharmaceutics. 1998;163:115-121. DOI: 10.1016/S0378-5173(97)00371-2

[37] Redenti E, Szente L, Szejtli J. Cyclodextrin complexes of salts of acidic drugs. Thermodynamic properties, structural features, and pharmaceutical applications. Journal of pharmaceutical sciences. 2001;90:979-986. DOI: 10.1002/jps.1050

Chapter 7

Modulating the Physicochemical Properties of Chitin and Chitosan as a Method of Obtaining New Biological Properties of Biodegradable Materials

Ilona Latańska, Piotr Rosiak, Paulina Paul, Witold Sujka and Beata Kolesińska

Abstract

Physical and chemical modifications of chitin and chitosan allow for obtaining new functional properties of the natural polymers. This is a particularly valuable feature for the design and manufacture of new materials for medical applications. Due to their wide and varied biological activity, chitin and chitosan materials are increasingly used as dressing materials with antibacterial and hemostatic properties and as materials accelerating the regeneration of damaged tissues because of stimulation of granulation tissue formation, re-epithelialization and reduction of the formation of scar tissue. In addition, chitosan derivatives have antifungal, antiviral, anticancer activity. The increasing use of chitin and chitosan also has a positive impact on the environment, as it is obtained as a result of chitin deacetylation, usually isolated from shellfish shells. The main source of chitin is waste coating of crustaceans. The annual natural reproducibility of chitin by biosynthesis is estimated at 2–3 billion tons. Our interest in the use of biodegradable biopolymers derived from chitin concerns the design, synthesis in laboratory scale, testing new material properties and the final implementation of new developments for industrial practice of new dressing materials useful in the treatment of bleeding wounds (haemostatic properties) as well as in the regeneration of wounds and ulcers of various etiologies. Examples of chitin-based dressing materials introduced by Tricomed SA are Medisorb R Ag, Medisorb R Membrane, Medisorb R Powder and Tromboguard®.

Keywords: chitin, chitosan, chemical modification, biological properties, dressing materials, manufacturing on an industrial scale

1. Introduction

Chitin and chitosan belong to the polymeric materials of natural origin from the polysaccharides group. The widely used polysaccharides of natural origin also include cellulose and derivatives of hyaluronic and alginic acid. Use for the production of medical devices (contact with the patient's body), makes them meet several requirements: they should maintain their physicochemical properties after treatment at elevated temperature during sterilization, after exposure to X-ray, detergents and aseptic. Polysaccharide biopolymers, like most polymeric materials, degrade after some time of use, so it is also important that their decomposition products do not cause inflammation, allergic or immune reactions or any other interactions with patients' bodies.

Chitin is a polysaccharide composed of N-acetylglucosamine residues linked by β-1,4-glycosidic bonds. Chitin is the main component of the fungal walls and the shells of arthropods (crustaceans, insects, and arachnids), but is also present in sponges, corals, and mollusks. However, for laboratory and industrial purposes, it is obtained mainly from marine invertebrates such as: crabs, shrimps, lobsters and krill. The properties of chitin depend on its origin. Chitin used in the production of medical devices must come from certified, controlled fisheries and must be properly purified. The methods of isolating chitin from natural sources are strictly dependent on the choice of the organism from which it is isolated. This polysaccharide is similar in structure to cellulose. It differs in the presence of an acetyl amide group (-$NHCOCH_3$) in place of one of the hydroxyl groups (**Figure 1**). The presence of this group means that there are much stronger intermolecular hydrogen bonds in chitin, which results in its greater mechanical strength compared to cellulose [1, 2].

Depending on the origin source, chitin can occur in three amorphous forms: α, β and γ [2, 3]. The most widespread is α chitin found in fungi, shells of crustaceans and krill, and the skeletons of insects. The β form, which can mainly be isolated from squids, is much less common. The differences in the crystal structure of both amorphous forms of chitin affect their processing capabilities. The ordered crystal structure of chitin limits its solubility in commonly used solvents, and thus, reduces its use in industry. α-Chitin is moderately soluble in aqueous thiourea solution, aqueous alkaline urea solution, 5% LiCl/DMAC, some ionic liquids, hexafluoroacetone, hexafluoro-2-propanol, methanesulfonic acid [4, 5]. The form of β-chitin, on the other hand, swells in water (forms a suspension) and is soluble in formic acid. Chitin has no cytotoxic effect *in vitro*, is physiologically inert, biodegradable, has antibacterial properties and is characterized by a high affinity to proteins. During its biodegradation in the wound environment, its oligomers and units are released. Most often, it is used in the form of gel, membranes, fibers, polymer films or is a component of polymer blends. Chitin activates macrophages, stimulates the proliferation of fibroblasts and influences vascularization [6–11].

Despite the very good biological properties of chitin, its practical use is moderate, which is related to its low solubility causing difficulties in its processing. Therefore, chitin is used as a raw material to obtain new derivatives with better performance parameters. In terms of practical use, the most important chitin derivatives are its esters and chitosan, which is a product of chitin deacetylation, which can be classified into the group of amino-polysaccharides.

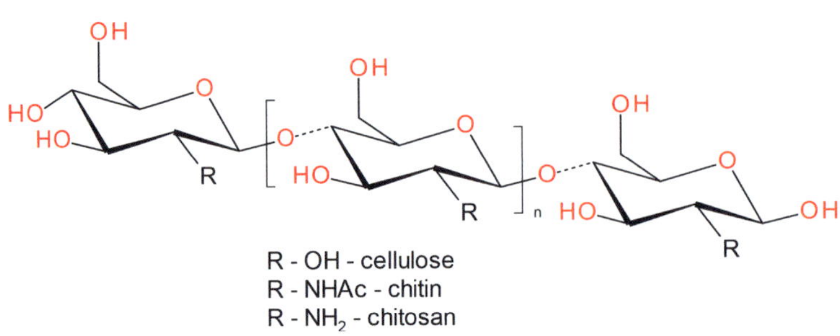

Figure 1.
Structural resemblance of cellulose, chitin and chitosan.

2. Chitin esters - materials with tailored functional properties

2.1 Chitin esters in dressing materials

The esterification of chitin hydroxyl groups allows to increase the utility potential of the polysaccharide by introducing various substituents, and thus, influencing the physical, chemical and biological properties of materials. The best known are chitin esters, in which the hydroxyl groups are esterified with one type of acylating reagent (presence of the same ester groups on both hydroxyl groups of chitin). Acetylated chitin derivatives (CH_3CO- substituent) are prepared with acetic anhydride in the presence of an acid catalyst. However, the physicochemical properties conditioning the processing of chitin acetate turned out to be unsatisfactory [12]. The use of a mixture of orthophosphoric acid and trifluoroacetic acid anhydride as a catalyst allowed to obtain a variety of chitin esters derived from: butyric acid, cyclopropanecarboxylic acid, cyclobutanecarboxylic acid, cyclopentanecarboxylic acid, cyclohexanecarboxylic acid and substituted benzoic acids. In the case of chitin butyrate, the process efficiency (DS (degree of substitution) included in the range 1.9–2.38) was dependent on the excess of butyric acid anhydride use [13–15]. Di-butyrylchitin (chitin di-butyrate, DBC) is an example of a chitin derivative soluble in typical organic solvents [16]. DBC is obtained by chitin esterification with butyric anhydride. Typically, it is a two-stage process. In the first step, chitin is purified from calcium salts with 2 M hydrochloric acid. The next stage is the process of proper esterification of purified chitin. The substrates of the reaction, apart from chitin, are butyric anhydride and the catalyst, which is most often chloric (VII) acid. The reaction is carried out in a heterogeneous system by adding powdered chitin in appropriate proportions to the reaction mixture consisting of butyric anhydride and chloric (VII) acid. The classical esterification process requires the use of substrates in a molar ratio of acid anhydride to N-acetylaminoglucose unit of 10: 1. It is also crucial to carry out the reaction at a temperature of 20°C. Increasing the reaction temperature to 40°C causes a rapid lowering of the molecular weight of the modified polymer. The catalyst concentration has a direct influence on the efficiency of the esterification reaction. The yield of the reaction is the higher when more concentrated chloric (VII) acid is used. However, it should be remembered that the use of too much concentrated chloric (VII) acid results in the macromolecule degradation. The esterification process is completed by adding diethyl ether to the reaction mixture. The isolated product is then heated with water to remove residual chloric (VII) acid. The product obtained in this way is treated for 24 hours with acetone, in which only di-butyrylchitin is dissolved. Then, the solution is concentrated to 5–6%. After the desired concentration is reached, the solution is poured into deionized water to precipitate the polymer, then the product is dried to obtain solid di-butyryl chitin. The above-described process of chitin esterification allows the conversion of free hydroxyl groups on the C3 and C6 carbon of the chitin into ester groups ($CH_3(CH_2)_2CO$- substituent). Di-butyrylchitin is composed of di-butyl-N-acetyl-amino-glucose units linked by 1,4-β-glycosidic bonds. The polymer is stabilized by hydrogen bonds between the polymer chains. Hydrogen bonds are formed with the participation of the hydrogen atom from the acetylamino group and the oxygen atom from the ester group. This kind of intermolecular interaction determines its good mechanical properties [12–15]. Di-butyrylchitin does not dissolve and does not swell in water, but it dissolves in many organic solvents such as: acetone, methanol, ethanol, tetrahydrofuran (THF), dimethylformamide (DMF), chloroform, methylene chloride and others. Di-butyryl chitin is not easily degraded, it is resistant to γ-radiation (possibility of radiation sterilization), while enzymatic degradation under the influence of lysozyme and CE econase occurs at a slow rate, which causes a slight change in molecular weight. Di-butyrylchitin with a molecular weight above 100000 Da has film-forming and fiber-forming properties [1, 2, 12–15]. Thus, obtaining DBC with the desired molecular weights directly determines its further processing capabilities (in particular electrospinning and leaching). The most important biological parameters of di-butyrylchitin are: prolongation of blood clotting time and good wettability. The use of DBC dressings has a positive effect on the granulation process (increasing the level of glycosaminoglycans in the wound), collagen cross-linking (generating more durable tissue), accelerating the wound healing process to form a healthy epidermis without scarring and protecting the wound against excessive moisture loss (optimal moist environment) [1, 2, 6–11]. In the course of treatment, the dressing slowly bio-degrades and resorbs until it disappears completely, which eliminates the painful act of changing it. The spontaneous, anti-pain effect of the dressing was also noted. DBC does not show cytotoxicity or irritation, it is a biocompatible polymer [9]. Di-butyrylchitin fibers are obtained by two methods: wet and dry-wet. The choice of the method used determines the structure of the obtained fibers. The fibers obtained in the wet-spinning process are less regular in shape, with a greater surface development than in the case of dry-wet spinning. DBC fibers produced by wet spinning are used as a raw material for the production of nonwovens. The technique of producing nonwovens from DBC depends on cutting the fibers into 6 cm long sections, from which the fleece is produced using a mechanical method on carding machines, and then the fibers in the fleece are joined by needling and calendaring [16, 17].

The dry-wet method of forming fibers from DBC hinges on preparing a spinning solution with a concentration of 15 to 25% in ethanol, heating it to 60°C and squeezing it through a spinning die. Then, the incompletely solidified fiber is introduced into a water bath, where it is completely solidified. The fiber is then wound onto drums, stretched, and dried. A microporous DBC fiber with a linear mass of 1.7 to 5.6 g is obtained, depending on the concentration of the spinning solution used. The fibers obtained by the dry method have an elongated and curved cross-sectional shape, similar to a croissant. The degree of crystallinity of the fibers determined in X-ray examinations is similar in both methods and amounts to approx. 19%, and the transverse dimension of the crystallites approx. 23 Å. It is also easy to obtain chitin materials (the so-called regenerated chitin) from these materials without damaging their macro-structure after a mild alkaline treatment. Fibers made of regenerated chitin and di-butyrylchitin do not induce cytotoxic, haemolytic or irritating effects and cause minimal local tissue reaction after implantation [17–19]. Di-butyrylchitin and regenerated chitin fibers can be used to obtain dry dressing materials, as well as materials for other biomedical purposes. DBC-based woven dressings are biodegradable within the wound and do not need to be replaced during use [16, 17]. Chitin di-pentanoate (chitin divalerate, Di-O-Valeryl-Chitin, DVCH) is also used for the production of commercially available dressings, where chitin is esterified with two valeryl groups ($CH_3(CH_2)_3CO$- substituent) at the 3 and 6 positions of N-acetylglucosamine units. The high degree of DVCH esterification was achieved by using a large excess of valeric anhydride used both as acylating agent and reaction medium, and perchloric acid as catalyst. It turned out to be optimal to conduct the reaction where the molar ratio of valeric acid anhydride to chitin was 7:1, which also facilitated thorough mixing of the components during the reaction and temperature control. The performance of the reaction under conditions of high homogeneity of the solution has a great influence on the quality of the product. Insufficient mixing of the solution during the acylation step led to a local temperature rise, uneven chitin acylation and ultimately resulted in products with varying degrees of esterification and higher polydispersity. Additionally, at elevated temperature it was observed reduction of the molecular weight of the biopolymer as a result of the acidic degradation of chitin that occurs in parallel with the acylation reaction in the presence of a strong acid. To obtain products with a high degree of esterification, 0.5 M perchloric acid was used (deacetylation of the N-acetyl group was not observed). The separation of the raw product from the reaction mixture takes place during the neutralization of the valeric acid excess with a 2.5% $NaHCO_3$ solution. The use of sodium bicarbonate as a weak base prevents deacetylation of the N-acetyl group. Depending on the reaction time and temperature, products with different parameters are obtained. The lower temperature leads to a product with a higher molecular weight. A longer reaction time increases the yield of the reaction, but is associated with a reduction in molecular weight due to acidic degradation of the polymer. The DVCH polydispersity index ranged from 1.47 to 2.06, suggesting a low molecular weight distribution. Due to the good solubility of DVCH in organic solvents such as acetone or ethanol, it is possible to prepare thin polymer layers by casting or porous structures by salt leaching. The DVCH shows a semi-ductile behavior and breaks when it exceeds the yield point. The stretching properties of DVCH films depend on the molecular weight. The modulus, yield stress, tensile stress as well as strain at break increase continuously with increasing DVCH molecular weight. The increase in the modulus with molecular weight results in higher mechanical strength of DVCH films. The elongation at break, although slightly increases with increasing molecular weight, remains low, not exceeding 4%. As a consequence, the higher DVCH molecular weight is, it behaves like a stiff plastic that can withstand relatively high stresses but does not withstand high elongation before breakage. Using the salt leach method, it is also possible to develop porous materials from DVCH. The structure of porous DVCH-based materials consists of a unified network of interconnected channels. This structure is characterized by a high content of open pores of various sizes. Two pore populations can be distinguished: large with a size in the range 150–780 μm (average pore size approx. 435 μm ± 168 μm) and small with a size in the range of 4–22 μm (average pore size 7.7 μm ± 3.3 μm [16–19]. Chitin divalerate is a technologically friendly biopolymer. The good solubility of DVCH in organic solvents (ethanol, DMAC, DMSO, acetone) due to the presence of hydrophobic valeryl groups in C-3 and C-6 positions enables its easy processing of its particles. The DVCH maintains the film-forming ability of chitin, so it can easily be used in the production of threads, films, foams and scaffolds, as well as non-woven fabrics. Biological data show that DVCH is not cytotoxic to fibroblasts and does not cause irritation or allergy to the skin of animals [20]. For the synthesis of chitin dihexanoate (DHCH) it is also possible to use appropriate acid anhydrides and methanesulfonic acid as a catalyst. In order to increase the homogeneity of the solution and better control the temperature in the process, by analogy to the synthesis of the valerate ester, an excess of acid anhydride and methanesulfonic acid are used, the mixture being the reagents and the reaction medium. Optimal methanesulfonic acid to chitin molar ratios are 16:1 and 10:1 for chitin di-hexanoate and chitin di-butyrate, respectively. This approach will result in a high degree of substitution of hydroxyl groups, equal to almost 2, and a low polydispersity. Moreover, under optimal conditions, no hydrolysis of the N-acetyl bond was observed. Good chitin solubility in methanesulfonic acid, even at low temperatures, allows the esterification process to be carried out under milder conditions. The key parameter is the intensity of agitation of the reaction suspension. Insufficient heat transfer due to poor mixing of the solution, similar to the synthesis of chitin di-pentanoate, leads to a lower degree of esterification, high polydispersity of the final product and a reduction in molecular weight. The neutralization process is carried out with a 4% sodium bicarbonate solution. The synthesis of DBC at a low temperature and short reaction time (temperature 0°C and 8°C) is ineffective due to the low reaction yields and possibly incomplete esterification of the chitin hydroxyl groups, resulting in the formation of a significant amount of insoluble gel fractions when dissolved in acetone prior to precipitation with water. For DHCH, it is preferable to use low synthesis temperatures (0°C and 8°C). The yield of DHCH synthesis was relatively high (above 70%), with the highest efficiency observed at 21°C (84 to 95%). Unfortunately, carrying out the synthesis of DHCH at 21°C resulted in a low molecular weight product. A trend analogous to that of chitin di-pentanoate was observed, indicating that the longer the reaction time, the higher the reaction performance and the lower the molecular weight of the obtained biopolymers. Although in DHCH the hydroxyl groups of chitin are substituted with longer alkyl chains than in DVCH or DBC, it has been found that DHCH retains good solubility in solvents such as ethanol, acetone, dichloromethane, 1,2-dichloroethane, N,N-dimethylformamide, N,N-dimethylacetamide and ethyl acetate and no solubility in water. Good solubility, filmogenic and fiber-forming properties of DHCH give greater possibilities of its processing (film casting, washing method, electrospinning method) compared to chitin alone. The mechanical properties of DHCH and DBC in the form of thin solid layers poured from solution were investigated in relation to their molecular weights. DHCH and DBC showed semi-continuous properties and cracked rapidly upon exceeding the plasticity point, similar to that observed for DVCH. The elongation at break was small and did not exceed 4%. For both biopolymers, their tensile properties correlate with the molecular weight. Parameters such as modulus of elasticity, stress at yield, as well as stress and strain at break, were found to increase with increasing DHCH and DBC molecular weight. Comparing the mechanical properties of DHCH, DBC and DVCH, it can be concluded that Young's modulus decreases with increasing chain length of the acyl group of chitin diesters (a similar relationship is observed for chitin monoesters, where only one hydroxyl group is acylated). Due to the good solubility of hydrophobic chitin diesters in organic solvents and their insolubility in water, it is possible to obtain porous structures based on DHCH and DBC by using the salt leaching method. The prepared porous materials are characterized by a united network of interconnected channels and a high degree of open porosity with pores of various sizes (pore size in the range 78–421 μm, average pore size 253 μm ± 93 μm) [21–24]. Due to its physic-chemical properties, DHCH can successfully replace or support materials based on di-butyrylchitin (e.g. in the form of mixtures of both biopolymers) and thus it can be used as a material for medical and pharmaceutical applications, especially in tissue engineering scaffolds and in healing wounds. The described procedure of chitin esterification to obtain products of high purity. Moreover, this method is universal (the possibility of preparation various chitin diesters) and is easy to produce and is not time-consuming [21]. Another method of chemical modification of chitin is esterification leading to carboxymethylchitin (CMCht, CM-chitin) [22, 23] or dicarboxymethylchitin using monochloroacetic or mono-chloropropionic acid followed by substitution of halogen with a hydroxyl group. This modification leads to the loss of the supramolecular structure of chitin and the formation of water-soluble derivatives [24].

2.2 Mixed chitin esters (co-esters)

The presence of two hydroxyl groups at the C-3 and C-6 positions of chitin allows the introduction of two different acyl substituents, leading to the formation of mixed esters (co-esters) of chitin. This possibility is also due to the differential reactivity of the hydroxyl groups linked to the primary and secondary carbon atoms in chitin. Thus, under ideal conditions, it is possible to obtain mixed esters containing the same molar fraction of different acyl groups in the modified material.

Different ester groups (e.g., butyric and acetic acid residues) are present in mixed esters in a single polysaccharide chain. The replacement of some large bulky butyl groups with much smaller acetate groups in one polysaccharide chain causes that in a condensed state, it becomes possible to obtain favorable conditions for the formation of intermolecular bonds of the hydrogen bridge type. This effect cannot be expected when using a mixture of two different biopolymers (e.g. chitin diacetate and chitin di-butyrate). Thus, the term mixed polymers is not the same as mixed chitin esters. In order to obtain a polymer mixture, it is necessary to use at least two chemically different polymers (**Figure 2**). In contrast mixed ester/co-ester of chitin contains only one component. It was found that the differences between chitin mixed esters (co-esters) and a mixture of two species can be observed in NMR spectra (^{1}H and ^{13}C) (**Figures 3** and **4**). The studies used 50:50 butyryl-acetyl-chitin co-polyester (**2**) (mixed ester), 90:10 butyryl-acetyl-chitin co-polyester (**3**) and a 1: 1 mixture of chitin di-butyrate (**1**) and chitin diacetate (**4**).

A comparative analysis of the ^{13}C-NMR spectra of the 180–150 ppm range characteristic for the chemical shifts of carbon in carboxylic acid derivatives showed that the distribution in the mixed ester of chitin **2** and **3** is different from the carbon signals of the 1: 1 mixture of polymers **1** and **4** (**Figure 3**). A similar result is observed in the range of 10–40 ppm, characteristic for carbons of aliphatic ester residues introduced as a result of esterification of chitin with acetic anhydride and butyric acid (**Figure 3**).

Comparative studies of ^{1}H-NMR spectra in the range of chemical shifts 2.5 ppm to 0.5 ppm also allowed to state that in the case of butyryl-acetyl chitin co-polyesters (samples **2** and **3**) the recorded signals are different than in the case of the 1: 1 mixture of polymers **1** and **4** (**Figure 4**).

The possibility of forming the intermolecular hydrogen bonds that stabilize butyryl-acetyl chitin co-polyester structure translates into fiber-forming properties, and thus the possibility of modulating the functional properties of the final materials and their use in the manufacture of new materials for medical use. In addition to stabilization by hydrogen bonds, it is also possible to create weak interactions based on hydrophobic interactions. The participation of such various weak interactions in the stabilization of materials may translate into their ability to interact with both hydrophobic and hydrophilic structures, which affects biological activity.

2.2.1 Chitin co-esters in dressing materials

Acetate-formate mixed chitin ester was obtained using formic acid, acetic anhydride and trifluoroacetic acid as a catalyst [25]. It turned out that this ester is slightly soluble in typical organic solvents. This is one of the reasons why this derivative has not found practical use, even though its biological properties are comparable to those of chitin. A similar situation was observed in the case of trifluoroacetate-formate derivatives of chitin [26].

Attempts to obtain a mixed butyric acetic ester of chitin by reaction using acetic and butyric anhydrides and methanesulfonic acid or trifluoroacetic acid as catalysts have been unsuccessful. The final product was a mixture of chitin acetate, chitin butyrate and the expected acetate-butyrate of chitin [27, 28].

The approach to synthesize mixed chitin esters using a mixture of trifluoroacetic acid and the corresponding organic acid as catalysts also led to the formation of mixtures of chitin monoesters and mixed esters (co-polyesters) of chitin. It has been shown that carrying out the reaction at the temperature of 70°C for a short time (30 min) under homogeneous conditions allows for obtaining co-polyesters: acetate-butyrate, acetate-hexanoate, acetate-octanoate and acetate-palmitate of chitin. The final co-polyesters have molecular weights ranging from

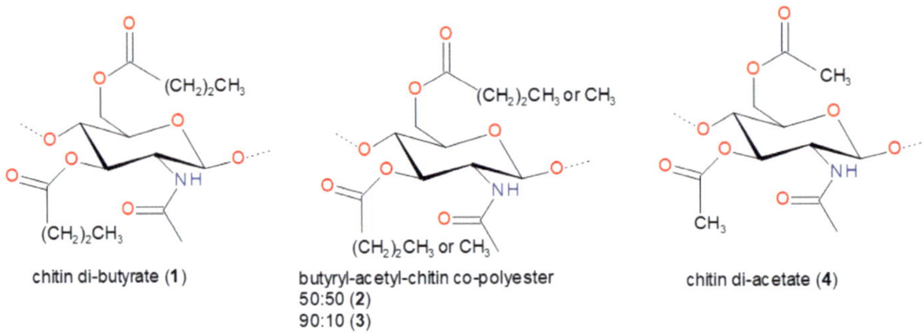

Figure 2.
Chemical structure of chitin di-acetate, chitin di-butyrate, butyryl-acetyl chitin co-polyester (mixture ester, co-ester).

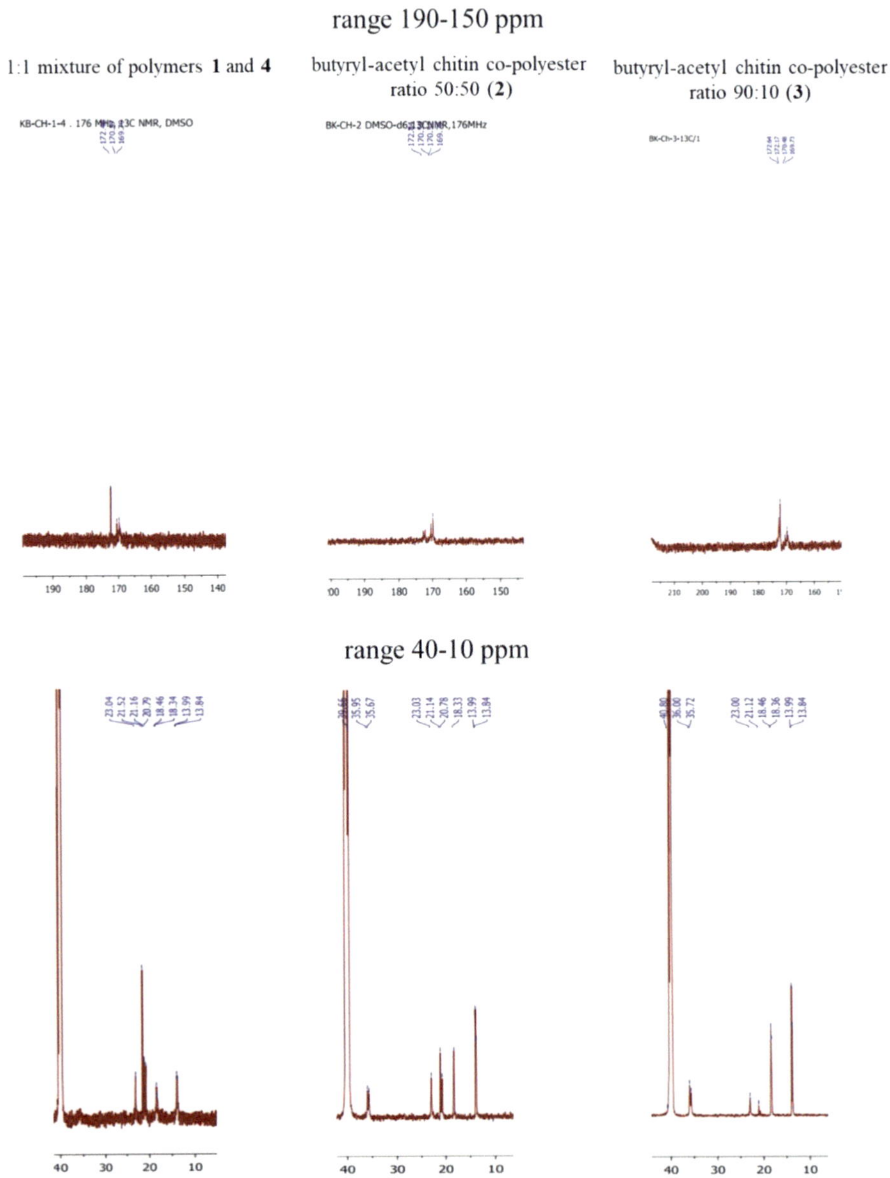

Figure 3.
Fragments of ^{13}C-NMR spectra of **1**, **2**, **3** *and* **4** *derivatives.*

30 to 150 kDa and the degree of esterification ranging from 1.0 to 2.0, depending on the raw materials used.

Another approach to obtain mixed butyryl-acetyl esters of chitin [29] is based on the use of butyric and acetic acid anhydrides and methanesulfonic acid as a catalyst. However, this method is not very friendly from the point of view of industrial stoppage. Thinking about the industrial synthesis of the butyryl-acetyl chitin derivative, it was necessary to establish reaction conditions that would eliminate the need to use methanesulfonic acid.

In the works on the development of a method for the production of the butyryl-acetyl chitin co-polyester on an industrial scale, it was necessary to develop, in the first stage, the conditions for the synthesis on a laboratory scale, which would later be transferred to an industrial scale. The esterification in laboratory conditions is carried out under heterogeneous conditions at 20-25°C, using chloric (VI) acid as a catalyst and a mixture of butyric and acetic acid anhydrides, used in a molar ratio of 90:10 [22, 23]. The product was obtained with a yield of 82 to 89% is soluble in DMF, DMSO and NMP, has a high molar mass (intrinsic viscosity of these products determined in DMF at the level of 2.0–2.05 dL/g) and a full degree of esterification. In the research on the development of a method of producing butyryl-acetyl chitin co-polyester in industrial conditions it was crucial to eliminate the possibility of formation an explosive mixture which may arise as a result of direct contact of acetic anhydride with perchloric acid. It turned out that the priority was to use an efficient cooling system so that the process temperature did not exceed 20°C. In laboratory conditions, it was sufficient to use an ice-water bath with NaCl (brine bath) and intensive mixing of the suspension. In laboratory conditions, diethyl ether is added to the slurry to remove excess unreacted anhydrides and formed carboxylic acids and the crude product is filtered off. The crude acetylation product is washed with water and dilute ammonia water, dried and finally dissolved in ethanol. The transfer of the conditions from the laboratory scale to the macro scale

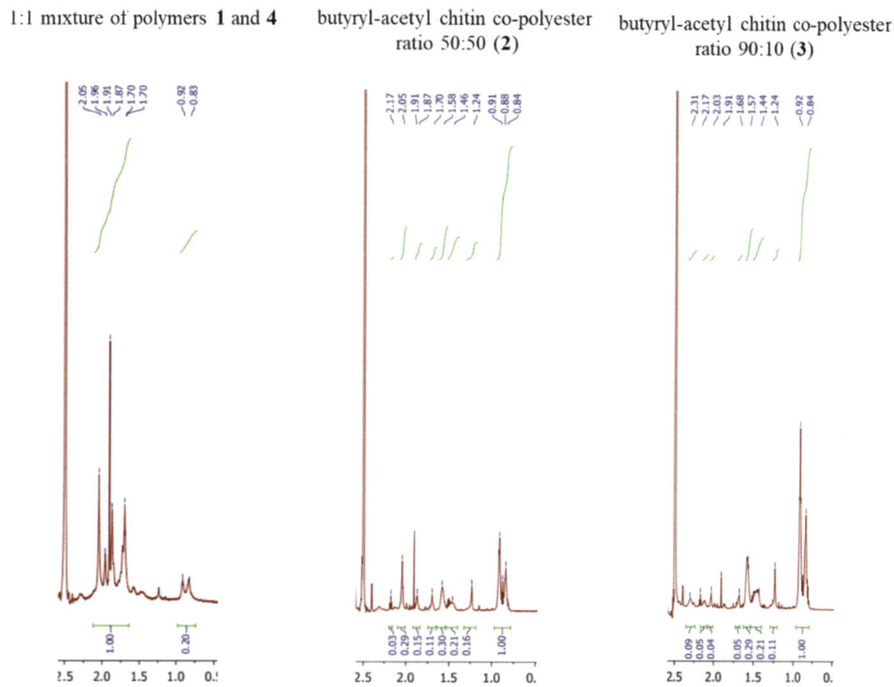

Figure 4.
Fragments of ^{1}H-NMR spectra of **1, 2, 3** *and* **4** *derivatives in the range 2.5–0.5 ppm.*

did not involve only increasing the amount of reagents and the size of the synthesizer. The key was to optimize the process conditions and the use of reagents that can be used in industrial conditions from the point of view of safety, environmental impact and economics. A reactor with a capacity of 60 dm^3 with an effective cooling system was used. A 3 kg chitin input was used for the synthesis. The remaining reagents (2 dm^3 of perchloric acid, 15 dm^3 of butyric anhydride and 1 dm^3 of acetic anhydride) were added in portions. The time required for all reagents to be introduced and for complete conversion was about 24 h. In place of diethyl ether, under industrial conditions, ethyl acetate was used to remove excess unreacted butyric and acetic anhydrides. In industrial conditions it was also necessary to replace the ammonia water to neutralize the acetic and butyric acid residues. It turned out that it is possible to use sodium carbonate for this purpose. Also, the step of separation of raw product required changes in the industrial process. In the synthesis under laboratory conditions, G4 Schott funnels were used for filtration. However, the use of this method on a large scale was ineffective. So suction filtration was used, the efficiency of which was 100 dm^3 per hour. The process efficiency on an industrial scale was comparable to that of a laboratory scale synthesis. The physical and chemical properties of the final products were also comparable. The conducted research guaranteed the reproducible and required parameters of the raw material for the production of new medical materials [22, 23]. **Figure 5** shows a set for the synthesis of butyryl-acetyl chitin co-polyester on an industrial scale.

The development of an efficient synthesis of the butyryl-acetyl chitin co-polyester (BAC 90/10) made it possible to demonstrate that the obtained chitin derivative has the ability to form fibers from a wet solution with a strength slightly above 20 cN/tex with high porosity. Fibers with a strength at this level can be the basis for the production of 3D polymer-fiber composites. BAC 90/10 fibers show a stronger predisposition to apatite crystallization; strong sorption tendencies of fibers allowing the possibility of local supersaturation favoring the nucleation of apatite. It has been also found that BAC 90/10 fibers degrade fast under *in vitro* conditions.

One application of the butyryl-acetyl chitin co-polyester (BAC 90/10) is its use to produce highly porous film materials [30].

The research work began with laboratory tests during which two methods of formation of porous materials were tested: (a) pouring a 5% ethanol BAC 90/10 solution on a layer of solid inorganic salt (porophor agent), which, after solidification, was exposed to water to wash out the porophor agent, and (b) the use of porophor

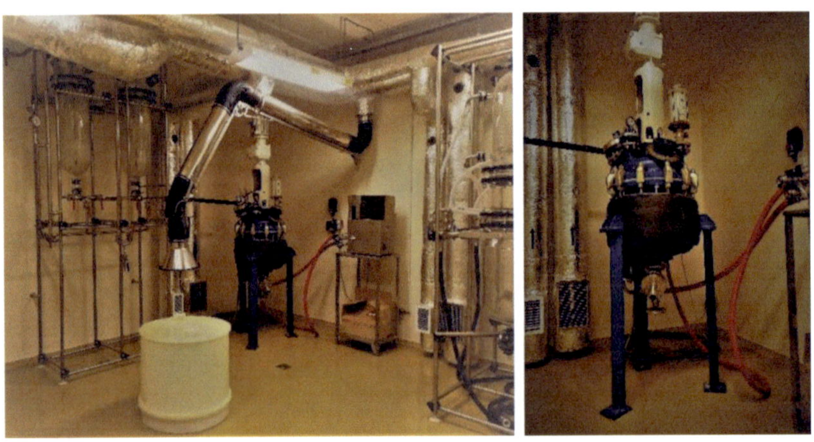

Figure 5.
Set for the synthesis of butyryl-acetyl chitin co-polyester on an industrial scale.

agent suspensions in BAC 90/10 solution, which was a mixture of solvents with a density close to the bulk density of the porophor agent. Various organic and inorganic salts (K_2CO_3, $KHCO_3$, $KHSO_4$, KNO_2, $(NH_4)_2CO_3$, $(NH_4)HCO_3$, $(NH_4)_2HPO_4$, $(NH_4)_2SO_4$, Na_2CO_3, $NaHCO_3$, Na_2HPO_4, $Na_2S_2O_3 \times 5H_2O$, NaCl, di-ammonium citrate, di-ammonium oxalate were tested. It was found that all tested salts can be used as porophors. However, the most optimal porophor agent, in terms of porosity (95–99%) and tensile strength of 5 cN, was NaCl.

Based on laboratory work, it was possible to start work on optimizing the production of porous dressing materials (Medisorb R, Medisorb R Ag). In the industrial method, solid NaCl as porophor agents and a 3% solution of BAC 90/10 in ethanol were used. The membrane was formed by pouring a 3% solution of BAC 90/10 onto a layer of porophor agent to produce a spongy structure. After drying, the membrane is rinsed in distilled water at 40°C until the porophor agent is washed off. The product is then dried at 80°C. After packing, the obtained membrane dressings are subjected to a process of radiation sterilization (in the case of the variant without the addition of an antibacterial substance). To obtain a silver-coated membrane, the membrane is sprayed with a suspension of metallic silver dispersed in water by means of a spray nozzle. The silver particles are evenly distributed in the suspension using an ultrasonic device. After drying and then packing, the product is subjected to radiation sterilization. The powder dressing is obtained by grinding the byturyl-acetyl chitin co-polyester, which is then sterilized by radiation [23, 31]. **Figure 6** shows a scheme for the preparation of porous dressings based on butyryl-acetyl chitin co-polyester.

Dressing materials obtained by leaching salt from butyryl-acetyl chitin co-polyester (BAC 90/10) and sodium chloride with a diameter of 0.16–0.40 nm and/or microsilver were characterized by a high degree of porosity, pore size in the range of 275–305 nm and a degree of crystallinity in the range of 27.2–27.4%. SEM

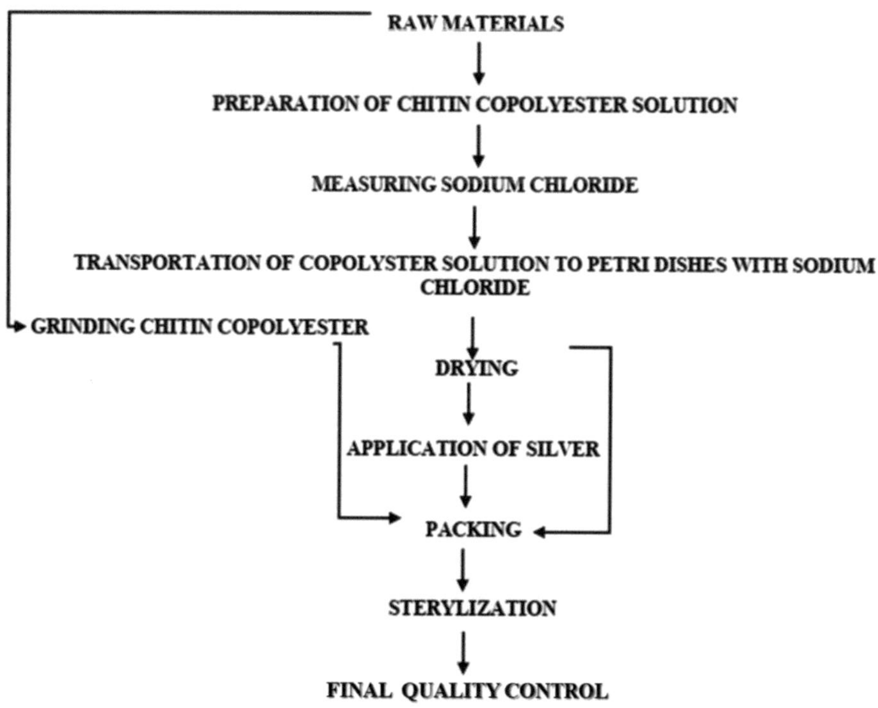

Figure 6.
A scheme for the preparation of porous dressings based on butyryl-acetyl chitin co-polyester.

tests confirmed the porous structure of pores, which are negative for the crystals of the inorganic porophor agent used. In addition, the pores are open pores, which increases the effectiveness of water adsorption. **Figure 7** shows microscopic picture of porous structures obtained by the porophor agent washout method.

Dressings made of butyryl-acetyl chitin co-polyester (**Figure 8**) are intended for wounds of various etiologies, including chronic wounds, where the healing process is disturbed by comorbidities. In order to accelerate the healing of deep wounds, a dressing in the form of a backfill has been designed. Wounds are often accompanied by a bacterial infection, therefore, apart from the dressing in the form of a membrane made of chitin co-polyester only, there is also a variant containing the addition of silver, showing a bactericidal effect in the wound environment. Silver may appear in various forms, however, it has been assumed that only the ionic form of silver has a bactericidal effect. Any other form of silver must be converted to its ionic form. Hence, metallic silver with a small particle size after oxidation and hydrolysis is characterized by the highest antibacterial activity. Silver in ionic form also has the ability to interact with proteins. It has been found that the ionic form of silver has a higher protein binding capacity compared to nanoparticles [32–36].

The presence of pores and microcapillaries in the structure of membrane dressings allows drainage of wound exudate. Dressings made on the basis of chitin co-polyesters are characterized by high biocompatibility. Biological tests confirmed the lack of cytotoxic, irritating and allergenic effects. These dressings are degraded

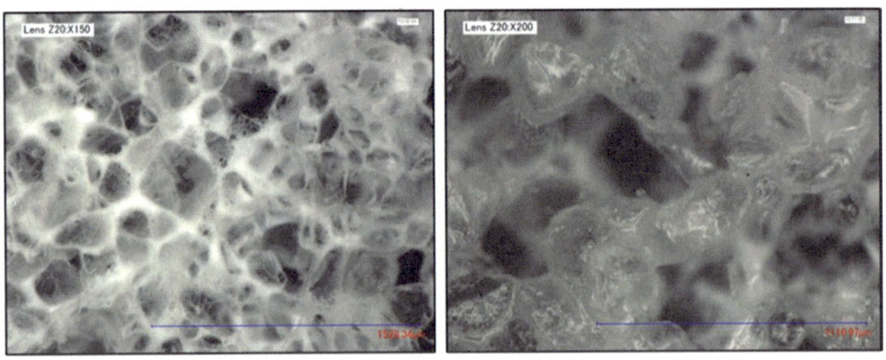

Figure 7.
Microscopic picture of porous structures obtained by the porophor agent washout method.

Figure 8.
Picture of porous structures obtained by the porophor agent washout method.

in the subcutaneous tissue and gradually become smaller. The dressing shortens and weakens the exudative phase, drains the wound and accelerates the productive phase. The epithelialization process under the butyryl-acetyl chitin co-polyester was completed faster compared to the control sample [37].

FTIR ATR analysis was made for samples of untreated Medisorb R dressings and material treated with fresh human plasma in order to test the dressing surface degradation and protein absorption on the dressing surface. Comparing spectra of samples treated with fresh human plasma and pure material, the decreasing of intensity of the vibration band of C=O at 1733 cm^{-1} in relation to the amide I signal at 1659 cm^{-1} was observed. It confirmed the sample surface degradation, which was connected to the hydrolysis of ester BAC 90/10 groups. In the *in vivo* tests, the dressings under macroscopic examination, in full thickness defects of subcutaneous tissue and skin caused wound healing with no inflammation, undergoing the most gradual biodegradation between weeks 4 and 8, and the observed differences were statistically significant [37].

The developed biodegradable dressings based on butyryl-acetyl chitin co-polyester were subjected to clinical evaluation using a wide range of patients. The use of dressings significantly accelerated the wound healing process caused by venous insufficiency and diabetes, also in patients whose healing process was disturbed by comorbidities. The improvement of the clinical condition of the wound depends on the individual patient and is most often observed after 30–60 days. The obtained results indicate that the tested dressings significantly reduce the time of wound healing. Medisorb R Ag is more effective than Medisorb R Membrane in the treatment of infected wounds. The powder form (Medisorb R Powder) allows the application of the dressing to deeper wounds. Thanks to their unique structure, dressings drain wound exudates beyond its environment, thus restoring the proper course of the cell reconstruction process. The ability to biodegrade in contact with the wound secretion eliminates the need to replace dressings, so the newly formed granulation tissue is not disturbed - cell reconstruction processes run smoothly [38].

3. Chitosan- raw materials obtained from chitin

Chitosan is obtained as a result of the hydrolysis of chitin N-acetylamide groups (partial deacetylation of chitin). The main advantage of chitosan is its much better solubility in aqueous acid solutions, especially organic acids. Chitin deacetylation by chemical or enzymatic methods allows for obtaining materials with various degrees of hydrolysis (**Figure 9**). However, it is assumed that at least 50% chitin deacetylation is necessary for the material to be classified as chitosan. The degree of deacetylation (DD) is defined as the ratio of the number of free NH_2 groups to the initial number of $NHCOCH_3$ groups present in chitin and is presented in the equation:

$$DD = \frac{N_{NH_2}}{N_{NH_2} + N_{NHCOCH_3}} \cdot 100\%$$

where N - the number of specific units (structural units) in the polymer.

The value of DD affects the biological and physicochemical properties of the polymer, such as solubility, swelling and stability, as well as reactivity.

Chitosan obtained by chemical (concentrated NaOH) or enzymatic (chitin deacetylase) deacetylation of chitin. The first step of preparation of chitosan

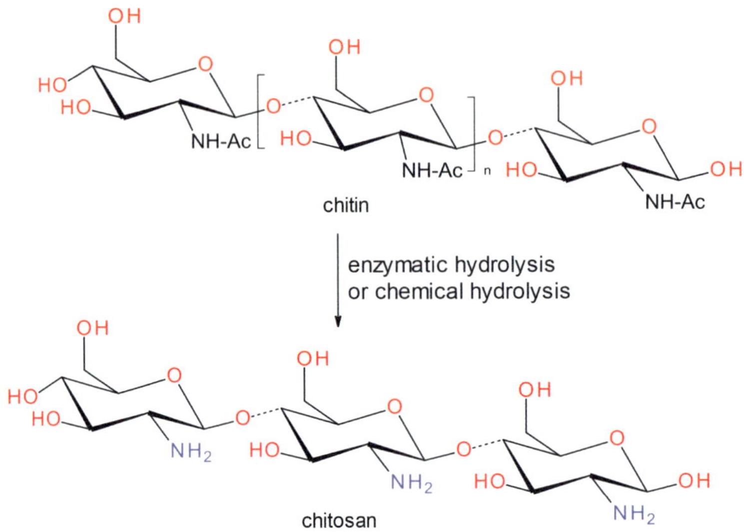

Figure 9.
Chitosan formation from chitin by chitin deacetylation.

is mechanical grinding of the raw chitin, and subsequent process of removing proteins, color compounds and inorganic salts takes place. The deproteinization process is usually performed with a dilute aqueous solution of sodium hydroxide at an elevated temperature [4, 5]. For protein removing also proteolytic enzymes were used [39, 40], but in the case of papain, trypsin and chymotrypsin, it was found that they did not completely remove the protein fraction. After deproteinization process, the residue is treated with dilute aqueous hydrochloric acid to dissolve the calcium carbonate. A similar effect can be obtained by using HCOOH, HNO_3, H_2SO_4 or EDTA [5]. The decolorization process is based on extraction with ethanol, acetone or treatment with oxidizing reagents ($KMnO_4$, NaOCl, H_2O_2). These activities allow to obtain chitin of required purity for its further transformation into chitosan. Chitosan from chitin obtained by chemical deacetylation is obtained at high temperature (above 100°C), under high pressure and in the presence of concentrated (40–50%) strong bases (usually NaOH). A typical industrial chitosan production process provides almost complete recovery of proteins, calcium oxide or calcium carbonate, carotenoid pigments and sodium acetate under the conditions of using sodium hydroxide as the deacetylating agent. However, this process has several disadvantages. It is not environmentally friendly as it consumes a large amount of energy and is also difficult to control leading to a heterogeneous product. The use of chitin deacetylase for the production of chitosan oligomers and polymers can potentially eliminate most of these drawbacks [41]. The advantages of enzymatic deacetylation are more pronounced in the processing of chitin oligomers, as these substrates are soluble in the aqueous medium and are therefore more susceptible to enzymes. The downside is the high cost of the enzyme and the requirement to use pre-processed chitin. The conditions of the chitin deacetylation process have a significant impact on the distribution of N-acetyl-D-glucosamine and D-glucosamine groups in the polysaccharide chain. Their location in the chain has a significant impact on the physicochemical properties of the material, such as crystallinity, solubility and susceptibility to degradation [3]. Depending on the final use, chitosan can be formed into a hydrogel, membranes, fibers, sponges and micro/nanoparticles [42].

4. Chitosan parameters

4.1 Physio-mechanical parameters of chitosan

Chitosan is a polysaccharide composed of randomly distributed acetylated and deacetylated units of D-glucosamine. Chitosan exists in five different crystal forms, four of which are hydrated and one is anhydrous. Microcrystalline chitosan is characterized by better biodegradability and bioactivity.

Most of the properties of chitosan depend on two parameters: degree of deacetylation and molecular weight distribution. Depending on the source and method of preparation, the deacetylation degree varies from 30 to 95%, and the molecular weight from 300 to over 1000 kDa [43]. The solubility of chitosan strongly depends on the deacetylation degree, which translates into the number of free amino groups. Chitosan is soluble in acidic solutions due to its susceptibility to protonation and formation of ammonium salts. It is soluble in acetic, formic, citric, lactic and hydrochloric acid and insoluble in most organic solvents. Chitosan, as a biodegradable polymer, is easily broken down by microorganisms into simple chemical compounds such as carbon dioxide (CO_2) and ammonia (NH_3). Like other biopolymers, it is susceptible to many chemical and physical factors leading to its degradation. The degradation process also depends on the degree of deacetylation and the molecular weight of the polymer [3, 5].

Chitosan has many valuable properties, such as: biocompatibility, biodegradability, non-toxicity, the ability to create polycations in an acidic environment, the possibility of modification, high affinity for metals, dyes and proteins, hydrophilicity, ability to create membranes and others [3, 5, 44]. These features make it applicable in medicine and pharmacy, in various industries, in environmental protection, including water treatment and separation processes. [5, 45, 46]. Chitosan also has a number of properties that limit its use in certain areas. It swells strongly in water (especially in an acidic environment), and in the swollen state it is characterized by low mechanical strength. The use of chitosan is also limited due to its high viscosity. The reduction of the viscosity of chitosan solutions can be achieved by increasing the deacetylation degree while reducing the molecular weight and increasing the temperature or ionic strength [5, 47]. The key problem with the use of chitosan is its susceptibility to external factors (humidity and temperature) and processing conditions (heating or sterilization), which can affect its structure and cause its degradation. Parameters such as molecular weight or the presence of impurities have a significant impact on the processing of chitosan [48]. This causes difficulties in maintaining the stability of chitosan (no changes in molecular weight) for a long time at room temperature [49]. The influence of many factors, such as increased temperature, the presence of strong acids, mechanical shear or radiation, on the molecular weight of chitosan was demonstrated. It is also believed that high molecular weight chitosan is more stable. The lack of reproducibility in the processing of chitosan is also due to significant differences in molecular weight, deacetylation degree and purity level depending on the source of the raw material. The level of chitosan purity may affect both biological properties, such as biodegradability or immunogenicity, as well as its solubility and stability [48, 50].

4.2 Biological parameters of chitosan

Chitosan is a non-toxic polysaccharide containing randomly distributed acetylated and deacetylated units of D-glucosamine. The results of many studies confirm the antibacterial effect of chitosan. The mechanism explaining this feature is unknown [51]. The antimicrobial activity of chitosan is strongly dependent on many factors, such as molecular weight [52], degree of deacetylation (DD), pH of the dissolving medium and its ionic strength. Stronger antibacterial activity was observed with a high degree of deacetylation [53] and a low molecular weight of chitosan [54]. The antibacterial activity of chitosan is also associated with the form of the polymer (hydrogels, membranes, dissolved form) and the presence of other compounds [55]. One of the factors responsible for the antibacterial activity of chitosan is its cationic nature. The positively charged ammonium groups of chitosan may interact with negatively charged components of the bacterial cell wall, causing damage to the cell membrane and destruction of bacterial cells (a mechanism proposed for high molecular weight chitosan) [56]. Ultimately, this causes the formation of an impermeable layer around the bacterial cell, affecting permeability and transport to the cell [57, 58]. It has been suggested that low molecular weight chitosan can penetrate bacterial cell walls and eventually enter the cytoplasm and bind to DNA affecting DNA transcription, mRNA synthesis and finally protein biosynthesis [59].

The difference in the hydrophilicity and the negative charge of the cell surface of the bacteria makes gram-negative bacteria interact more strongly with chitosan, resulting in higher antibacterial activity against them.

The antibacterial activity of chitosan or its derivatives on gram-negative bacteria has been demonstrated for various strains: *Escherichia coli*, *E. coli* K88, *E. coli* ATCC 25922, *E. coli* O157, *Pseudomonas aeruginosa*, *Proteus mirabilis*, *Salmonella enteritidis*, *S. choleraesuis* ATCC 50020, *S. typhimurium*, *S. typhimurium* ATCC 50013, *Enterobacter aerogenes*, *Listeria monocytogenes* [60–64]. The antibacterial activity of chitosan or its derivatives on gram-positive bacteria has been demonstrated for: *Staphylococcus aureus*, *S. aureus* ATCC 25923, *Corynebacterium*, *Staphylococcus epidermidis*, *Enterococcus faecalis*, *Bacillus cereus*, *Bacillus megaterium*. This indicates a broad spectrum of chitosan activity. It was also found that the effectiveness of chitosan binding to the bacterial cell wall depends on the pH value. At low pH, chitosan shows better adsorption to the bacterial cell wall due to the increased positive charge of protonated amino groups [65–67].

The antifungal activity of chitosan also depends on its molecular weight and degree of acetylation. It was found that chitosan shows antifungal activity against selected phytopathogenic fungi *Penicillium* sp. in citrus [68], *Botrytis cinerea* in cucumber [69], *Phytophthora infestans* [70], *Alternaria solani* and *Fusarium oxysporum* [71]. Chitosan is also active against fungal species pathogenic to humans, while being non-toxic to human cells. The antifungal activity of chitosan or its derivatives has been demonstrated against: *Candida albicans*, *Candida parapsilosis*, *Candida krusei*, *Candida glabrata*, *Penicillium digitatum*, *Penicillium italicum*, *Fusarium proliferatum*, *Hamigera avellanea*, *Aspergillus fumigatus*, *Rhizopus stolonifer*, *Cryptus stolofattans*. The suggested mechanism of action is to create a permeable layer by chitosan which disrupts fungal growth [72–75]. It is believed to be related to the activation of defense processes, including chitinse accumulation, synthesis of proteinase inhibitors, callus synthesis and the lignification process [76].

The antiviral activity of chitosan derivatives is also suggested. The research focuses mainly on HIV. Peptide-chitosan conjugates (GlnMetTrp-chitosan and TrpMetGln-chitosan) show a protective effect on C8166 cells by counteracting the cytolytic effects of the HIV-1RF strain. These derivatives have the ability to suppress HIV-induced syncytium formation and reduce HIV load without inhibiting HIV-1 reverse transcriptase and protease *in vitro* [77]. Sulfated low molecular weight chitosan derivatives inhibit HIV-1 replication, HIV-1 induced syncytium formation, lytic activity and p24 antigen production. These derivatives are believed to influence the binding of HIV-1 gp120 to the CD4 cell surface receptor [78]. In comparative studies, chitosan conjugates with thioglycolic acid, lactic acid, PEG and antiviral drugs, significantly higher efficacy was observed compared to the use of the drug alone [79–85].

Recent studies have shown that chitosan and its derivatives exhibit anti-tumor activity in both *in vitro* and *in vivo* models. It was found that chitosan derivatives increased the secretion of interleukin-1 and 2, influencing the maturation and infiltration of cytotoxic T lymphocytes [86]. The results have been confirmed in an *in vivo* study [87]. Moreover, a direct cytotoxic effect on neoplastic cells was found by inducing apoptosis. For A375, SKMEL28 and RPMI7951 cancer cell lines it was found to reduce cell adhesion, inhibit proliferation, inhibit specific caspases, regulate the activity of Bax as well as Bcl-2 and Bcl-XL, induce CD95 receptor expression through greater susceptibility to chitosan-induced apoptosis by FasL [88]. In the case of the PC3 A549, HepG2 and LCC cell lines, in the presence of chitosan, inhibition of the growth of neoplastic cells and inhibition of MMP-9 expression, hampering of cells in the S phase and reduction of the rate of DNA synthesis, p21 and PCNA regulation were found [89]. HepG2 and LCC xenografts in a mouse model showed inhibition of tumor growth and a reduction in the number of metastatic colonies at the dose of 500 mg/kg [90]. Carboxymethylated chitosan [91] inhibited hydrogen peroxide-induced apoptosis in Schwann cells by reducing caspase-3, −9 and Bax activity and increasing Bcl-2 activity.

Various biological properties of chitosan also include good adhesion to cells, macrophage activation, stimulation of fibroblast proliferation, stimulation of cytokinin production, stimulation of type IV collagen synthesis, promoting angiogenesis processes, haemostatic properties [92–95]. Moreover, it has a positive effect on granulation and epithelialization, and reduces scar formation. It is believed that many of the listed biological activities of chitosan are related to its unique feature, namely its cationic nature. Chitosan molecules with a positive charge interact with negatively charged erythrocytes and thrombocytes activating the extrinsic coagulation, effectively stopping bleeding.

Chemical modification of chitosan allows to modulate the biological activity of chitosan, for example, heparin inactivation or antiviral activity. Chitosan can be in the form of: gel, sponge, fiber or porous composition with ceramics, collagen or gelatin. Chitosan is used as a component of wound healing dressings, while in the case of scaffolds, it is usually used with other natural polymers (hyaluronic acid, alginic acid, poly-L-lactidic acid, elastin, collagen, gelatin) or additives (hydroxyapatite, calcium phosphate, ceramic components) [95–97].

Chitosan increases the inflow of phagocytic cells (segmented granulocytes and macrophages) to the site of infection, stimulates the migration and proliferation of endothelial cells and fibroblasts. The effect of chitosan on the proliferation of fibroblasts depends on the degree of deacetylation and molecular weight. Forms with a higher degree of deacetylation and lower molecular weight stimulate fibroblast proliferation to a greater extent [98–123]. Chitosan is widely researched for its use in bone and cartilage reconstruction. It has the ability to create porous structures, which makes it used in tissue engineering, orthopedics and bone regeneration. It has also been used in drug delivery systems or therapeutic substances (DNA plasmids, siRNA, nanosilver), for the production of surgical sutures, wound healing dressings and artificial internal organs [124–150].

An interesting chitosan feature is also its ability to bind with mucus and cross epithelial barriers, so that its use as an adjuvant or auxiliary adjuvant in vaccines is considered. It is also included among the auxiliary substances that enable the preparation of various forms of drugs with specific properties.

It is an excellent metal ion complexing agent. This parameter is useful due to the immobilization of metal ions with antibacterial activity and enabling their controlled release, depending on the needs [97].

Chitosan can also be an environmentally friendly agent used to obtain textiles with antibacterial properties. Attempts were made to introduce chitosan powder into cotton and polyester-cotton fabric. Chitosan was introduced after the fabric surface was activated by 20% NaOH. The performed studies confirmed that chitosan is well implemented in fabrics made of a cotton and polyester/cotton blend [151].

4.3 Chitosan in dressing materials

Due to its physicochemical and biological properties, chitosan and its derivatives are considered to be versatile biomaterials with various biological activities [152–159].

Chitosan and its derivatives as materials with antimicrobial activity and low immunogenicity are widely used in wound healing. They provide a three-dimensional matrix for tissue growth, activate macrophages and stimulate cell proliferation [160]. Chitosan improves the activity of polymorphonuclear leukocytes, macrophages and fibroblasts, which increase granulation and organization of repaired tissues [161]. Its degradation to N-acetyl-β-D-glucosamine stimulates the proliferation of fibroblasts, supports regular collagen deposition, and also stimulates the synthesis of hyaluronic acid at the wound site. These properties accelerate healing and prevent scarring [162]. The development of chitosan formation in the form of nanofibers with the assumed adhesive properties allowed to obtain a material useful for the creation of dressing materials [163]. Chitosan nanofibers obtained by electrospinning method are porous, have high tensile strength, large surface area combined with an ideal rate of water vapor and oxygen transfer. They are also compatible with stem cells derived from adipose tissue, which is beneficial for wound healing [164, 165].

A characteristic feature of chitosan dressings is their ability to effectively control bleeding [166]. The most important element of hemostasis is blood clotting, which leads to the formation of a clot consisting mainly of the fibrin network and platelets embedded in it. This process prevents further loss of fluid and electrolytes from the wound and reduces contamination of the wound. There is erythema around the wound, swelling, pain and locally increased temperature. Inflammation widens local blood vessels, which facilitates the penetration of macrophage cells and fibroblasts into the wound, which cleanse the wound of tissue residues, vascular clots and pathogenic bacteria. In the next phase of healing, fibroblasts synthesize collagen and other proteins needed to build and regenerate connective tissue and rebuild damaged blood vessels. In the course of scar formation, type III collagen fibers transform into type I collagen until they reach the balance characteristic of healthy skin and are necessary to restore skin continuity. The final remodeling process leads to a significant increase in the mechanical strength of the wound. The haemostatic effect of chitosan has been clearly documented. Chitosan in the form of a non-woven fabric has a positive effect on each stage of wound healing. The unique features of chitosan include: macrophage activation, stimulation of fibroblast proliferation, absorption of growth factors, stimulation of cytokinin production, stimulation of type IV collagen synthesis, support for angiogenesis processes, antibacterial and hemostatic properties. The positive effect of chitosan on granulation tissue, epidermis and reduction of scar formation has been proven. Like chitin, chitosan is susceptible to enzymatic biodegradation which produces biologically active oligosaccharides. The positively charged chitosan molecules react with negatively charged erythrocytes and thrombocytes to activate the external clotting pathway and effectively block bleeding. At the same time, chitosan can serve as a carrier of specific medicinal substances (DNA plasmids, siRNA, nanosilver particles), which enhance its positive effect on the healing process. Chitosan has also been found to significantly increase the adhesion and aggregation of platelets in the process of hemostasis [167, 168].

Currently, there are many chitosan materials available on the market that are used to heal wounds in patients undergoing plastic surgery [169], skin grafting [170, 171] and endoscopic sinus surgery [172]. Chitosan-containing materials in the form of nonwovens, nanofibers, composites, films and sponges are: HemCon®, GuardaCare®, ChitoFlex®, ChitoGauze®, Celox™ Granules, Celox™ Gauze, Chito-Seal™, Clo-SurPLUS PAD Tegasorb™, Tegaderm™ ChiGel, ChitopackC®, and TraumaStat™ [173–176].

Haemostatic dressings also include Tromboguard® - a multi-layer dressing made of three layers: semi-permeable polyurethane foil, hydrophilic polyurethane foam, and a layer containing chitosan. The film layer protects the dressing against seepage, allowing the wound environment to remain moist, ensuring optimal air permeability to its interior and creating a barrier against external factors. Polyurethane foam is a load-bearing layer and has strong absorbent properties thanks to the "pore-in-pore" structure. The polyurethane layer is responsible for storing exudate and keeping it outside the wound surface, ensuring adequate wound moisture. Additionally, it is a layer that protects the wound against mechanical damage.

The active layer, which is created by a unique composition of chitosan and alginates, activates the blood coagulation process, significantly reducing bleeding time. By reacting on the wound surface with erythro- and thrombocytes, chitosan significantly shortens the bleeding time. Calcium alginate accelerates the natural clotting process, and sodium alginate - by absorbing wound discharge - creates a layer of gel on the surface of the dressing that prevents it from sticking to the wound. Alginates are resorbable, non-toxic, non-carcinogenic, non-allergic and haemostatic [177]. When used as dressing materials, it is important that during contact with the wound, part of the alginate dressing passes in the form of a gel, which prevents the wound surface from drying out, and thus creates the possibility of creating a favorable, moist environment within the skin lesion [178]. At the same time, hemostatic properties result in a faster wound healing process and allow for more effective scarring. Patients also benefit from using these dressings to reduce pain when changing them. A significant advantage of using alginate-containing dressings is the elimination of the dressing sticking to the wound and high absorbency.

The Tromboguard® dressing (**Figure 10**) is used to stop bleeding in the case of: traumatic wounds, postoperative wounds, skin graft collection sites in surgery and reconstructive surgery - including combustiology, wounds requiring emergency care, gunshot and puncture wounds, wounds resulting from traffic accidents. It is characterized by a quick hemostatic effect (stops bleeding in 3 minutes), an antibacterial effect inside the product (protecting the dressing against the growth of microorganisms), and effective blood absorption even under pressure. It is not irritating, sensitizing and cytotoxic.

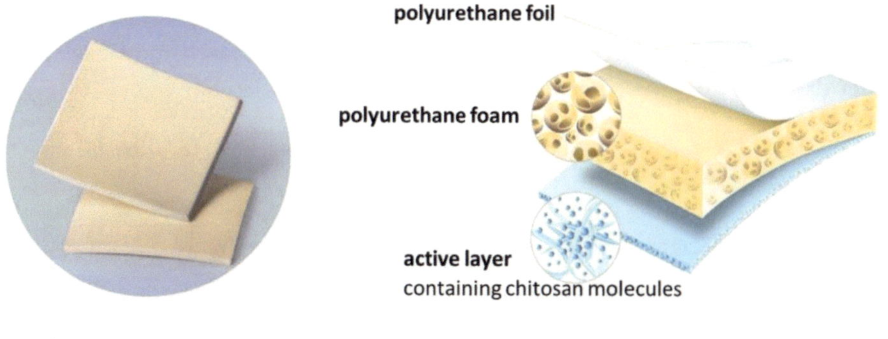

Figure 10.
Tromboguard® dressing structure.

Tests of operational parameters: tensile strength, the ability to adapt to the injury site or the transmission of moisture vapors have shown that this dressing has a tensile strength (for porous materials) of min. 75 kPa (according to PN-EN ISO 1798), which corresponds to the value recommended for dressing materials, and vapor permeability (transmission of moisture vapor) of min. 400 $g/m^3/24h$.

The results of clinical trials have demonstrated the high haemostatic efficacy of Tromboguard®. The high effectiveness and durability of the antihaemorrhagic effect was confirmed 24 hours after application, which allowed the introduction of an absorbent foam dressing [179] and a three-layer hemostatic dressing to the market [96, 97].

5. Conclusion

Chitin and its ester derivatives, as well as chitosan obtained as a result of chitin deacetylation, have many valuable chemical, physical and biological properties that determine their use in many areas, also in medicine.

The widest use of chitin and its derivatives is observed in biomedical sciences, in particular: in dressing materials (active dressings), active substance carriers (drugs and growth factors), in tissue engineering (cell scaffolds - scaffolds, mainly orthopedics) and in regenerative medicine (stem cell differentiation). Chitin accelerates the wound healing process by having a beneficial effect on processes such as angiogenesis, granulation, epithelialization and scar formation, which play a key role in the physiological wound healing process. It increases the inflow of phagocytic cells (segmented granulocytes and macrophages) to the site of infection, stimulates the migration and proliferation of endothelial cells and fibroblasts. Chitin derivative dressings are considered to be very effective medical devices in the healing of difficult-to-heal wounds [6–11].

The results of clinical trials with dressings based on butyryl-acetyl chitin co-polyesters have also shown their high effectiveness in healing wounds of various etiologies, mainly those caused by chronic venous insufficiency and diabetes. Their use leads to a reduction in the ulcer area and its depth. These dressings were assessed as having a high safety profile [38].

On the other hand, the results of clinical trials with chitosan dressings showed high effectiveness and durability of the anti-haemorrhagic effect. These studies also confirmed the safety of the dressing [96]. The antibacterial test confirmed that the dressing is bactericidal. Thus, there are currently many different hemostatic dressings based on chitosan on the market.

Acknowledgements

This research was funded by the National Centre for Research and Development under Project POIR.04.01.02-00-0004/17.

Conflict of interest

The authors declare no conflict of interest.

Author details

Ilona Latańska[1], Piotr Rosiak[1], Paulina Paul[1], Witold Sujka[1] and Beata Kolesińska[2*]

1 Tricomed S.A., Łódź, Poland

2 Lodz University of Technology, Faculty of Chemistry, Institute of Organic Chemistry, Lodz, Poland

*Address all correspondence to: beata.kolesinska@p.lodz.pl

References

[1] Du Y, Zhao Y, Dai S, Yang B. Preparation of water-soluble chitosan from shrimp shell and its antibacterial activity. Innov. Food Sci. Emerg. Technol. 2009; 10: 103-107. DOI: 10.1016/j.ifset.2008.07.004.

[2] Pillai CKS, Paul W, Sharma CP. Chitin and chitosan polymers: Chemistry, solubility and fiber formation. Prog. Polym. Sci. 2009; 34: 641-678. DOI: 10.1016/j.progpolymsci.2009.04.001.

[3] Mucha M. Chitosan: a versatile polymer from renewable sources, WNT, Warsaw, 2010.

[4] No HK, Meyers SP, Lee KS. Isolation and characterization of chitin from crawfish shell waste. J. Agric. Food Chem. 1989; 37: 575-579. DOI: 10.1021/jf00087a001.

[5] Struszczyk H. Mint: Chitin and Chitosan, Part I. Properties and Production. Polimery. 2002; 47: 316-325.

[6] Aranaz I, Mengíbar M, Harris R, Paños I, Miralles B, Acosta N, Galed G, Heras Á. Functional Characterization of Chitin and Chitosan. Curr. Chem. Biol. 2009; 3: 203-230. DOI: 10.2174/2212796810903020203.

[7] Park BK, Kim MM. Applications of Chitin and Its Derivatives in Biological Medicine. Int. J. Mol. Sci. 2010; 11: 5152-5164. DOI: 10.3390/ijms11125152.

[8] Kumirska J, Weinhold MX, Thöming J, Stepnowski P. Biomedical Activity of Chitin/Chitosan Based Materials -Influence of Physicochemical Properties Apart from Molecular Weight and Degree of N-Acetylation. Polymers. 2011; 3:1875-1901.

[9] Yadav RP, Chauhan MK. Pharmaceutical Diversity of Chitin and Chitosan: A Review. Int. J. Pharm. Sci. Res. 2017; 2: 6-11.

[10] Tripathi K, Singh A. Chitin, chitosan and their pharmacological activities: a review. Int. J. Pharm. Sci. Res. 2018; 9: 2626-2635.

[11] Zargar V, Asghari M, Dashti A. A review on chitin and chitosan polymers: structure, chemistry, solubility, derivatives and applications. Chem. Bio. Eng. Rev. 2015; 2: 1-24. DOI: 10.1002/cben.201400025.

[12] Sugimoto M, Kawahara M, Teramoto Y, Nishio Y. Synthesis of acyl chitin derivatives and miscibility characterization of their blends with poly (ε-caprolactone). Carbohydr. Polym. 2010; 79: 948-954. DOI: 10.1016/j.carbpol.2009.10.014.

[13] Bhatt LR, Kim BM, Hyun K, Kang KH, Lu C, Chai KY. Preparation of chitin butyrate by using phosphoryl mixed anhydride system. Carbohydr. Res. 2011; 346: 691-694. DOI: 10.1016/j.carres.2011.01.033.

[14] Bhatt LR, Kim BM, An CY, Lu CC, Chung YS, Soung MG, Park SH, Chai KY. Synthesis of chitin cycloalkyl ester derivatives and their physical properties. Carbohydr. Res. 2010; 345: 2102-2106. DOI: 10.1016/j.carres.2010.07.017.

[15] Bhatt LR, Kim BM, Hyun K, Kwak GB, Lee CH, Chai KY. Preparation and characterization of chitin benzoic acid esters. Molecules. 2011; 16: 3029-3036. DOI: 10.3390/molecules16043029.

[16] Szosland L, East GC. The Dry Spinning of Dibutyrylchitin Fibers. J. Appl. Polym. Sci. 1995; 58: 2459-2466. DOI: 10.1002/app.1995.070581313.

[17] Szosland L, Krucinska I, Cisło R, Paluch D, Staniszewska-Kus J, Solski L, Szymonowicz M. Synthesis of dibutyrylchitin and preparation of new textiles made from dibutyrylchitin. Fibres Text. East. Eur. 2001; 9: 54-57.

[18] Paluch D, Pielka S, Szosland L, Staniszewska-Kus J, Szymonowicz M, Solski L, Zywicka B. Biological investigation of the regenerated chitin fibres. Eng. Biomater. 2000; 12: 17-22.

[19] Chilarski A, Szosland L, Krucinska I, Kiekens P, Blasinska A, Schoukens G, Cislo R, Szumilewicz J. Novel Dressing Materials Accelerating Wound Healing Made from Dibutyrylchitin. Fibers Text. East. Eur. 2007; 15: 77-81.

[20] Skołucka-Szary K, Ramięga A, Piaskowska W, Janicki B, Grala M, Rieske P, Stoczyńska-Fidelus E, Piaskowski S. Chitin dipentanoate as the new technologically usable biomaterial. Mater. Sci. Eng. C Mater. Biol. Appl. 2015;55: 50-60. DOI: 10.1016/j. msec.2015.05.051.

[21] Skołucka-Szary K, Ramięga A, Piaskowska W, Janicki B, Grala M, Rieske P, Bartczak Z, Piaskowski S. Synthesis and physicochemical characterization of chitin dihexanoate - A new biocompatible chitin derivative - In comparison to chitin dibutyrate. Mater. Sci. Eng. C Mater. Biol. Appl. 2016; 60: 489-502. DOI: 10.1016/j. msec.2015.11.078.

[22] Draczyński Z, Boguń M, Sujka W, Kolesińska B. An industrial scale synthesis of biodegradable soluble in organic solvents butyric-acetic chitin co-polyesters. Adv. Polym. Technol. 2018; 37: 3210-3221. DOI: 10.1002/ adv.22090

[23] Draczynski Z, Kolesinska B, Latanska I, Sujka W. Preparation Method of Porous Dressing Materials Based on Butyric-Acetic Chitin Co-Polyesters. Materials. 2018; 11: 2359. DOI: 10.3390/ma11122359.

[24] Tokura S, Nishi N, Tsutsumi A, Somorin O. Studies on chitin VIII. Some properties of water soluble chitin derivatives. Polym. J. 1983; 1:, 485-489. DOI:

[25] Tokura S, Nishi N, Noguchi J. Studies on Chitin. III. Preparation of Chitin Fibers. Polym. J. 1979; 11: 781-786.

[26] Tokura S, Nishi N, Somorin O, Noguchi J. Studies on chitin. IV. Preparation of acetylchitin fibers. Polym. J. 1980; 12: 695-700. DOI: 10.1295/polymj.12.695.

[27] Van Luyen D, Rossbach V. Mixed esters of chitin. J. Appl. Polym. Sci. 1995; 55: 679-685. DOI: 10.1002/ app.1995.070550504.

[28] Bourne EJ, Stacey M, Tatlow JC, Tedder JM. Studies on trifluoroacetic acid. Part I. Trifluoroacetic anhydride as a promoter of ester formation between hydroxy-compounds and carboxylic acids. J. Chem. Soc.1949; 2976-2979. DOI: 10.1039/JR9490002976.

[29] Van Luyen D, Rossbach V. Mixed esters of chitin. J. App. Pol. Sci. 1995; 55: 679-685. (DOI: 10.1002/ app.1995.070550504.

[30] Latańska I, Kolesińska B, Draczyński Z, Sujka W. The use of chitin and chitosan in manufacturing dressing materials. Prog. Chem. Appl. Chitin Deriv. 2020; XXV: 16-36. DOI: 10.15259/PCACD.25.002.

[31] Draczyński Z, Szosland L, Janowska G, Sujka W, Rogaczewska A. Chitin ester dressing and method of producing chitin ester dressing. P.404249 of 07.06.2013.

[32] Maramu N, Rao BS. A Study on Stabilization of Polymers Against Radiation. RRJoPHY. 2018; 6:18-20.

[33] Tipnis NP, Burgess DJ. Sterilization of implantable polymer-based medical devices: A review. Int. J. Pharm. 2018; 544: 455-460. DOI: 10.1016/j. ijpharm.2017.12.003.

[34] Alariqi SAS, Pratheep Kumar A, Rao BSM, Tevtia AK, Singh RP.

Stabilization of γ-sterilized biomedical polyolefins by synergistic mixtures of oligomeric stabilizers. Polym. Degrad. Stab. 2006; 91: 2451-2464. DOI:10.1016/j.polymdegradstab. 2006.03.010.

[35] Shamshad A, Basfar AA. Radiation resistant polypropylene blended with mobilizer: antioxidants and nucleating agent. Radiat. Phys. Chem. 2000; 57: 447-450. DOI: 10.1016/ S0969-806X(99)00412-0.

[36] Przybytniak G, Mirkowski K, Rafalski A, Nowicki A, Legocka I, Zimek Z. Effect of hindered amine light stabilizers on the resistance of polypropylene towards ionizing radiation. Nukleonika. 2005; 50: 153-159.

[37] Sujka W, Draczynski Z, Kolesińska B, Latańska I, Jastrzębski Z, Rybak Z, Żywicka B. Influence of porous dressings based on butyric-acetic chitin copolymer on biological process in vitro and in vivo. Materials 2019; 12: 970. DOI: 10.3390/ma12060970.

[38] Latanska I, Kozera–Żywczyk A, Paluchowska EB, Owczarek W, Kaszuba A, Noweta M, Tazbir J, Kolesinska B, Draczyński Z, Sujka W. Characteristic features of wound dressings based on butyric-acetic chitin copolyesters results of clinical trial, Materials. 2019; 12: 4170. DOI: 10.3390/ ma12244170.

[39] Gagne N, Simpson BK. Use of proteolytic enzymes to facilitate the recovery of chitin from shrimp wastes. Food Biotechnol. 1993; 7: 253-263. DOI:

[40] Gildberg A, Stenberg E. A new process for advanced utilisation of shrimp waste. Process Biochem. 2001; 36: 809-812. DOI: 10.1016/ S0032-9592(00)00278-8.

[41] Tsigos I, Martinou A, Kafetzopoulos D, Bouriotis V. Chitin deacetylases: new, versatile tools in biotechnology. Trends Biotechnol. 2000; 18: 305-312. DOI: 10.1016/ s0167-7799(00)01462-1.

[42] Jayakumar R, Prabaharan M, Kumar PS, Nair SV, Tamura H. Biomaterials based on chitin and chitosan in wound dressing applications. Biotechnol. Adv. 2011; 29: 322-337. DOI: 10.1016/j.biotechadv.2011.01.005.

[43] Olteanu CE. Applications of functionalized chitosan. Scientific Study & Research. 2007; VIII(3): 227-256. ISSN 1582-540X.

[44] Rinaudo M. Chitin and chitosan: properties and applications. Prog. Polym. Sci. 2006; 31: 603-632. DOI: 10.1016/j.progpolymsci.2006.06.001.

[45] Chakrabarty T, Kumar M, Shahi VK. Chitosan based membranes for separation, pervaporation and fuel cell applications: Recent developments. Biopolymers. 2010: 10: 201-226. DOI: 10.5772/10263.

[46] Kim SK. (Ed.) Chitin, chitosan, oligosaccharides and their derivatives: biological activities and applications. CRC Press, 2010.

[47] Mironov AV, Vikhoreva GA, Kil'deeva NR, Uspenskii SA. Reasons for unstable viscous properties of chitosan solutions in acetic acid. Polym. Sci. Ser. B. 2007; 49: 15-17. DOI: 10.1134/ S1560090407010046.

[48] Szymańska E, Winnicka K. Stability of chitosan—a challenge for pharmaceutical and biomedical applications. Mar. Drugs. 2015; 13: 1819-1846. DOI: 10.3390/md13041819.

[49] Furuike T, Komoto D, Hashimoto H, Tamura H. Preparation of chitosan hydrogel and its solubility in organic acids. Int. J. Biol. Macromol. 2017; 104: 1620-1625. DOI: 10.1016/j. ijbiomac.2017.02.099.

[50] Khanmohammadi M, Elmizadeh H, Ghasemi K. Investigation of size and morphology of chitosan nanoparticles used in drug delivery system employing chemometric technique. Iran. J. Pharm. Sci. 2015; 14:, 665-675.

[51] Cheung RCF, Ng TB, Wong JH, Chan WY. Chitosan: An update on potential biomedical and pharmaceutical applications. Mar. Drugs. 2015;13: 5156-5186. DOI: 10.3390/md13085156.

[52] No HK, Park NY, Lee SH, Meyers SP. Antibacterial activity of chitosans and chitosan oligomers with different molecular weights. Int. J. Food Microbiol. 2002; 74: 65-72. DOI: 10.1016/s0168-1605(01)00717-6.

[53] Kumar ABV, Varadaraj M, Gowda LR, Tharanathan RN. Characterization of chito-oligosaccharides prepared by chitosanolysis with the aid of papain and Pronase, and their bactericidal action against *Bacillus cereus* and *Escherichia coli*. Biochem. J. 2005; 391: 167-175. DOI: 10.1042/BJ20050093.

[54] Jeon YJ, Park PJ, Kim SK. Antimicrobial effect of chitooligosaccharides produced by bioreactor. Carbohydr. Polym. 2001; 44: 71-76. DOI: 10.1016/s0144-8617(00)00200-9.

[55] Dai T, Tanaka M, Huang YY, Hamblin MR. Chitosan preparations for wounds and burns: antimicrobial and wound-healing effects. Expert Rev. Anti-Infect. Ther. 2011; 9: 857-879. DOI: 10.1586/eri.11.59.

[56] Sahariah P, Masson M. Antimicrobial Chitosan and Chitosan Derivatives: A Review of the Structure–Activity Relationship. Biomacromolecules. 2017; 18: 3846-3868. DOI: 10.1021/acs.biomac.7b01058.

[57] Zheng LY, Zhu JF. Study on antimicrobial activity of chitosan with different molecular weights. Carbohydr. Polym. 2003; 54: 527-530. DOI: 10.1016/j.carbpol.2003.07.009.

[58] Muzzarelli R, Tarsi R, Filippini O, Giovanetti E, Biagini G, Varaldo PE. Antimicrobial properties of N-carboxybutyl chitosan. Antimicrob. Agents Chemother. 1990; 34: 2019-2023. DOI: 10.1128/aac.34.10.2019.

[59] Sudarshan NR, Hoover DG, Knorr D. Antibacterial action of chitosan. Food Biotechnol. 1992; 6: 257-272. DOI: 10.1080/08905439209549838.

[60] Liu N, Chen XG, Park HJ, Liu CG, Liu CS, Meng XH, Yu LJ. Effect of MW and concentration of chitosan on antibacterial activity of *Escherichia coli*. Carbohydr. Polym. 2006; 64: 60-65. DOI: 10.1016/j.carbpol.2005.10.028.

[61] Kong M, Chen XG, Xing K, Park HJ. Antimicrobial properties of chitosan and mode of action: A state of the art review. Int. J. Food Microbiol. 2010;144: 51-63. DOI: 10.1016/j.ijfoodmicro.2010.09.012.

[62] Park SC, Nam JP, Kim JH, Kim YM, Nah JW, Jang MK. Antimicrobial action of water-soluble beta-chitosan against clinical multi-drug resistant bacteria. Int. J. Mol. Sci. 2015; 16: 7995-8007. DOI: 10.3390/ijms16047995.

[63] Sahariah P, Benediktssdottir BE, Hjalmarsdottir MA, Sigurjonsson OE, Sorensen KK, Thygesen MB, Jensen KJ, Masson M. Impact of chain length on antibacterial activity and hemocompatibility of quaternary N-alkyl and N,N-dialkyl chitosan derivatives. Biomacromolecules. 2015; 16: 1449-1460. DOI: 10.1021/acs.biomac.5b00163.

[64] Sarhan WA, Azzazy HM. High concentration honey chitosan electrospun nanofibers: Biocompatibility and antibacterial effects. Carbohydr. Polym. 2015;

122: 135-143. DOI: 10.1016/j.carbpol.2014.12.051.

[65] Chung YC, Wang HL, Chen YM, Li SL. Effect of abiotic factors on the antibacterial activity of chitosan against waterborne pathogens. Bioresour. Technol. 2003: 88: 179-184. DOI: 10.1016/s0960-8524(03)00002-6.

[66] Rhoades J, Roller S. Antimicrobial actions of degraded and native chitosan against spoilage organisms in laboratory media and foods. Appl. Environ. Microbiol. 2000; 66: 80-86. DOI: 10.1128/aem.66.1.80-86.2000.

[67] Younes I, Sellimi S, Rinaudo M, Jellouli K, Nasri M. Influence of acetylation degree and molecular weight of homogeneous chitosans on antibacterial and antifungal activities. Int. J. Food Microbiol. 2014; 185: 57-63. DOI: 10.1016/j.ijfoodmicro.2014.04.029.

[68] Tayel AA, Moussa SH, Salem MF, Mazrou KE, El-Tras WF. Control of citrus molds using bioactive coatings incorporated with fungal chitosan/plant extracts composite. J. Sci. Food Agric. 2016; 96; 1306-1312. DOI: 10.1002/jsfa.7223.

[69] Ben-Shalom N, Ardi R, Pinto R, Aki C, Fallik E. Controlling gray mould caused by Botrytis cinerea in cucumber plants by means of chitosan. Crop. Prot. 2003; 22: 285-290. DOI: 10.1016/S0261-2194(02)00149-7.

[70] Atia MMM, Buchenauer H, Aly AZ, Abou-Zaid MI. Antifungal activity of chitosan against *Phytophthora infestans* and activation of defence mechanisms in tomato to late blight. Biol. Agric. Hortic. 2005; 23: 175-197. DOI: 10.1080/01448765.2005.9755319.

[71] Saharan V, Sharma G, Yadav M, Choudhary MK, Sharma SS, Pal A, Raliya R, Biswas P. Synthesis and in vitro antifungal efficacy of Cu-chitosan nanoparticles against pathogenic fungi of tomato. Int. J. Biol. Macromol. 2015; 75: 346-353. DOI: 10.1016/j.ijbiomac.2015.01.027.

[72] El Ghaouth A, Arul J, Grenier J, Asselin A. Antifungal activity of chitosan on two postharvest pathogens of strawberry fruits. Phytopathology. 1992; 82: 398-402. DOI: 10.1016/S0953-7562(09)80447-4.

[73] Wang LS, Wang CY, Yang CH, Hsieh CL, Chen SY, Shen CY, Wang JJ, Huang KS. Synthesis and anti-fungal effect of silver nanoparticles-chitosan composite particles. Int. J. Nanomed. 2015; 10: 2685-2696. DOI: 10.2147/IJN.S77410.

[74] Lopez-Moya F, Colom-Valiente MF, Martinez-Peinado P, Martinez-Lopez JE, Puelles E, Sempere-Ortells JM, Lopez-Llorca LV. Carbon and nitrogen limitation increase chitosan antifungal activity in Neurospora crassa and fungal human pathogens. Fungal Biol. 2015; 119: 154-169. DOI: 10.1016/j.funbio.2014.12.003.

[75] Gabriel JS, Tiera MJ, Tiera VA. Synthesis, characterization, and antifungal activities of amphiphilic derivatives of diethylaminoethyl chitosan against Aspergillus flavus. J. Agric. Food Chem. 2015; 63: 5725-5731. DOI: 10.1021/acs.jafc.5b00278.

[76] Bai RK, Huang MY, Jiang YY. Selective permeabilities of chitosan-acetic acid complex membrane and chitosan-polymer complex membranes for oxygen and carbon dioxide. Polym. Bull. 1988; 20: 83-88. DOI: 10.1007/BF00262253.

[77] Karagozlu MZ, Karadeniz F, Kim SK. Anti-HIV activities of novel synthetic peptide conjugated chitosan oligomers. Int. J. Biol. Macromol. 2014; 66: 260-266. DOI: 10.1016/j.ijbiomac.2014.02.020.

[78] Artan M, Karadeniz F, Karagozlu MZ, Kim MM, Kim SK. Anti-HIV-1 activity of low molecular weight sulfated chitooligosaccharides. Carbohydr. Res. 2010; 345: 656-662. DOI: 10.1016/j.carres.2009.12.017.

[79] Meng J, Zhang T, Agrahari V, Ezoulin MJ, Youan BB. Comparative biophysical properties of tenofovir-loaded, thiolated and nonthiolated chitosan nanoparticles intended for HIV prevention. Nanomedicine. 2014; 9: 1595-1612. DOI: 10.2217/nnm.13.136.

[80] Aghasadeghi MR, Heidari H, Sadat SM, Irani S, Amini S, Siadat SD, Fazlhashemy ME, Zabihollahi R, Atyabi SM, Momen SB. Lamivudine-PEGylated chitosan: A novel effective nanosized antiretroviral agent. Curr. HIV Res. 2013; 11: 309-320. DOI: 10.2174/1570162x113119990043.

[81] Khan AB, Thakur RS. Formulation and evaluation of mucoadhesive microspheres of tenofovir disoproxil fumarate for intravaginal use. Curr. Drug Deliv. 2014; 11: 112-122. DOI: 10.2174/15672018113106660064.

[82] Ramana LN, Sharma S, Sethuraman S, Ranga U, Krishnan UM. Evaluation of chitosan nanoformulations as potent anti-HIV therapeutic systems. Biochim. Biophys. Acta. 2014; 1840: 476-484. DOI: 10.1016/j.bbagen.2013.10.002.

[83] Belletti D, Tosi G, Forni F, Gamberini MC, Baraldi C, Vandelli MA, Ruozi B. Chemico-physical investigation of tenofovir loaded polymeric nanoparticles. Int. J. Pharm. 2012; 436: 753-763. DOI: 10.1016/j.ijpharm.2012.07.070.

[84] Meng J, Sturgis TF, Youan BB. Engineering tenofovir loaded chitosan nanoparticles to maximize microbicide mucoadhesion. Eur. J. Pharm. Sci. 2011; 44: 57-67. DOI: 10.1016/j.ejps.2011.06.007.

[85] Yang L, Chen L, Zeng R, Li C, Qiao R, Hu L, Li Z. Synthesis, nanosizing and in vitro drug release of a novel anti-HIV polymeric prodrug: Chitosan-O-isopropyl-5′-O-d4T monophosphate conjugate. Bioorg. Med. Chem. 2010; 18: 117-123. DOI: 10.1016/j.bmc.2009.11.013.

[86] Tokoro A, Tatewaki N, Suzuki K, Mikami T, Suzuki S, Suzuki M. Growth-inhibitory effect of hexa-N-acetylchitohexaose and chitohexaose against Meth-A solid tumor. Chem. Pharm. Bull. 1988; 36: 784-790. DOI: 10.1248/cpb.36.784.

[87] Lin SY, Chan HY, Shen FH, Chen MH, Wang YJ, Yu CK. Chitosan prevents the development of AOM-induced aberrant crypt foci in mice and suppressed the proliferation of AGS cells by inhibiting DNA synthesis. J. Cell Biochem. 200; 100: 1573-1580. DOI: 10.1002/jcb.21152.

[88] Gibot L, Chabaud S, Bouhout S, Bolduc S, Auger FA, Moulin VJ. Anticancer properties of chitosan on human melanoma are cell line dependent. Int. J. Biol. Macromol. 2015; 72: 370-379. DOI: 10.1016/j.ijbiomac.2014.08.033.

[89] Jiang Z, Han B, L, H, Li X, Yang Y, Liu W. Preparation and anti-tumor metastasis of carboxymethyl chitosan. Carbohydr. Polym. 2015; 125: 53-60. DOI: 10.1016/j.carbpol.2015.02.039.

[90] Park JK, Chung MJ, Choi HN, Park YI. Effects of the molecular weight and the degree of deacetylation of chitosan oligosaccharides on antitumor activity. Int. J. Mol. Sci. 2011; 12: 266-277. DOI: 10.3390/ijms12010266.

[91] He B, Tao HY, Liu SQ. Neuroprotective effects of carboxymethylated chitosan on hydrogen peroxide induced apoptosis in Schwann cells. Eur. J. Pharmacol.

2014; 740: 127-134. DOI: 10.1016/j.ejphar.2014.07.008.

[92] Kozen BJ, Kircher SJ. An alternate hemostatic dressing: comparison of CELOX, HemCon and quick Clot. J. Emerg. Med. 2005; 15: 74-81. DOI: 10.1111/j.1553-2712.2007.00009.x

[93] Gegel BT, Austin PN, Johnson AD. An evidence-based review of the use of combat gauze (QuikClot) for hemorrhage control. AANA J. 2013; 81 453-458.

[94] Boateng J, Catanzano O. Advanced therapeutic dressings for effective wound healing-a review. J. Pharm. Sci. 2015; 104: 3653-3680. DOI: 10.1002/jps.24610.

[95] Paul P, Kolesińska B, Sujka W. Chitosan and its derivatives - biomaterials with diverse biological activity for manifold applications. Mini Rev. Med. Chem. 2019; 19: 737-750. DOI: 10.2174/1389557519666190112142735.

[96] Witkowski W, Surowiecka-Pastewka A, Sujka W, Matras-Michalska J, Bielarska A, Stępniak M. PMCF evaluation of efficiency, safety and silver ion secretion from the TROMBOGUARD hemostatic first aid tactical dressing. Lekarz Wojskowy. 2015;93: 301-314.

[97] Modrzejewska Z, Rogacki G, Sujka W, Zarzycki R. Sorption of copper by chitosan hydrogel: Kinetics and equilibrium. Chem Eng Process. 2016; 109: 104-113. DOI: 10.1016/j.cep.2016.08.014.

[98] Domard A, Domard M. Chitosan: Structure properties relationship and biomedical applications. Biomater. 2002; 9: 187-212.

[99] Younes I, Rinaudo M. Chitin and chitosan preparation from marine sources. Structure, properties and applications Mar. Drugs. 2015; 13: 1133-1174. DOI: 10.3390/md13031133.

[100] D'Ayala GG, Malinconico M, Laurienzo P. Marine derived polysaccharides for biomedical applications: chemical modification approaches. Molecules. 2008; 13: 2069-2106. DOI: 10.3390/molecules13092069.

[101] Chandran VS, Amritha TS, Rajalekshmi G, Pandimadevi M. Potential wound healing materials from the natural polymers -a review. Int. J. Pharm. Bio. Sci. 2015; 6: (B) 1365-1389.

[102] Baldrick P. The safety of chitosan as a pharmaceutical excipients. Regul. Toxicol. Pharmacol. 2009; 56: 290-299. DOI: 10.1016/j.yrtph.2009.09.015.

[103] Dai T, Tanaka M, Huang Y, Hamblin MR. Chitosan preparations for wounds and burns: antimicrobial and wound-healing effects. Expert Rev. Anti. Infect. Ther. 2011; 9: 857-879. DOI: 10.1586/eri.11.59.

[104] Bhattarai N, Gunn J, Zhang M. Chitosan-based hydrogels for controlled, localized drug delivery. Adv. Drug Deliv. Rev. 2010; 62: 83-99. DOI: 10.1016/j.addr.2009.07.019.

[105] Kang HK, Seo CH. The effects of marine carbohydrates and glycosylated compounds on human health. Int. J. Mol. Sci. 2015; 16: 6018-6056. DOI: 10.3390/ijms16036018.

[106] Ong SY, Wu J, Moochhala SM, Tan MH, Lu J. Development of a chitosan-based wound dressing with improved hemostatic and antimicrobial properties. Biomaterials. 2008; 29: 4323-4332. DOI: 10.1016/j.biomaterials.2008.07.034.

[107] Davydova VN, Kalitnik A. Cytokine-inducing and anti-inflammatory activity of chitosan and its low-molecular derivative. Appl. Biochem. Microbiol. 2016; 52: 476-482. DOI: 0.1134/S0003683816050070.

[108] Friedman AJ, Phan J, Schairer DO, Champer J, Qin M, Pirouz A,

Blecher-Paz K, Oren A, Liu PT, Modlin RL, Kim J. Antimicrobial and anti-inflammatory activity of chitosan-alginate nanoparticles: A targeted therapy for cutaneous pathogens. J. Invest. Dermatol. 2013; 133: 1231-1239. DOI: 10.1038/jid.2012.399.

[109] Jeon YJ, Kim SK. Antitumor activity of chitosan oligosaccharides produced in ultrafiltration membrane reactor system. J. Microbiol. Biotechnol. 2002; 12: 503-507.

[110] Papineau A, Hoover M, Knorr DG, Farkas, DF. Antimicrobial effect of water soluble chitosans with high hydrostatic pressure. Food Biotechnol. 1991; 5: 45-57. DOI: 10.1080/08905439109549790.

[111] Sudarshan NR, Hoover DG, Knorr D. Antibacterial action of chitosan. Food Biotechnol. 1992; 6: 257-272. DOI: 10.1080/08905439209549838.

[112] Raafat D, Haas KA, Sahl HG. Insights into the mode of action of chitosan as an antibacterial compound. Appl. Environ. Microbiol. 2008; 74: 3764-3773. DOI: 10.1128/AEM.00453-08.

[113] Devlieghere F, Vermeulen A, Debevere, J. Chitosan: antimicrobial activity, interactions with food components and applicability as a coating on fruit and vegetables. Food Microbiol. 2004; 21: 703-714. DOI: 10.1016/j.fm.2004.02.008.

[114] Fang S, Li W, Shih CF. Antifungal activity of chitosan and its preservative effect on low-sugar Candied Kumquat. Food Prot. 1994; 57: 136-140. DOI: 10.4315/0362-028x-57.2.136.

[115] Chung YC, Chen CY. Antibacterial characteristic and activity of acid soluble chitosan. Bioresource Technol. 2008; 99: 2806-2814. DOI: 10.1016/j.biortech.2007.06.044.

[116] Helander IM, Lassila NEL, Ahvenainen R, Rhoades J, Roller S. Chitosan disrupts the barrier properties of the outer membrane of gram-negative bacteria. Int. J. Food Microbiol. 2001; 30: 235-244. DOI: 10.1016/s0168-1605(01)00609-2.

[117] Másson M, Holappa J, Hjálmarsdóttir M, Rúnarsson Ö, Nevalainen VT, Järvinen T. Antimicrobial activity of piperazine derivatives of chitosan. Carbohydr. Polym. 2008; 74: 566-571. DOI: 10.1016/j.carbpol.2008.04.010.

[118] Yalpani M, Johnson F, Robinson LE. Antimicrobial activity of some chitosan derivatives, in: Brine CJ, Sandford PA, Zikakis JP. (Eds), Advances in Chitin and Chitosan, Elsevier, London, 2001, 543-548.

[119] Sebti I, Carnet A, Pantiez A, Grelier S, Coma VJ. Chitosan polymer as bioactive coating and film against Aspergillus niger contamination. Food Sci. 2005; 70: 100-104. DOI: 10.1111/j.1365-2621.2005.tb07098.x.

[120] Agrawal K. Chitosan as classic bio-polymer, a review. IJPLS 2010; 1: 369-372.

[121] Majekodunmi SO. Current development of extraction, characterization and evaluation of properties of chitosan and its use in medicine and pharmaceutical industry. Am. J. Polym. Sci. 2016; 6: 86-91. DOI: 10.5923/j.ajps.20160603.04.

[122] Selvaral S, Karthikeyan J, Saravanakumar N. Chitosan loaded microspheres as an ocular delivery system for acyclovir. Int. J. Pharm. Pharm. Sci. 1996; 4: 125-132.

[123] Kong M, Chen XG, Xing K, Park HJ. Antimicrobial properties of chitosan and mode of action: a state of the art review. Int. J. Food Microbiol. 2010; 144: 51-63. DOI: 10.1016/j.ijfoodmicro.2010.09.012.

[124] Li Y, Ju D. The application, neurotoxicity, and related mechanism of cationic polymers. Academic Press 2017; 285-329.

[125] Zhang W, Zhang J, Jiang Q, Xia W. The hypolipidemic activity of chitosan nanopowder prepared by ultrafine milling. Carbohydr. Polym. 2013; 95: 487-491. DOI: 10.1016/j.carbpol.2013.02.037.

[126] Chen D, Hu B, Huang C. Chitosan modified ordered mesoporous silica as micro-column packing materials for on-line flow injection-inductively coupled plasma optical emission spectrometry determination of trace heavy metals in environmental water samples. Talanta. 2009; 78: 491-497. DOI: 10.1016/j.talanta.2008.11.046.

[127] Gomaa YA, Darwish IA, Boraei NA, El-Khordagui LK. Formulation of wax oxybenzone microparticles using a factorial approach. J. Microencapsul. 2010; 27: 628-639. DOI: 10.3109/02652048.2010.506580.

[128] Muramatsu K, Oba K, Mukai D, Hasegawa K, Masuda S, Yoshihara Y. Subacute systemic toxicity assessment of β-tricalcium phosphate/carboxymethyl-chitin composite implanted in rat femur. J. Mater. Sci. Mater. Med. 2007; 18: 513-522.

[129] Madhumathi K, Kumar S, Kavya K, Furuike T, Tamura H, Nair S, Jayakumar R. Preparation and characterization of novel β-chitin-hydroxyapatite composite membranes for tissue engineering applications. Int. J. Biol. Macromol. 2009; 45: 289-292. DOI: 10.1016/j.ijbiomac.2008.09.013.

[130] Agnihotri SA, Mallikarjuna NN, Aminabhavi TM: Recent advances on chitosan-based micro- and nano particles in drug delivery. J. Control Release. 2004; 100: 5-28. DOI: 10.1016/j.jconrel.2004.08.010.

[131] Dai M, Zheng X, Xu X, Kong X, Li X, Guo G, Qian Z. Chitosan-alginate sponge: preparation and application in curcumin delivery for dermal wound healing in rat. Biomed. Res. Int. 2009; 1-8. DOI: 10.1155/2009/595126.

[132] Momin M, Kurhade S, Khanekar P, Mhatre S. Novel biodegradable hydrogel sponge containing curcumin and honey for wound healing. J. Wound Care. 2016; 25: 364-372. DOI: 10.12968/jowc.2016.25.6.364.

[133] Vinová J, Vavíková E: Chitosan derivatives with antimicrobial, antitumour and antioxidant activities - a review. Curr. Pharm. Des. 2011; 17: 3596-3607. DOI: 10.2174/138161211798194468.

[134] Rajasree R, Rahate KP. An overview on various modifications of chitosan and its applications. Int. J. Pharm. Sci. Res. 2013; 4: 4175-4193.

[135] Tan YL, Liu CG. Self-aggregated nanoparticles from linoleic acid modified carboxymethyl chitosan: Synthesis, characterization and application in-vitro. Colloids Surf. B. Biointerfaces. 2009; 69: 178-182. DOI: 10.1016/j.colsurfb.2008.11.026.

[136] Ahmed EM. Hydrogel: Preparation, characterization and applications: A review. J. Adv. Res. 2015; 6: 105-121. DOI: 10.1016/j.jare.2013.07.006.

[137] Huo M, Zhang Y , Zhou J. Synthesis and characterization of low-toxic amphiphilic chitosan derivatives and their application as micelle carrier for antitumour drug. Int. J. Pharm. 2010; 394: 162-173. DOI: 10.1016/j.ijpharm.2010.05.001.

[138] Fang L, Jianing L. Anti-tumor activity of paclitaxel-loaded chitosan nanoparticles: An in vitro study. Mater. Sci. Eng. C.2009; 29: 2392-2397. DOI: 10.1016/j.msec.2009.07.001.

[139] Vongchan P. Sajomsang W, Subyen D, Kongtawelert P. Anticoagulant activity of a sulfated chitosan. Carbohydr. Res. 2002; 337: 1239-1242. DOI: 10.1016/s0008-6215(02)00098-8.

[140] Olivier B, Jean YR. Glucosamine and chondroitin sulfate as therapeutic agents for knee and hip osteo-arthritis. Drugs Aging. 2007; 24: 573-580. DOI: 10.2165/00002512-200724070-00005.

[141] Prasad RS. Preparation, characterization and anti- inflammatory activity of chitosan stabilized silver nano particles. Research J. Pharma. Dosage Forms and Tech 2013; 5: 161-167.

[142] Zhang JL. Effects of chitosans physico-chemical properties on binding capacities of lipid and bile salts in- vitro. Chin. Food Sci. 2008; 29: 45-49.

[143] Liu JN, Zhang JL, Maezaki Y, Tsuji K, Nakagawa Y, Kawai Y, Akimoto M, Tsugita T. Hypocholesterolemic effects of different chitosan samples in-vitro and in-vivo. Food Chcm. 2008; 107: 419-425.

[144] Huimin Q, Jiwen S. The antihyperlipidemic mechanism of high sulfate content Ulvan in rats. Mar. Drugs. 2015; 13: 407-3421. DOI: 10.3390/md13063407.

[145] Pan H, Yang Q, Huang G, Ding C, Cao P, Huang L, Xiao T, Guo J, Su Z. Hypolipidemic effects of chitosan and its derivatives in hyperlipidemic rats induced by a high-fat diet. Food Nutr. Res. 2016; 60: 31137. DOI: 10.3402/fnr.v60.31137.

[146] Kerch G. The Potential of chitosan and its derivatives in prevention and treatment of age-related diseases. Mar. Drugs. 2015; 13: 2158-2182. DOI: 10.3390/md13042158.

[147] Patil Kumar M, Dash D. Chitosan: a versatile bio-polymer for various medical applications. J. Sci. Eng. Res. 2013; 4: 1-16.

[148] Davydova VN, Nagorskaia VP, Gorbach VI, Kalitnik AA, Reunov AV, Solov'eva TF, Ermak IM. Chitosan antiviral activity dependence on structure and deploymerization method. Prikl. Biokhim. Mikrobiol. 2011; 47: 113-118.

[149] George P, Nikolaos B. Swelling studies and in-vitro release of verapamil from calcium alginate and calcium alginate chitosan beads. Int. J. Pharm. 2006; 3231: 34-42. DOI: 10.1016/j.ijpharm.2006.05.054.

[150] Park PJ, Je JY, Kim SK. Angiotensin I converting enzyme (ACE) inhibitory activity of hetero chitooligosaccharides prepared from partially different deacetylated chitosans. J. Agric. Food Chem. 2003; 51: 4930-4934. DOI: 10.1021/jf0340557.

[151] Grgac SF, Tarbuk A, Dekanić T, Sujka W, Draczyński Z. The chitosan implementation into cotton and poliester/cotton blend fabrics, , Materials. 2020; 13: 1616. DOI: 10.3390/ma13071616.

[152] Kyzas GZ, Bikiaris DN. Recent modifications of chitosan for adsorption applications: a critical and systematic review. Mar. Drugs, 2015; 13: 312-337. DOI: 10.3390/md13010312.

[153] Moeini A, Pedram P, Makvandi P, Malinconico M, d'Ayala GG. Wound healing and antimicrobial effect of active secondary metabolites in chitosan-based wound dressings: A review. Carbohydr. Polym. 2020; 233: 115839. DOI: 10.1016/j.carbpol.2020.115839.

[154] Bagher Z, Ehterami A, Safdel MH, Khastar H, Semiarif H, Asefnejad A, Mohammad S, Mehdi D, Salehi MM. Wound healing with alginate/chitosan hydrogel containing hesperidin in rat

model. J. Drug Deliv. Sci. Technol. 2020; 55: 101379. DOI: 10.1016/j.jddst.2019.101379.

[155] Ruiz GAM, Corrales HFZ. Chitosan, Chitosan Derivatives and their Biomedical Applications In Biological Activities and Application of Marine Polysaccharides. InTech, 2017; 87-106. DOI: 10.5772/66527.

[156] Pokhrel S, Yadav PN, Adhikari R. Applications of chitin and chitosan in industry and medical science: a review. Nep. J. Sci. Technol. 2015; 16: 99-104. DOI: 10.3126/njst.v16i1.14363.

[157] Silva SS, Mano JF, Reis RL. Ionic liquids in the processing and chemical modification of chitin and chitosan for biomedical applications. Green Chem. 2017; 19: 1208-1220. DOI: 10.1039/C6GC02827F.

[158] Zhang J, Xia W, Liu P, Cheng Q, Tahi T, Gu W, Li B. Chitosan modification and pharmaceutical/biomedical applications. Mar. Drugs. 2010; 8: 1962-1987. DOI: 10.3390/md8071962.

[159] Elieh-Ali-Komi D, Hamblin MR. Chitin and chitosan: production and application of versatile biomedical nanomaterials. Int. J. Adv. Res.. 2016; 4: 411-427.

[160] Jayasree RS, Rathinam K, Sharma CP. Development of artificial skin (Template) and influence of different types of sterilization procedures on wound healing pattern in rabbits and guinea pigs. J. Biomater. Appl. 1995; 10: 144-162. DOI: 10.1177/088532829501000205.

[161] Ueno H, Mori T, Fujinaga T. Topical formulations and wound healing applications of chitosan. Adv. Drug Deliv. Rev. 2001; 52: 105-115. DOI: 10.1016/s0169-409x(01)00189-2.

[162] Muzzarelli RA, Mattioli-Belmonte M, Pugnaloni A, Biagini G. Biochemistry, histology and clinical uses of chitins and chitosans in wound healing. EXS. 1999; 87: 251-264. DOI: 10.1007/978-3-0348-8757-1_18.

[163] Azuma K, Izumi R, Osaki T, Ifuku S, Morimoto M, Saimoto H, Minami S, Okamoto Y. Chitin, chitosan, and its derivatives for wound healing: Old and new materials. J. Funct. Biomater. 2015; 6: 104-142. DOI: 10.3390/jfb6010104.

[164] Naseri N, Algan C, Jacobs V, John M, Oksman K, Mathew AP. Electrospun chitosan-based nanocomposite mats reinforced with chitin nanocrystals for wound dressing. Carbohydr. Polym. 2014; 109: 7-15. DOI: 10.1016/j.carbpol.2014.03.031.

[165] Guo R, Xu S, Ma L, Huang A, Gao C. Enhanced angiogenesis of gene-activated dermal equivalent for treatment of full thickness incisional wounds in a porcine model. Biomaterials. 2010; 31: 7308-7320. DOI: 10.1016/j.biomaterials.2010.06.013.

[166] Ishihara M, Obara K, Nakamura S, Fujita M, Masuoka K, Kanatani Y, Takase B, Hattori H, Morimoto Y, Ishihara M. Chitosan hydrogel as a drug delivery carrier to control angiogenesis. J. Artif. Organs. 2006; 9: 8-16. DOI: 10.1007/s10047-005-0313-0.

[167] Chou TC, Fu E, Wu CJ, Yeh JH. Chitosan enhances platelet adhesion and aggregation. Biochem. Biophys. Res. Commun. 2003; 302: 480-483. DOI: 10.1016/s0006-291x(03)00173-6.

[168] Okamoto Y, Yano R, Miyatake K, Tomohiro I, Shigemasa Y, Minami S. Effects of chitin and chitosan on blood coagulation. Carbohydr. Polym. 2003; 53: 337-342. DOI: 10.1016/S0144-8617(03)00076-6.

[169] Biagini G, Bertani A, Muzzarelli R, Damadei A, DiBenedetto G, Belligolli A, Riccotti G, Zucchini C,

Rizzoli C. Wound management with N-carboxybutyl chitosan. Biomaterials. 1991; 12: 281-286. DOI: 10.1016/0142-9612(91)90035-9.

[170] Stone CA, Wright H, Devaraj VS, Clarke T, Powell R. Healing at skin graft donor sites dressed with chitosan. Br. J. Plast. Surg. 2000; 53: 601-606. DOI: 10.1054/bjps.2000.3412.

[171] Azad AK, Sermsintham N, Chandrkrachang S, Stevens WF. Chitosan membrane as a wound-healing dressing: Characterization and clinical application. J. Biomed. Mater. Res. B Appl. Biomater. 2004; 69: 216-222. DOI: 10.1002/jbm.b.30000.

[172] Valentine R, Athanasiadis T, Moratti S, Hanton L, Robinson S, Wormald PJ. The efficacy of a novel chitosan gel on hemostasis and wound healing after endoscopic sinus surgery. Am. J. Rhinol. Allergy. 2010; 24: 70-75. DOI: 10.2500/ajra.2010.24.3422.

[173] Bennett BL, Littlejohn LF, Kheirabadi BS, Butler FK, Kotwal RS, Dubick MA, Bailey JA. Management of external hemorrhage in tactical combat casualty care: Chitosan-based hemostatic gauze dressings-TCCC guidelines-change 13-05. J. Spec. Oper. Med. 2014; 14: 40-57.

[174] Hatamabadi HR, Zarchi FA, Kariman H, Dolatabadi AA, Tabatabaey A, Amini A. Celox-coated gauze for the treatment of civilian penetrating trauma: a randomized clinical trial. Trauma Monthly. 2015; 20: e23862. DOI: 10.5812/traumamon.23862.

[175] Nguyen N, Hasan S, Caufield L, Ling FS, Narins CR. Randomized controlled trial of topical hemostasis pad use for achieving vascular hemostasis following percutaneous coronary intervention. Catheter. Cardiovasc. Interv. 2007; 69: 801-807. DOI: 10.1002/ccd.21024.

[176] Weng MH. The effect of protective treatment in reducing pressure ulcers for non-invasive ventilation patients. Intensive Crit. Care Nurs. 2008; 24: 295-259. DOI: 10.1016/j.iccn.2007.11.005.

[177] Augst AD, Kong HJ, Mooney DJ. Alginate Hydrogels as Biomaterials. Macromol. Biosci. 2006; 6: 623-633. DOI: 10.1002/mabi.200600069.

[178] Lee KY, Mooney DJ. Alginate: properties and biomedical applications. Prog. Polym. Sci. 2012; 37: 106-126. DOI: 10.1016/j.progpolymsci.2011.06.003.

[179] Rogaczewska A, Pluta A, Gąsiorowski T, Sujka W, Szymczyk P. Opatrunek przyspieszający ziarninowanie, sposób wytwarzania opatrunku przyspieszającego ziarninowanie. Patent P.391310 of 24.05.2010.

Section 5

Electrical Applications

Chapter 8

Proton Conductivity in Chitin System

Takashi Kawabata

Abstract

We have created and researched fuel cells using biomaterials as next-generation low-environmental-load energy. As is well known, chitin is a biomass that is discharged in large quantities as a marine product processing waste. We have focused on the chitin and have been studying the production of fuel cells and proton conductivity using it. It was revealed that chitin can be used as an electrolyte membrane for fuel cells under humidified conditions and becomes a proton conductor. It was found that the presence of water molecules is important for the appearance of proton conduction in chitin system. This study presents the utility value of chitin in new fields and provides insight into the proton conduction mechanism of chitin-based biomaterials.

Keywords: Chitin, chitosan, fuel cell, proton conductor, electrolyte

1. Introduction

In the olden days of biomaterials, prostheses and medical alternatives have been mainly developed. However, there are reports that biomaterials play a wider range of roles today and can be used in the field of electronic devices such as sensors. In other words, biomaterials are not limited to medical materials, but are becoming more valuable than materials such as plastics and metals that we usually use.

We have fabricated fuel cells based on chitin and investigated its proton conductivity in chitin systems. It is well known that the chitin is superior biomass emitted from marine products and is obtained from crabs and shrimp shells. It is also famous that chitin has excellent biocompatibility and can be easily decomposed in the environment. For a long time, most research on chitin focused on biocompatibility and ion adsorption capacity. For example, Malette *et al.* have studied the curative effect of chitosan on the vulnerary [1]. Sandford *et al.* reported the useful substituent effect of chitosan to skin [2, 3]. Nair and Madhavan have suggested the method for the elimination of Hg in solution using chitosan, and Peniche-covas *et al.* have investigated the efficiency of adsorption of Hg [4, 5]. Recent studies seem to focus specifically on biocompatibile medical application and biomass. Romana *et al.* have suggested that chitin-PLA laminated composite becomes a candidate for medical applications such as implants [6]. Mohamoud *et al.* have shown that insects can be used as an alternative low-cost chitin source, and bio-convert chitin directly to ethanol by using strain of *M. circinelloides* [7].

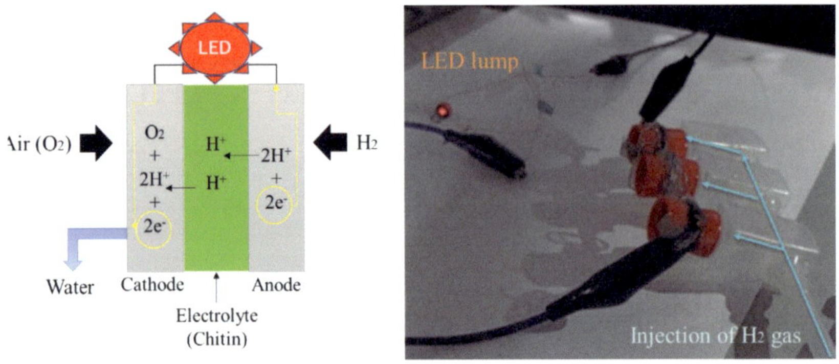

Figure 1.
Schematic diagram of fuel cell used for demonstration (left) and photograph of turning up LED lamp by chitin fuel cell (right).

However, there were few reports in the field of energy such as using chitin in fuel cells. Biomaterials such as DNA, protein and polysaccharide are abundant in nature, and they are disassembled in environment by microbial. Active use of biomaterials is expected to have less environmental impact and manufacturing costs than chemical processes.

We have revealed that the chitin is proton conductor and available for electrolyte of fuel cells (**Figure 1**) [8]. Moreover, it was found that appearance of proton conductivity in chitin demand water molecules, and the acetyl group plays important role in injection water molecule into chitin. These suggestions are basis on relationship between results of impedance measurement and water content measurement with humidified condition. In appearance of proton conductivity in chitin, it is considered that one more important factor is exist of amino acetyl group. Effects of amino acetyl group have been revealed by comparing to proton conductivity in chitosan which is basic structure of chitin. Considering proton conduction system of chemical polymer Nafion® which is used the most for fuel cell, it is deduced that the amino and acetyl group in chitin involves forming hydration supporting proton transport. Although, it is found that power density and proton conductivity in chitin are lower than the Nafion®.

Therefore, in order to improve proton conductivity, there is room for further investigation of the relationship between the appearance of proton conductivity in chitin and water molecules. These understandings are expected to present the necessary and important factors for applying polysaccharides with little change in basic structure to electrolyte membranes.

2. Experimental

2.1 Sample preparation

Chitin films were prepared with the purified chitin of under 5% of deacetylation degree, obtained from crab (Sugino Machine Limited). This purified 2%(w/v) chitin slurry was well dispersed in distilled water and the chitin sheets were prepared by suction filtration using Teflon membrane filter (ADVANTEC, Co.) [8–10]. **Figure 2** shows the photograph of the chitin film. The thickness of the film is approximately 0.07 mm.

The chitin fiber specimens were prepared by purifying chitin obtained from the tendon of crab's legs. Based on the article of Prosky *et al.*, the purification

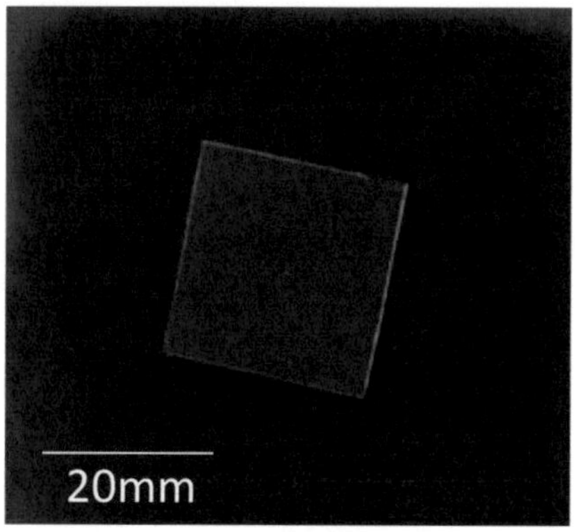

Figure 2.
Photograph of chitin sheet [11].

method was performed using a group of enzymes including α-amylase and protease obtained from *Streptomyces griseus* (Wako Pure Chemical Industries, Ltd.) [8–10]. The oriented chitosan was prepared by oriented chitin to deacetylation treatment with 25%(w/v) sodium hydroxide under reflux conditions for 5 hours.

2.2 Fabrication of the fuel cell based on chitin and operating condition

Figure 3 shows the shape of the fuel cell based on the chitin electrolyte. As show in **Figure 3**, the chitin electrolyte was inserted between Pt-C electrodes (anode and cathode). The current was collected from the current collector plates. The hydrogen and oxygen gases were introduced from the up and down sides of the fuel cell, respectively. In the fuel gas flow, the relative humidity, temperature and gas-flow ratio were controlled by the humidified gas-flow control system of Auto PEM (Toyo Corporation) at room temperature. The H_2 gas flow rate and the air flow rate are 0.1 L/min and 0.25 L/min, respectively [8].

2.3 Impedance and water contents measurements

The water contents were measured from the relative humidity dependence of the weight of chitin using the electronic analytical balance (OHAUS Inc.) and the number of water molecules per a chitin molecule was calculated from the obtained water contents and molecular weights of water and mono-chitin [8]. The water content *n* was calculated using the following equation,

$$n = \frac{(w - d) / Mw}{d / Mc} \tag{1}$$

This time, *w* and *d* show each weight of wet and dry sample. *Mw* is molecular weight of water. *Mc* is molecular weight of mono-chitin or mono-chitosan.

The measurement of electrical conductivity was carried out using precision LCR meter (E4980A, Agilent Technologies Inc.). The relative humidity and temperature were con-trolled by the humidified gas-flow control system (Auto PEM). In the impedance measurement, the electrical conductivities perpendicular to the surface and parallel to the surface in chitin sheet were measured. In the case of chitin fiber specimens, impedance measurements were performed for specimens along the fiber direction and normal to the fiber direction, respectively [8–10].

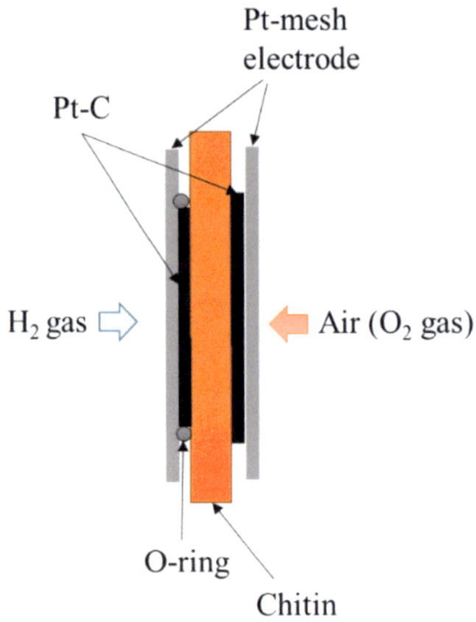

Figure 3.
Fuel cell based on the chitin.

3. Result and discussion

3.1 Power density in chitin and chitosan fuel cells

So far, we have revealed that power density in the fuel cell based on chitin or chitosan [8]. **Figure 4** shows *i-V* characteristics of the fuel based on the chitin. As shown in **Figure 4**, chitin electrolyte shows typical *i-V* curve and is also a polysaccharide material, but it has a high output, an open circuit voltage of 0.76 V and a power density of 1.35 mW/cm^2. The red dot in **Figure 4** shows the *i-V* characteristics when injecting unhumidified H_2 gas, but it can be seen that the current obtained is very low. Chitosan did the same test, but its maximum power density was 0.032 mW/cm^2, which was lower than that of chitin. These results indicate that chitin and chitosan become the proton conductor with humidified condition. Moreover, it is expected that these differences are due to difference between chitin and chitosan, that is, the deacetylation degree of the chitin system.

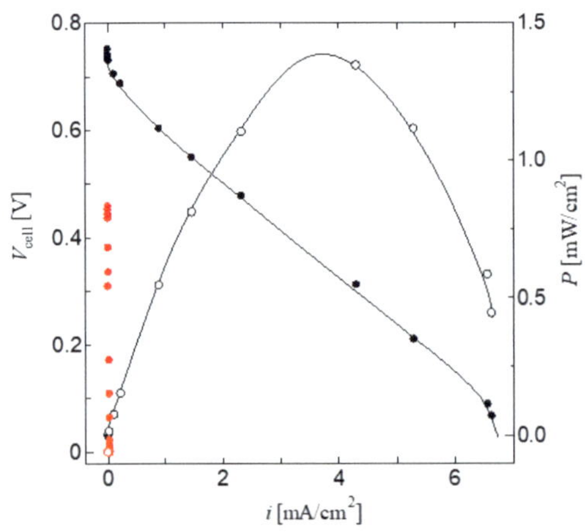

Figure 4.
i-V characteristics of fuel cell based on chitin electrolyte. Red and black dots show relation between open circuit voltage and current density. Black and red circle show power density.

3.2 Proton conductivity in chitin system

Regarding the proton conductivity in chitin and chitosan, which have already revealed, these values in sheet specimens show approximately $10^{-4} \sim 10^{-1}$ S/m [1]. The proton conductivity of the chitin system sheet increases monotonically with increasing humidity, and the value of chitin is tens of times higher, especially when the relative humidity is changed from 60% to 100%. In the case of chitosan, its value changes approximately 10 times higher. Further, we focused on acetyl group in chitin, and have revealed that relationship between proton conductivity in chitin system and the deacetylation by using Fourier transform infrared spectrometer (FT-IR). **Figure 5** shows the FTIR spectrum of chitin used in the experiment [10]. It was found that the acetyl group plays important role for appearance of proton conductivity in chitin system because the degree of deacetylation gave conductivity of anomalous behavior [10].

Furthermore, we have approached the proton conductivity in chitin system by using measurement of water contents and comparing chemical component. First, we have used sheet and fiber specimen of chitin, and measured the degree of swelling by microscopic observation after immersion experiment in water. As a result, it was confirmed that the sheet specimen was isotropically swelled in both the cross section and the in-plane, and the fiber specimen was anisotropically swelled by approximately 20% in both the fiber cross section and the fiber direction. From the above, it was clear that the chitin system introduced water molecules, so we made a comparison based on the water content and chemical component. The results obtained by these researches indicated that injection of water molecule and existence of acetyl group for promotion it, that is necessary to appearance of the proton conductivity in chitin system. However, it has not been completely understood that water molecule how to behave in chitin system yet.

3.3 Percolation conductivity in chitin

In order to clarify the role of water molecules in chitin, we obtained the volume fraction of water molecules with respect to chitin molecules from the results of the relative humidity dependence of the hydration number, and investigated the relationship with proton conductivity. Yamada *et al.* measure the resistance value that changes by gradually increasing the volume fraction of the conductor material in the non-conductor material [12]. As a result, Yamada *et al.* report that the resistance

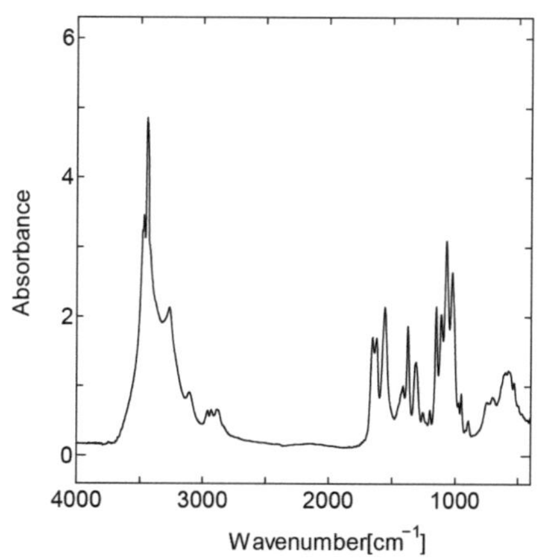

Figure 5.
FT-IR spectra of chitin sample [10].

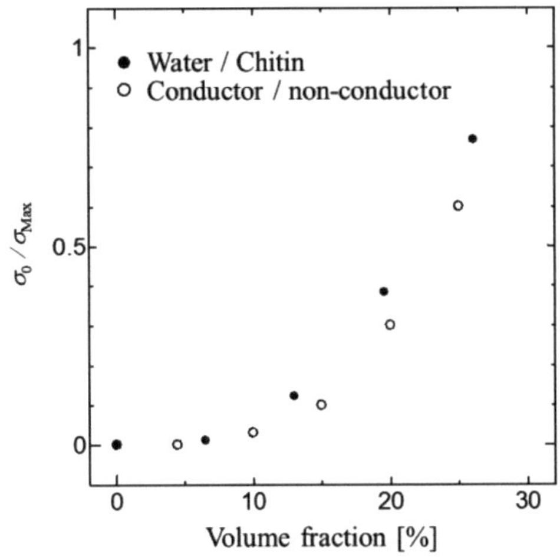

Figure 6.
Relationship between volume fraction of water molecules and percolation conduction in chitin [11].

value becomes constant at a volume fraction of about 30%. It is well known that the crystal structures of chitin and chitosan have been clarified for a long time and many reports have been published [13–19]. Sawada *et al.* report on the crystal structure of β-chitin dihydrate. We have found from water content experiments of chitin that each constituent unit has two water molecules, and that the proton conductivity is saturated during this dihydrate formation [8, 9]. Moreover, it has already confirmed that XRD diffraction pattern of our chitin sample is consistent with pattern of general chitin. Based on these results, the volume fraction of water molecules contained in the crystal lattice was estimated with reference to the report by Sawada *et al* [14–16]. **Figure 6** shows the relationship between the volume fraction of water molecules in chitin hydrate and the proton conductivity. The relationship between the volume fraction and the conductivity of the conductor in the non-conductor of **Figure 6** is a value calculated from the resistance value reported by Yamada *et al.* As can be seen in **Figure 6**, water molecules in chitin behave very much like conductors in non-conductors. In other words, this result indicates that water molecules in chitin function as a proton transport pathway.

3.4 Activation energy and proton pathway

Investigation of proton transport pathways in chitin is important for the future development and development of polysaccharide electrolyte membranes. So far, it has been shown that chitin has proton conductivity, and it is important that the introduction of water molecules behaves like a conductor for the proton conductivity. In addition, since the structure of chitin hydrate has been clarified, we approached the proton conduction pathway of chitin by measuring impedance using an oriented sample. As a result, it was confirmed that chitin and chitosan have orientation dependence of proton conductivity [8, 9]. From this result, the temperature dependence of the proton conductivity of the chitin system was investigated. **Figure 7** shows the proton conductivity in the chitin fiber direction when only the temperature factor is changed while maintaining a constant wet weight. As shown in **Figure 7**, the relationship between the reciprocal of temperature and the proton conductivity of chitin shows an Arrhenius-like linear change. This result indicates that the proton conductivity of chitin has thermal activity. **Table 1** shows the activation energy of proton conductivity derived from the Arrhenius equation

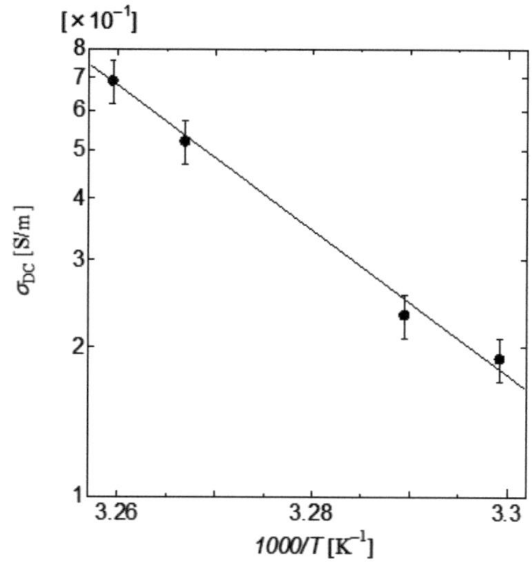

Figure 7.
Arrhenius plot of proton conductivity in chitin [11].

	Fiber direction (eV)	Fiber vertical direction (eV)
Chitin	0.55	0.30
Chitosan	0.57	0.58

Table 1.
Activation energy of proton conduction in the chitin system.

in each fiber direction of chitin and chitosan. Arrhenius equation is represented by following equation:

$$\sigma = \sigma_0 \exp\left(-\frac{\Delta E}{kT}\right) \quad (2)$$

Here, σ and σ_0 are proton conductivity and proton conductivity when frequency is 0, k and T are boltzmann constant and temperature, ΔE is activation energy.

As shown in **Table 1**, the activation energy of chitin-based proton conduction changed depending on the fiber direction in the case of chitin, and no change was observed in the case of chitosan. In general, a decrease in activation energy indicates a decrease in energy required for proton transport, and is therefore expected to contribute to the realization of high proton conductivity. Considering this, it is considered that the decrease in activation energy of chitin in the vertical direction of the fiber is appropriate. However, in the oriented sample, the activation energy of chitin in each fiber direction seems to be inconsistent, considering that the proton conduction of chitin in the fiber direction is the highest. Since the activation energy is not the only element required for proton conduction, the following equation:

$$\sigma = zne\mu \quad (3)$$

Here, z and n are ion valence and number of proton transport pathway, e and μ are charge density and mobility. Since the charge of the proton is +1 and z = 1, the charge concentration e is that the amount of water of crystallization is e, assuming that the proton conduction is through the water of crystallization of the chitin crystal. Furthermore, from Eq. 3, considering that the mobility μ is related to the activation energy, the value of the proton conductivity σ on the left side is determined, so that the high proton conductivity in the fiber direction of the oriented chitin is determined. It is suggested that it is brought about by the number n of proton transport pathways. Taking these things into consideration, we gained insight into the relationship between the crystal structure of chitin and chitosan hydrates and the proton transport pathway from the hydrogen bond distance.

Figure 8 shows the hydrogen bonds formed in chitin hydrate. In **Figure 8**, each color line shows hydrogen bonds, pink is about 2.6 Å of water-chitin molecule, yellow is 2.98 Å, and light blue is between chitin and chitin. Further, broken line shows the hydrogen bond formed along the fiber direction of chitin, and the rigid line shows the hydrogen bond formed in the direction perpendicular to the fiber. As shown in **Figure 8**, among the hydrogen bonds formed in chitin hydrate, the hydrogen bonds between the water molecule and chitin are formed at a distance of about 2.6 Å in the fiber vertical direction (a-axis direction). On the other hand, hydrogen bonds with a distance of 2.6 Å and 2.98 Å are alternately formed in the fiber direction. **Figure 9** shows the results in the case of chitosan. This result suggests that the proton transport pathway of chitosan is mediated by the hydrogen bond of 3.0 Å, which is the yellow dotted line in **Figure 9**, which is common in both the fiber direction and the fiber vertical direction. From these results, it is considered that the relationship between the proton conductivity of chitin and chitosan and the activation energy is due to the hydrogen bond distance of approximately 3.0 Å, which is common to both samples, as a bottleneck. It is expected that the high proton conductivity generated in the fiber direction of oriented chitin will be realized by increasing the number of pathways.

From the above results, the proton conduction of chitin having high proton conductivity is expected as shown in **Figure 10**. Proton conduction in chitin is considered to be realized by the Grosus mechanism in consideration of the relationship between the result of percolation conduction and the hydration structure. In the proton conduction, the crystal water becomes an oxonium ion by the proton, and the proton is passed to the adjacent water molecule by repeating the breaking and rearrangement of the hydrogen bond. In addition, as shown in **Figure 10**, it is considered that the high proton conductivity of chitin in the fiber direction was caused by the increase in the number of pathways due to the hydrogen bond of the bottleneck and the vertical path with low activation energy. On the other hand,

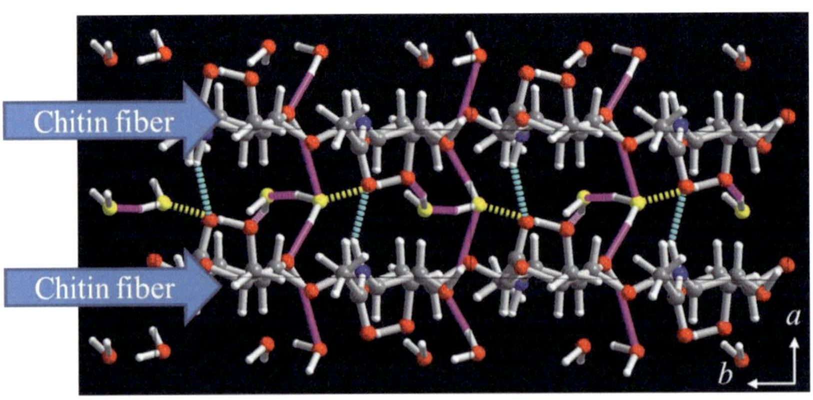

Figure 8.
a-b plane of the crystal structure of chitin hydrate [11, 14–16]. White, gray, blue and red balls show hydrogen, carbon, nitrogen and oxygen.

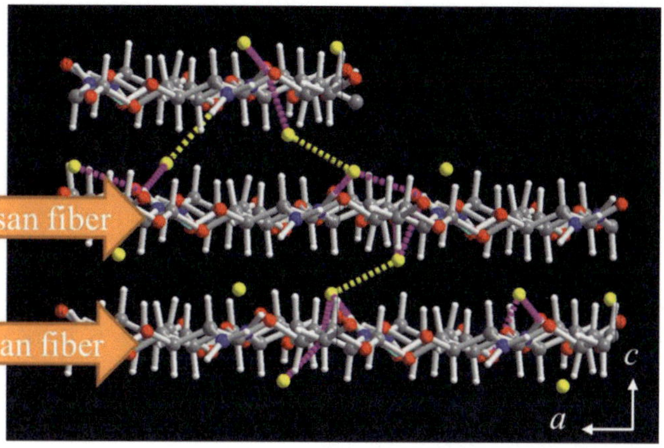

Figure 9.
a-c plane of the crystal structure of chitosan hydrate [11, 17–19]. White, gray, blue and red balls show hydrogen, carbon, nitrogen and oxygen. Yellow ball shows oxygen derived from water molecule.

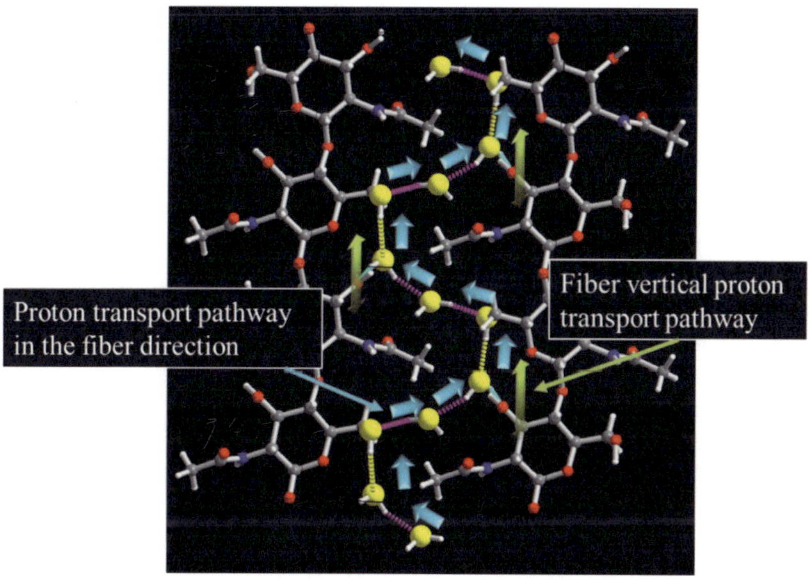

Figure 10.
Schematic diagram of proton transport pathway in chitin [11, 14–16].

the decrease in the proton conductivity in the vertical direction of the fibers of the oriented chitin can pass through the low activation energy pathway in the vertical direction, but the bottleneck is difficult to utilize, so that the number of pathways is reduced.

3.5 Creation of fuel cells using oriented chitin

The high proton conductivity of chitin appeared in the oriented sample and in the fiber direction. From this result, fuel cell using an oriented chitin sample was prepared. **Figure 11** shows schematic diagram of the fuel cell when oriented chitin is used and its *i-V* characteristics. As shown in **Figure 11**, the fuel cell using oriented chitin showed a much higher power density than the sheet sample, and a maximum of 33 mW/cm^2 was obtained. In other words, chitin indicates that the output can be

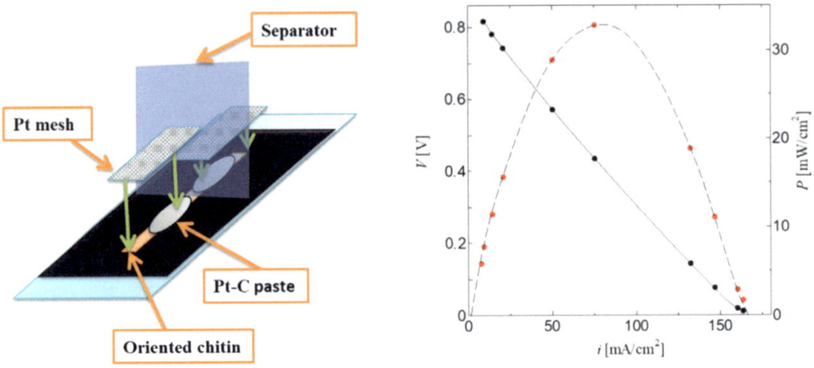

Figure 11.
Schematic diagram of fuel cell using oriented chitin (left) and its i-V characteristics (right) [11].

improved by nearly 24 times by using the structure. In the future, it is expected that the key to practical use will be the improvement of output by chemical modification and the development of technology for producing oriented chitin at low cost.

4. Conclusions

Since chitin functions as an electrolyte membrane for fuel cells, it has been clarified that it is a proton conductor. It was found that the presence of water molecules is very important for the appearance of proton conductivity in chitin, and that these water molecules behave like conductors in chitin hydrate to conduct protons. Furthermore, it was suggested that the orientation dependence of the proton conductivity appearing in the oriented chitin is due to the bottleneck caused by the different hydrogen bond distances with the water molecules formed in the chitin depending on the position. It was also indicated that the high proton conductivity of oriented chitin appears by increasing the number of pathways by utilizing the bottleneck and the pathway in the fiber vertical direction, which is the low activation energy pathway. When a fuel cell using the fiber direction of oriented chitin having high proton conductivity was prepared, an improvement in proton conductivity of nearly 24 times was observed. In the future, chitin orientation methods and chemical approaches are expected to provide a foothold for more practical polysaccharide-based electrolyte membranes and to clarify new value in the energy field.

Acknowledgements

Professor Yasumitsu Matsuo of Setsunan University provided appropriate advice and research facilities in carrying out this research. I'm really thankful to you.

Conflict of interest

The authors declare no conflict of interest.

Author details

Takashi Kawabata
Faculty of Science and Engineering, Setsunan University, Neyagawa, Osaka, Japan

*Address all correspondence to: takashi.kawabata@setsunan.ac.jp

References

[1] Malettas W.G, Quingley H.J. and Adickes E.D. Chitin in Nature andTechnology. Plenum Press, New York, 1986:435.

[2] Olsen R, Schwartzmiller D, Weppner W. and Winandy R. Biomedical Application of Chitin and Its Derivates, in Chitin and Chitosan. Sources, Chemistry, Biochemistry, Physical Properties and Applications. Elsevier Applied Science, NewYork. 1989

[3] Sandford P.A. and Stinnes A. Biomedical Applications of High Purity Chitosan; Physical, Chemical and Bioactive Properties. ACS Symposium Series. 1991:467:430.

[4] Nair K.G.R. and Madhavan P. Chitosan for Removal of Mercury from Water. Fishery Technology. 1984:21:109-112.

[5] Peniche-Covas C, Alwarez L.W. and Arguelles-Monal W. The Adsorption of Mercuric Ions by Chitosan. Journal of Applied Polymer Science. 1992:46:1147-1150. https://doi.org/10.1002/app.1992.070460703

[6] Romana N, Shanta B, Taslim U.R, Sanjida A, Rumana A.J, Papia H, Mohammed M.R, Preparation of Chitin-PLA laminated composite for implantable application. Bioactive Materials. 2017:2:199-207. DOI: 10.1016/j.bioactmat.2017.09.003.

[7] Mahmoud K, Eslam A, Sulaiman A.A, Amany S.K, Magda H.R, Nevin A.I Exploring Simplified Methods for Insect Chitin Extraction and Application as a Potential Alternative Bioethanol Resource. Insects. 2020:11:1-14. DOI:10.3390/insects11110788

[8] Takashi K, Yasumitsu M, Chitin Based Fuel Cell and Its Proton Conductivity. Materials Sciences and Applications. 2018:9:779-789. DOI: 10.4236/msa.2018.910056

[9] Takashi K, Yasumitsu M, Role of acetyl group on proton conductivity in chitin system. Journal of Materiomics. 2019:5:258-263. DOI:10.1016/j.jmat.2019.02.010

[10] Takashi K, Yasumitsu M, Anomalous Proton Conductivity in Chitin-Chitosan Mixed Compounds. Materials Sciences and Applications. 2020:11:1-11. DOI: 10.4236/msa.2020.111001

[11] Takashi K, Research on Fabrication of fuel cells using chitin-based materials and its proton conduction mechanism. Doctoral dissertation. 2020

[12] Hitoshi Y, Mechanism of physical characteristics of conductive composite materials based on percolation theory [Internet]. 2002. Available from: https://www.nichias.co.jp/research/technique/pdf/333/parcoration.pdf

[13] Mogilevskaya E. L, Akopova T. A, Zelenetskii A. N, Ozerin A. N, *Polymer Sience Sir* A. 2006:48:116-123.

[14] Sawada D, Kimura S, Wada M, Nishiyama Y. Chitin and chitosan research. 2010:16:226.

[15] Kobayashi K, Kimura S, Togawa E, Wada M, Kuga S. Carbo hydrate Polymers. 2010:80:491-497.

[16] Kobayashi K, Kimura S, Togawa E, Wada M. Carbohydrate Polymers. 2010:79: 882-889.

[17] Okuyama K, Noguchi K, Miyazawa T. Macromolecules. 1997:30: 5849-5855.

[18] Sawada D, Kimura S, Wada M, Nishiyama Y. Chitin and chitosan research. 2011:17:164-165.

[19] Naito K P, Sawada D, Wada M, Nishiyama Y. Chitin and chitosan research. 2011:17:258.

Section 6

Filtration Membranes

Chapter 9

Characterization of Chitosan Membrane Modified with Silane-Coupled Nanosilica for Polymer Electrolyte

Ella Kusumastuti, Fadila Mauliani, F. Widhi Mahatmanti, Jumaeri, Lukman Atmaja and Nurul Widiastuti

Abstract

The electrolyte membrane currently being developed is limited to materials that are toxic and expensive. Chitosan as a natural organic polymer supports modification to produce the desired physical and chemical properties, one of which is as solid electrolyte. In the presence of functional groups on chitosan, it is possible to modify it with nanosilica as inorganic filler to improve its characteristics. Incorporation of chitosan matrix with of silane-coupled nanosilica in nanosilica:silane ratio (w/w) are 1:0; 1:0.25; 1:0.50; 1:1; 1:1.50; and 1:2. Evaluation on their properties are both quantitatively (water uptake, tensile strength, proton conductivity, methanol permeability, and selectivity) and qualitatively (functional groups, morphology, topography, and thermal properties). The results show that silane addition to the chitosan-nanosilica membrane in nanosilica:silane as 1:0.50 achieve the best characteristics for polymer electrolyte. The results of functional groups, morphology and topography analysis on selected membranes show that optimum silane addition provides the hydrogen interaction between chitosan matrix and silane-coupled nanosilica so there is an enhanced in membrane properties for electrolyte membrane.

Keywords: chitosan matrix modification, polymer electrolyte, silane-coupled nanosilica

1. Introduction

People's dependence on energy from fossil fuels is now getting higher, so energy resource supply is day by day decreasing. One of the alternative energy sources that is potential to overcome that problem is fuel cell [1]. A fuel cell is an electrochemical device that converts chemical energy into electrical energy continuously. The use of fuel cells is expected to reduce people's reliance on fossil fuels and reduce the damage to the atmosphere due to emissions [2]. Therefore, fuel cell is a promising alternative energy that is environmentally friendly.

The main component of the fuel cell is polymer electrolyte membrane (PEM) [2]. Since the role of polymer electrolyte membrane in fuel cell performance is very important, the study of technological development of polymer electrolyte membrane is needed [3]. Currently, the electrolyte membrane that most widely used in industry is Nafion 117 membrane or Perfluoro sulfonic acid [1]. As a polymer electrolyte membrane in Direct Methanol Fuel Cell (DMFC), the Nafion 117 membrane has very good ability to deliver proton with good chemical stability, but it has a weakness of methanol cross over which is signed by the high methanol permeability and low operating temperature (60–120°C) [1]. Another drawback is its poisonous nature due to the elemental fluorine content and its high price. Therefore, several studies have been conducted to obtain membranes that environmentally friendly and have better capabilities than Nafion 117 membrane at an economical cost [4].

Chitosan is one alternative polymer matrix that is potential for replacing the Nafion 117 membrane. Chitosan is an environmentally friendly biopolymer, produced by utilizing marine waste such as shrimp, crabs, lobsters, and fish shells. Chitosan is easily generated through deacetylation process of chitin by strong alkali or prepared from fungal cell walls via fermentation technology [5]. Chitosan is inexpensive, hydrophilic, has a functional group in the backbone that can be modified according to the desired characteristics and has low methanol permeability [6]. Chitosan has a free amine group that can be protonated and has a hydroxy group, so it can be categorized as natural polycation. These two groups make it possible to modify chitosan to produce the desired physical and chemical properties [7]. However, according to [8], chitosan membrane still has low proton conductivity that is 1.74×10^{-2} S/cm, so this value is lower than the proton conductivity of Nafion 117 membrane that is 5.66×10^{-2} S/cm.

To improve the performance of chitosan-based membrane matrix, some efforts have been done to modify chitosan by combining with other such as silica based materials [8, 9]. Silica based materials are chosen for several reasons. They can reduce hydrophilicity degree of the chitosan main chain that is hydrophilic due to the presence of free amino acid groups and hydroxides on its carbon atom [9]. Silica materials can absorb methanol on the surface of chitosan membrane so that most of the methanol does not pass through the membrane. The addition of inorganic additives such as zeolites and montmorillonites which able to act as molecular filters will provide tetrahedral silica that can cover the pores in the membrane so that the transfer of methanol through the membrane is very small [8, 10]. It is reported that silica addition into the polymer matrix can reduce crystallinity of the polymer and increase mechanical strengths such as water resistance, stretch strength, and tensile strength [11]. The ability of silica to combine with chitosan is very limited because of the different hydrophilicity properties of the two materials. Therefore, silica in the form of nanoparticles was developed to create strong composites.

Tetraethylorthosilicate (TEOS) or $Si(OC_2H_5)_4$ is a source of silica materials which is widely used because it is easy to do purification, the reaction rate is slow and it can be controlled [12]. TEOS is even easily converted into nanosilica particles by reacting with water. The hydrolysis reaction occurs in the sol–gel process. TEOS is commonly used as a crosslinking agent in inorganic polymers synthesis because of its ability to form Si-O-Si chains. The sol–gel method has been extensively developed in the surface modification of silica particles because it will form a more reactive silica precursor with the formation of Si-OH groups [13]. The weakness is that silica is insoluble in chitosan solution, so the dope solution formed from nanosilica and chitosan is not homogeneous [14] because physical mixing causes nanosilica dispersion in the chitosan matrix to be less homogeneous and weak interface interactions of the two material surfaces.

To improve the compatibility of silica particles and chitosan, it is needed a coupling agent [15]. The addition of 10% of GPTMS (3-glycidyloxypropyl trimethoxysilane) on chitosan/zeolite-β membrane produces methanol permeability of 2.20×10^{-7} cm^2/s and proton conductivity of 1.31×10^{-2} S/cm [8]. The electrolyte membrane investigation of SPAEK-C with 10% GPTMS addition produces proton conductivity of $(1.5 \pm 0.1) \times 10^{-2}$ S/cm and methanol permeability of 2.99×10^{-7} cm^2/s [12, 16].

Therefore, in this study, the addition of silane (GPTMS) is done as a coupling agent on the chitosan-nanosilica membrane. This coupling agent is selected because it has epoxy groups that can react with a free amine group in chitosan, so nanosilica can also be bounded strongly with chitosan (12, 15). Nanosilica is chosen because of its small size and wide surface so that making it possible to penetrate the polymer matrix easily.

The use of GPTMS as a coupling agent and nanosilica particles (from TEOS) as fillers in the chitosan matrix for polymer electrolyte membrane makes this study different from previous studies. The use of GPTMS as a coupling agent on nanosilica particles was carried out by [17]. Silica precursors used were silica fume and the investigation carried out was nanosilica variations on chitosan and did not investigate silane variation effect on membranes. The membranes produced for biopolymer applications which have high thermal resistance. Chitosan modified with silica (TEOS) and GPTMS has been coated on a cotton fabrics in [18] results clear transparent thin layer on cotton surface. While [19] has combined chitosan with silica and GPTMS for tissue engineering application. For dehydration of ethanol application, chitosan-silica study has been performed by [20, 21]. For fuel cell application, chitosan have been combined with silica-based material [5, 8, 10, 22, 23]. The result of [5, 10] used GPTMS-montmorillonite as filler in chitosan membrane, while research [8] used GPTMS-modified zeolite, [22] used sulfonated polyaniline/ silica and [23] used silica/ sulfonated polyether-ether ketone as filler in chitosan.

2. Experiments and characterization

2.1 Materials

The materials are tetraethyl orthosilicate (TEOS 99%, ρ = 0.94 L/kg Merck), absolute ethanol (C_2H_5OH 99–100%, ρ = 0.79 L/kg Merck), ammonia (NH_3 25%, ρ = 0.90 L/kg Merck), aqua demineralization, chitosan from CV Ocean Fresh Bandung (deacetylation degree = 82.7%, solubility in 1% acetic acid ≥99%, molecular weight = 8.78 kDa), acetic acid (CH_3COOH 100%, ρ = 1.05 L/kg Merck), methanol (CH_3OH 99.9%, ρ = 0.79 L/kg Merck), dimethyl formamide (DMF $HCON(CH_3)_2$ 99.80%, ρ = 0.94 L/kg Merck) and silane (3-glycidyloxypropyl trimethoxysilane) (GPTMS $C_9H_{20}O_5Si$ ≥ 98%, ρ = 1.07 L/kg Sigma-Aldrich).

2.2 Nanosilica preparation

Nanosilica synthesis from TEOS by sol–gel method was adopted from [8]. A total of 2.27 mL of TEOS was dissolved in 46 mL absolute ethanol at a speed of 600 rpm at room temperature for 15 minutes. Then ammonia was added dropwise until pH 10. After 1 hour, 1 mL distilled water was then added dropwise into the solution. The stirring process was continued for 6 hours. Then the solution was allowed to stand for 24 hours during the aging process. After the aging process, the obtained gel was then roasted at a temperature of 80°C for 24 hours. The crystals were crushed into powder in further calcined for 2 hours at the temperature of 600°C. The resulting crystals were sieved at 230 mesh.

2.3 Silane-coupled nanosilica preparation

The coupling process of nanosilica and silane was done by developing [12] method. A total of 0.03 grams of nanosilica and 0.0075 grams of 3-glycidoxypropyl trimethoxy silane (GPTMS) (nanosilica:silane = 1:0.25) were dissolved in 0.3 mL dimethylformamide (DMF) at room temperature, then stirred using magnetic stirrer for 6 hours. Then the homogeneous solution was put into a beaker and heated in an oven at 60° C for 24 hours. After that, it was heated at 100° C for 1 hour and at 120° C for 2 hours. Drying was carried out at 155° C for 2 hours. The obtained solids were soaked with 1 M HCl solution at 80° C for 24 hours until hydrolysis and condensation occur in the solution. The resulting solids were crushed and sieved with 230 mesh sieves. The final produced powder is nanosilica filler that had been modified with silane coupling agent. Furthermore, the coupling process was also carried out at the ratio of nanosilica: silane = 1:0; 1:0.50; 1:1.0; 1:1.5 and 1:2.0.

2.4 Membrane synthesis by modifying chitosan with silane-coupled nanosilica

The chitosan membrane synthesis with silane-coupled nanosilica addition by developing [8, 24] method. First, a total of silane-coupled nanosilica (w:w) with nanosilica:silane respectively 1:0; 1:0.25; 1:0.50; 1:1.0; 1:1.5 and 1:2.0 was dissolved in each 50 mL of 2% acetic acid and was stirred at a temperature of 60°C for 7 hours. Secondly, as many as 1 g of chitosan was dissolved in 50 mL of 2% acetic acid at room temperature for 4 hours. Then the first solution and second solution were mixed and stirred for 4 hours at the temperature of 60°C. The homogeneous solution formed is called dope solution that was then poured in the acrylic mold of 20 x 20 cm and was dried in an oven at the temperature of 60°C for 21 hours that produced dry membrane. The process of membrane formation used the phase inversion method.

2.5 Membrane characterization

The characterization of chitosan-nanosilica membrane with silane addition was performed by water uptake measurement, membrane tensile strength analysis (by tensile test equipment Strograph VG 10-E Toyoseiki), methanol permeability (by diffusion cell method), proton conductivity analysis (by EIS or Electrochemical Impedance Spectroscopy Autolab PG STAT 128 N Instrument) [2], membrane selectivity determination, functional group analysis (by FT-IR PRESTIGE-21 Shimadzu), membrane morphology analysis (by SEM FEI Inspect S50), membrane topography analysis (by AFM Bruker N8 Neos 5.5 IF367), and thermal degradation analysis (by TGA Mettler Star SW 10.00).

Water uptake is the ability of a membrane to absorb water for 24 hours, so it is performed by weighing the mass of absorbed water (w_2) and comparing it with the mass of membrane (w_1) in a percentage format as expressed in Eq. (1) adopted from [10, 25–27]. In determining the water uptake, demineralized water which is free of metal ions is used.

$$\text{Water uptake} = \frac{w_2 - w_1}{w_1} \times 100\% \tag{1}$$

The mechanical strength test is generally measured by tensile strength test, in which material sample of a certain size is exerted with force and pulled until it breaks. Eq. (2) until Eq. (4) express the equation for determining tensile strength, elongation break and modulus young adopted from [10, 28]. Tensile strength test performed according to ASTM D882 at room temperature at a constant cross head speed of 10 mm min^{-1} and 100 N load cells. The samples were dumbbell-shaped with gauge dimensions of 15 mm × 3 mm × 0.22 mm. Eq. (2) until Eq. (4) describe that σ is tensile strength, ε is elongation break, E is modulus young, P is the force applied to the specimen, A is the specimen cross-sectional area (cm^2), ΔL is the length increase (cm) and L_0 is the initial length of the specimen (cm).

$$\sigma = \frac{P}{A} \tag{2}$$

$$\varepsilon = \frac{\Delta L}{L_0} \tag{3}$$

$$E = \frac{\sigma}{\varepsilon} \tag{4}$$

Proton conductivity measured by Electrochemical Impedance Spectroscopy as Eq. (5) adopted from [5, 25–27]. Where σ is the proton conductivity (S cm^{-1}), d is the membrane thickness (cm), Rb is the bulk resistance value (Ω) and A is the membrane area in the sample (cm^2) [2].

$$\sigma = \frac{d}{R_b \times A} \tag{5}$$

Methanol permeability of the membrane is measured by counting the methanol concentration passing the membrane for the diffusion of methanol in progress [16, 26]. The methanol permeability test is carried out using a permeation measurement cell that has two identical compartments. Compartment A is filled with methanol solution in deionized water, and compartment B is filled with deionized water. The principle used cell diffusion between the two compartments. The solution in both compartments is stirred until it is homogeneous so that the diffusion process runs well [14]. Methanol permeability was calculated from the methanol concentration versus permeation time curve. The methanol concentration in compartment B (C_B) is obtained from the Eq. (6) adopted from [25, 27].

$$C_B = \frac{A}{V}\frac{DK}{L}C_A\left(t - t_0\right) \tag{6}$$

Where C_A is the methanol concentration in compartment A, A and L are the polymer membrane area and thickness, D and K are the methanol diffusivity and the partition coefficient between the membrane and solution, V is solution volume in compartment B and t is permeation time. The result of DK or P is the methanol permeability of membrane (DK = P), and t_0 is also called the time lag associated with diffusivity, $t_0 = L^2 / 6D$ [29].

Membrane selectivity is comparison between proton conductivity and methanol permeability [8] expressed in Eq. (7). Where β is membrane selectivity, σ is proton conductivity, and P is methanol permeability.

$$\beta = \frac{\sigma}{P} \tag{7}$$

The chitosan membrane to be analyzed for its functional groups by FT-IR PRESTIGE-21 Shimadzu was taken as much as 0.1–0.2 g. In addition, KBr powder of 0.5–1.0 g was also prepared. The two solids were mixed and grinded until smooth, then the mixture powder was made into pellets with a hydraulic press and the measurement analysis was carried out with a wavelength between 4000 and 400 cm^{-1} [30, 31].

The membrane surface morphology was observed using SEM FEI Inspect S50. The membrane (1x1 cm^2) was firstly coated using gold so that it could be detected by the device.

Membrane topography was observed in three dimensions and two dimensions using AFM Bruker N8 Neos 5.5 IF367. The membrane was taken several parts then put on the tip and detected by the device at certain distance.

Thermal stability analysis using TGA Mettler Star SW 10.00 was carried out on the selected membrane specimens that had the best and worst properties therefore it represented all variations. Thermal stability analysis data was recorded in nitrogen atmosphere at every 10°C/minute heating rate at 30–500°C temperatures.

3. Results and discussions

3.1 Synthesis of silane-coupled Nanosilica and its incorporation into chitosan matrix

Figure 1 illustrates the interaction model between chitosan, nanosilica, and silane in membranes. At the final stage of silylation process that is epoxy groups deformation (ring-opening) on silane organofunctional group end chain, silane form two hydroxyl groups (–OH) which interacts with the amine group (–NH_2) and and hydroxyl groups (-OH) in chitosan to form hydrogen bonds that could form a good interface interaction of chitosan and silane-coupled nanosilica network. Thus forming groups Si–OH via hydrogen bonds [5, 10] and siloxane groups (Si–O–Si) via oxane bonds [32].

3.2 The results characterization of chitosan membrane modified with silane-coupled nanosilica

3.2.1 Water uptake of chitosan-nanosilica membrane with silane addition

The water uptake determines how much water is absorbed by the membrane, so the PEM membrane must be able to hold water because the proton will be transported along the water channel created in the membrane polymer matrix. Thus, the high water uptake is favorable for PEM high-performance to facilitate great numbers of protons hopping and diffusion through the membrane [5, 14]. **Figure 2** shows all chitosan-nanosilica membranes with silane addition in this research potentially can be used as a DMFC membrane because it has water uptake value <50% as mentioned by [33]. The water on the membrane serves as proton transport medium and needed as the mobile phase to facilitate proton conductivity but if is it too high it will damage the membrane easily and lower their mechanical properties [33]. In **Figure 2**, it appears that silane addition to nanosilica has an impact on water uptake, which is an increase in water uptake on nanosilica:silane variations of 1:0, 1:0.25 to 1:0.5, but decrease on nanosilica:silane variations of 1:1 to 1:2. The most optimum membrane is variation nanosilica:silane 1:0.50 with water uptake is 37.52%, while chitosan-nanosilica membrane without silane reaches 32.16% water uptake.

The addition of silane to nanosilica increases the active Si-OH groups as shown in **Figure 1** (silane-coupled nanosilica) which will cause increased water absorption of the membrane. The addition of more silane that is silane 1:1 to 1:2 addition causes excess silane to interact with silica. Silane itself is hydrophobic so that if it is added to hydrophobic nanosilica [23] it will increase water absorption to chitosan membrane. This fact is following the results of the [14] study that pure chitosan

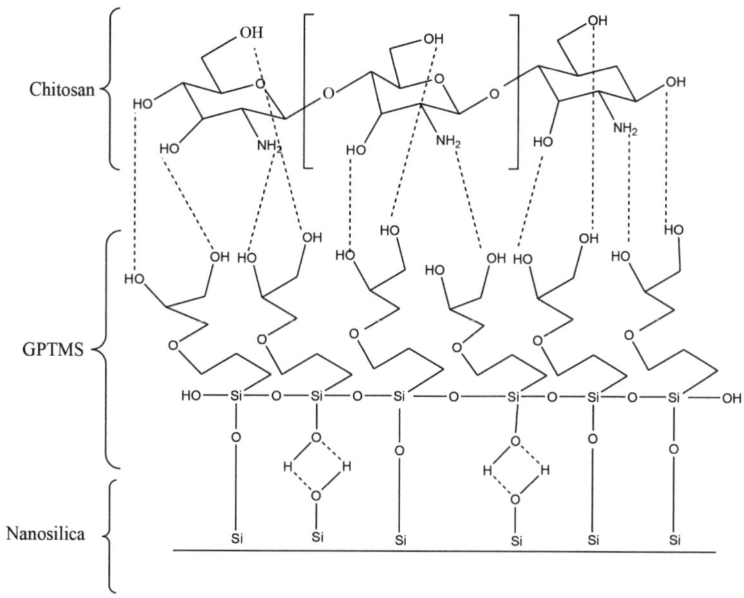

Figure 1.
Illustration of nanosilica-silane and chitosan interaction in membrane.

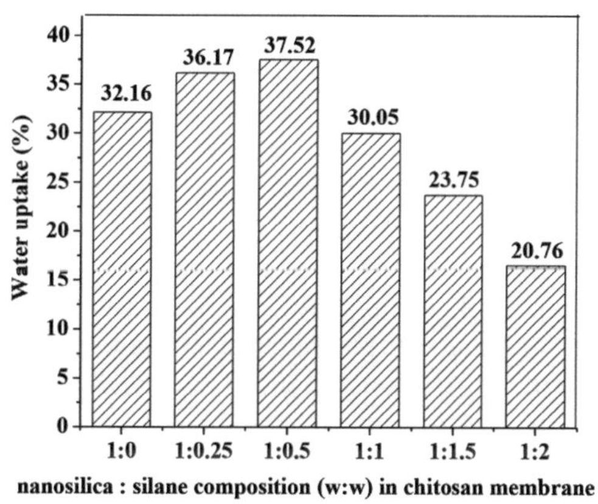

Figure 2.
The relationship between water swelling value (%) and nanosilica:silane composition (w:w).

membrane has 40.66% water uptake and increases along with nanosilica increase addition and water uptake increase along with silane addition in line with the results of [12] study. Nafion 117 has 18,3% in water uptake [12], so the water uptake membranes produced in this study have bigger than Nafion 117.

3.2.2 Mechanical properties of chitosan-nanosilica membrane with silane addition

Figure 3 explains the relations between silane addition in chitosan-nanosilica membrane and their mechanical properties, that is tensile strength, break and elasticity (modulus young) elongation. The tensile strength value of nanosilica:silane variations of 1:0 to 1:0.5 increases with the increasing silane composition that is 4.7, 5.1, 11.8 MPa respectively. It shows that the coupling agent presence causes strong interaction through hydrogen bonding between chitosan and nanosilica so that the tensile strength of the resulting membrane also increases. However, there

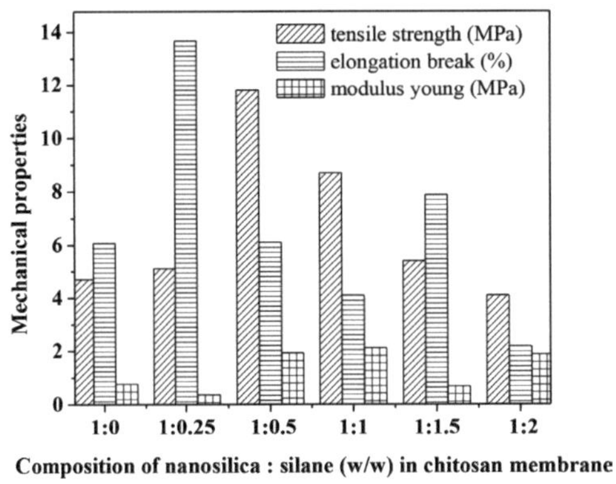

Figure 3.
The relationship of silane addition in chitosan-nanosilica membrane to their mechanical properties.

is a decrease in the silane variation of 1:1 to 1:2 (excess) that is 8.7, 5.4, and 4.1 MPa respectively. This happens because excess silane causes an imbalance in the interaction between silane and nanosilica. The effect of silane addition to nanosilica gives various levels of elongation at break on the membrane that are 6.07, 13.67, 6.1, 4.1, 7.88, and 2.18% respectively. Membrane elasticity is determined by the magnitude of modulus young. The fact is that silane addition can increase modulus young value of nanosilica:silane variation membrane 1:0.25 to 1:1, and decrease significantly in 1:1.5, then increase again in silane addition variation 1:2.

A membrane is said to be good and has the best mechanical performance seen from the high tensile strength values and low elongation at break values [22, 34], so modulus young values which expected is high. In this research chitosan-nanosilica membrane with nanosilica:silane is 1:0.5 reach optimum mechanical properties. Silane addition to nanosilica increases physical interaction between nanosilica and chitosan. Hydrogen bonds formed between hydroxyl groups in polysiloxane with amine and ether groups in chitosan as described in **Figure 1**. Strong interaction between nanosilica and chitosan causes high physical and mechanical strength including tensile strength, elongation and modulus young.

3.2.3 Proton conductivity, methanol permeability, and selectivity of chitosan-nanosilica membrane with silane addition

Proton conductivity of chitosan-nanosilica silane addition membrane can be seen in **Figure 4(a)**. The membrane proton conductivity is measured by using EIS to determine impedance of the membrane. **Figure 4(a)** explains that the proton conductivity increases in line with silane addition to a certain point then decreases. Chitosan-nanosilica with the addition of silane membrane nanosilica:silane variation 1:0.50 has the highest proton conductivity value than another variation of silane addition. It shows that silane composition addition is optimal for interacting with amine group on the chitosan matrix and form a polysiloxane network on nanosilica addition of silane. Excess silane addition (>50%) on the variation nanosilica:silane 1:1, 1:1.50 and 1:2 can cause proton conductivity value become low. This is because the hydrogen bonding that occurs between the silane with chitosan matrix and polysiloxane network is already saturated, so the ability to transport proton facilitate is less than optimal. **Figure 4(a)** shows the highest proton conductivity values obtained at chitosan-nanosilica addition of silane membrane variations nanosilica:silane 1:0.50 is 7.89 x 10^{-4} S/cm. This fact is consistent with the previous fact obtained in the water uptake test. The membrane with the highest

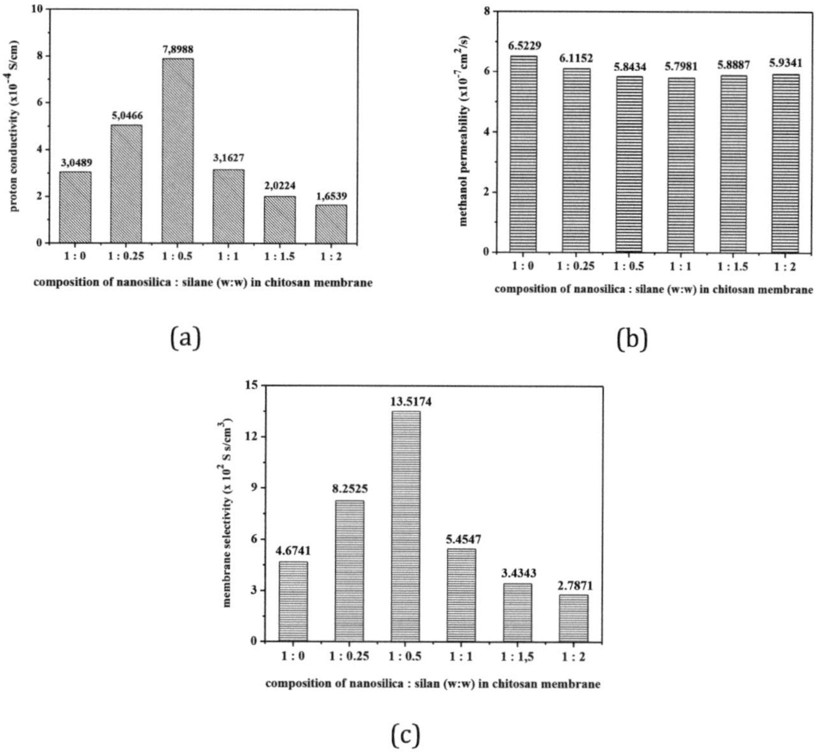

Figure 4.
The relationship of silane addition in chitosan-nanosilica membrane to their (a) proton conductivity, (b) methanol permeability, (c) membrane selectivity.

water uptake has the highest proton conductivity as well. This is true as water in a proton exchange medium, so the higher the water uptake the higher the proton conductivity.

The proton conductivity value obtained by chitosan-nanosilica membrane by silane addition of nanosilica:silane 1:0.5 is 7.89 x 10^{-4} S/cm. This value is smaller than the Nafion 117 proton conductivity membrane which is 5.66 x 10^{-2} S/cm [8]. According to [35], this is related to the Nafion 117 structure. There are many Fluor atoms (F) which have large electronegativity value so that Nafion 117 can be more easily forming hydrogen bonds with water, so it is more easily to absorb water that is needed as a proton transport medium, so the conductivity of the Nafion 117 proton is greater. Although the proton conductivity value of chitosan-nanosilica membrane of nanosilica:silane = 1:0.5 in the study is smaller than Nafion 117, but this membrane can still be used as PEM for applications in DMFC. The membrane can still deliver protons even though they are slow. The proton conductivity value obtained is above the minimum requirement, which is must be bigger than 1 × 10^{-5} S/cm [35].

Figure 4(b) shows that membrane methanol permeability is affected by silane addition. Methanol permeability occurs due to methanol transport in aggregate pores that is further dependent on the pores volume in the membrane. There are two types of pore in the polymer membrane: network pores or ionic clusters and aggregate pores. Proton transport occurs through two types of pore, while the mass transport (methanol, water, and gas) occurs only through the porous aggregate [36]. Network pores is a small cavity between the polymer chains that is responsible for proton conduction. Aggregate pores are large cavities that cover polymer aggregates which cause mass transport (methanol) [37].

The decrease in methanol membrane permeability as shown in **Figure 4(b)** from the composition of nanosilica:silane 1:0 to 1:1 is caused by properties adhesion and interface interaction (hydrogen bonds) of chitosan-nanosilica with the silane addition is stronger [15] than chitosan-nanosilica membranes without silane addition (1:0). According to [29], the decrease in hydrophilicity in this case is due to the silane nature which is able to balance the hydrophilic and hydrophobic nature of an organic or inorganic material. The addition of nanosilica which has been carried out by silane can close pores on the chitosan membrane through strong interactions between amines in chitosan with polysiloxane in nanosilica so that most of the methanol does not pass through the membrane.

Chitosan-nanosilica membrane with silane addition in nanosilica:silane variation 1:1.5 and 1:2 there is an increase in methanol permeability value that is 5.8887×10^{-7} cm^2/s and 5.9341×10^{-7} cm^2/s respectively. This is due to the composition of silane addition cannot interact perfectly with chitosan matrix so the adhesion force decreases and the chitosan matrix interaction with silane nanosilica addition is weak. The highest methanol permeability value was obtained on the chitosan-nanosilica membrane without silane is equal to 6.5229×10^{-7} cm^2/s. The addition of silane chitosan nanosilica membrane that has good performance in methanol permeability value followed by high proton conductivity value is the chitosan-nanosilica membrane addition of nanosilica:silane 1:0.5 variation which is 5.8434×10^{-7} cm^2/s. This is also supported by the proton conductivity value obtained by 7.8988×10^{-4} S/cm. Chitosan-nanosilica membrane with silane addition nanosilica variation: silane 1:0.5 has the lowest methanol permeability value when compared to Nafion 117 membrane that is 1.01×10^{-6} S/cm [16]. For DMFC applications, it is expected that membranes with small methanol permeability. Small permeability prevents leaks and avoids methanol cross over.

Figure 4(c) explains that the highest membrane selectivity is obtained on the chitosan-nanosilica membrane with silane addition in nanosilica:silane variation 1:0.5. Membrane selectivity is a parameter that connects proton conductivity with methanol permeability. For DMFC applications, the desired membrane is a membrane with high conductivity and low methanol permeability, so the determination of membrane selectivity uses Eq. (7) [8]. Through this membrane selectivity determination, the facts show that the chitosan-nanosilica membrane with 1:0.5 silane addition is the membrane with the best performance as PEM for DMFC applications. Membrane selectivity aims to evaluate membrane performance capability based on high proton conductivity values and low methanol permeability values. The membrane selectivity is influenced by modulus young value as mechanical properties of the membrane. It can be explained that the higher membrane selectivity, the higher modulus young so that the membrane physical properties are in harmony with the membrane chemical properties as polymer electrolyte membrane (PEM). The results of membrane synthesis in this study are supported by high modulus young values, high proton conductivity values and low methanol permeability.

When compared to Nafion 117, chitosan-nanosilica membrane with silane addition in nanosilica:silane variation 1:0.5 has lower membrane selectivity value. It shows that chitosan-nanosilica membrane with silane nanosilica variation: silane 1:0.5 addition has the ability under Nafion 117, but this membrane can still be used as PEM for DMFC applications, it means that the membrane can deliver protons even though it runs slowly but it can reduce the entry of methanol fuel through the membrane supported by methanol permeability obtained that is smaller than Nafion 117 membrane.

3.2.4 Functional groups analysis of chitosan-nanosilica membrane with silane addition

Figure 5 shows FT-IR spectra of nanosilica and chitosan-nanosilica membrane with silane addition. The FT-IR spectrum of chitosan on the membrane is characterized by the presence of —CN groups (amide groups) in all variations of the chitosan-nanosilica membrane with the addition of silane at the wavenumbers 1627.92 cm^{-1} and 1635.64 cm^{-1} [38]. The wavenumbers shift in the FTIR spectrum of chitosan indicates an interaction between the amide group in the chitosan matrix and the polysiloxane network in the nanosilica addition of silane through hydrogen bonds. The absorption around the wave number 1060–1084 cm^{-1} indicates the —CO group (ketone group) [38]. The —CO group (ketone group) which is a characteristic of the polysaccharide appears at the wave number 1080.14 cm^{-1}. The -CH bonds in $-NHCOCH_3$ appear at wave numbers 2854.65 cm^{-1} and 1404.18 cm^{-1} [39, 40]. This absorption in assigned to the $–^{+}N–H$ bonds, indicating that the chitosan could be interacting with another groups by coulombic forces [40].

In FT-IR spectrum of nanoslilca in **Figure 5**, the main functional group analysis results is the emergence of Si–O–Si groups on nanosilica and chitosan-nanosilica membrane with silane addition. Absorption around wave numbers 1100–1200 cm^{-1} and 460–480 cm^{-1} indicates the presence of the Si–O–Si group (siloxane group) [39]. This absorption band (Si-O-Si groups) did not appear in the chitosan membrane spectra but appear in chitosan-nanosilica membrane. This sharp increase in absorption band intensity indicates the presence of stretching vibrations of Si–O–Si (siloxane group) which is getting stronger. According to [24], it indicates that some of the siloxane groups (Si–O–Si) of polysiloxane network in silane addition and nanosilica have interacted strongly with chitosan matrix. Absorption band at wave number 894.97 cm^{-1} indicates the presence of the Si–OH groups (silanol groups) [38]. The Si–OH groups (silanol groups) show hydrogen bonds existence between silanol groups in nanosilica-silane with amide groups in the chitosan matrix. It is also seen in **Figure 5** that there are absorption bands around wave number 3400–3500 cm^{-1} which indicate the –OH (hydroxyl)

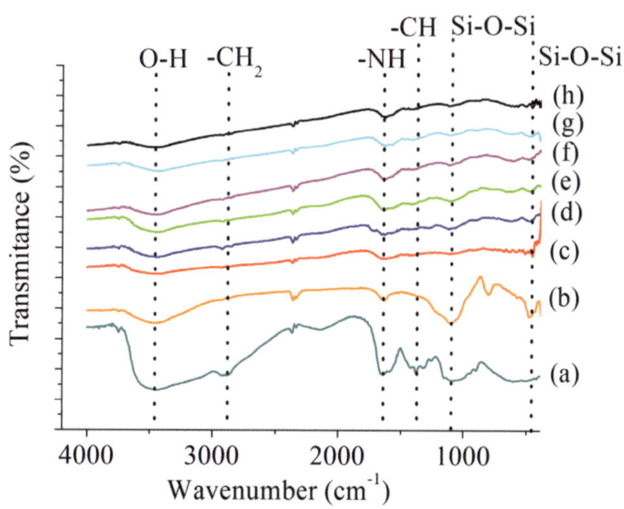

Figure 5.
FT-IR spectra of chitosan-nanosilica membranes with variation of silane addition. (a) Chitosan; (b)Nanosilica; (c) chitosan membrane with nanosilica:silane = 1:0; (d) 1:0,25; (e) 1:0,05; (f) 1:1; (g) 1:1,5; (h) 1:2.

groups presence. A sharp intensity increase occurs in the vibration stretching –OH (hydroxyl group). According to [39], it indicates that some hydroxyl groups (–OH) on the chitosan matrix have interacted strongly with the polysiloxane network in nanosilica-silane through hydrogen bonds. Another important absorption band is at 1644–1637 cm^{-1} which indicates the bending vibration H-O-H indicates the vibration of water molecules bound to the inorganic framework.

Variation of nanosilica:silane membrane at 1:0.50 is the best membrane that has the highest value of membrane selectivity and has the best mechanical properties. FT-IR spectra of the chitosan-nanosilica membrane with silane addition in nanosilica:silane membrane at 1:0.50 as in **Figure 5** shows that absorption band at wavenumber 3448.72 cm^{-1} is stretching vibration absorption –OH groups more sharply when compared with other variation. It shows that the hydrogen bonds between the chitosan matrixes with this silane addition on nanosilica more easily formed than the nanosilica without silane addition. The amine group on the chitosan is easier to form hydrogen bonds with epoxy groups on the silane. Besides, there are new absorption wavenumbers around 900–912 cm^{-1} which indicate the presence of epoxy groups ($-C_2H_3O$) in silane [15, 21], and from 2900 to 2960 cm^{-1} indicating $-CH_2$ groups of silane compound [4] in. It can be seen in **Figure 8** that absorption at wavenumber 902.69 cm^{-1} indicates epoxy groups presence of organofunctional group silane compound and absorption at wavenumber 2924.09 cm^{-1} that indicates the stretching vibration $-CH_2$ groups. Absorption at wavenumber 2924.09 cm^{-1} is absorption of the bridge group $-CH_2$ alkyl silane compound and it is not from chitosan.

3.2.5 Morphology analysis of chitosan-nanosilica membrane with silane addition

Figure 6 shows membrane morphology of chitosan, chitosan-nanosilica without silane addition (variation 1:0) and chitosan-nanosilica with silane addition. **Figure 6(a)** shows a homogenous surface of chitosan membrane, while in **Figure 6(b-d)** show heterogeneous surface of chitosan matrix containing nanosilica particles. Besides, there is fairly large agglomeration in **Figure 6(b)** as chitosan-nanosilica without silane. It suggests that the interaction between the chitosan matrix with nanosilica is poor without silane as coupling agent. Good interaction will produce homogeneous surface morphology [21].

Figure 6(c) shows membrane morphology of chitosan-nanosilica membrane with silane addition variation 1:0.50 (the best membrane) that shows good interaction between chitosan and nanosilica with silane addition. This is due to the occurrence of attachment between nanosilica and silane as coupling agents with relatively small amounts so that they can spread evenly with chitosan matrix, eventually being able to cover the pores of membrane well. Excessive agglomeration occurs in chitosan-nanosilica membranes with nanosilica:silane 1:2.0 composition (**Figure 6(d)**) due to excessive amounts of silane that interfere effective attachment process, so that the morphology resulting membrane appears to be lumps and not homogenous. But, other techniques are necessary to support this facts. A good alternative is X-ray photoelectron spectroscopy (XPS), specifically high-resolution XPS of C, O, N, and Si.

3.2.6 Membrane topographic analysis using AFM

Membrane topographic analysis using AFM quantitatively is shown in **Table 1**. Topographic analysis with backward amplitude results in surface morphology of surface roughness (S_a) and root mean square (RMS) roughness or S_q, as well as the highest surface height (H_{max}) and lowest surface (H_{min}) as in [41–45] studies. In

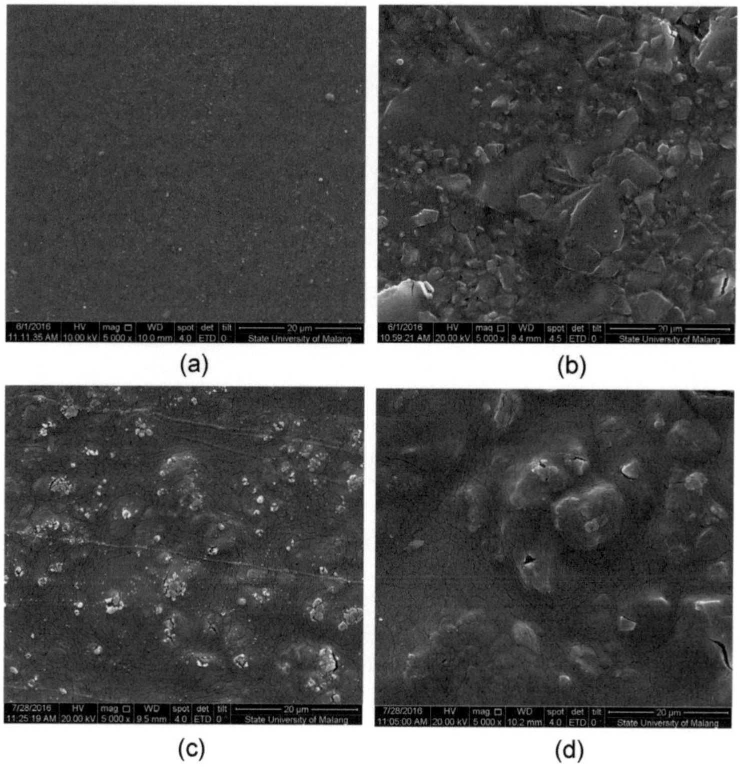

Figure 6.
Morphology surface (5000 magnification) by SEM of (a) chitosan; (b) chitosan-nanosilica membrane with nanosilica:silane = 1:0, (c) 1:0.5, and (d) 1:2.0.

Chitosan-nanosilica Membranes	S_a mean (nm)	S_q mean (nm)	H_{max} (nm)	H_{min} (nm)
Nanosilica:silane = 1:0	0.956	1.327	43.000	−28.000
Nanosilica:silane 1 = 1:0.5	2.308	3.713	121.667	−83.333
Nanosilica:silane 1 = 1:2.0	1.473	1.857	82.667	−66.333

Table 1.
Results of quantitative AFM analysis on chitosan-nanosilica membranes with Silane addition.

Table 1 the surface roughness parameter values on chitosan-nanosilica membrane without silane showing the value of S_a, S_q, H_{max}, and H_{min} have the lowest value. The addition of 50% GPTMS silane or in composition 1:0.5 as the coupling agent raises the value of the surface roughness parameter even at this point reaching the highest value. These peaks indicate the optimum interaction with the 50% silane addition to chitosan-nanosilica. It proves that silane addition as the coupling agent between chitosan and nanosilica at the optimum amount raises the value of average roughness or S_a and root mean square roughness S_q, as well as the highest surface height (H_{max}) and lowest surface (H_{min}) and then decreases at next addition. The next point is the addition of 100% silane or in composition of 1:2.0 the surface roughness parameter shows a decrease that is, the excess silane causes reduced interaction between chitosan and nanosilica.

The qualitative results of membrane topography are shown in **Figure 7**. The AFM analysis provides topographic information of chitosan-nanosilica composite membranes with GPTMS (silane) addition as coupling agent in two-dimensional

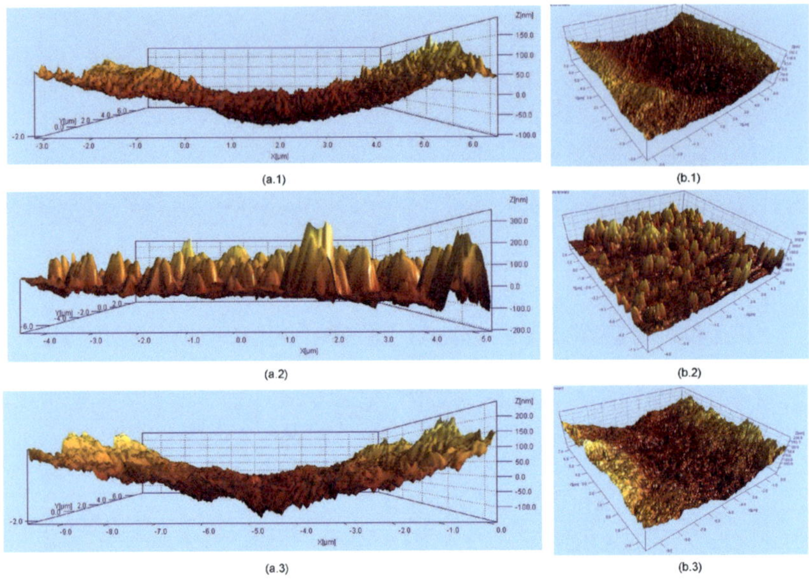

Figure 7.
The AFM analysis of chitosan-nanosilica composite membranes with GPTMS silane coupling agent addition in (a) two-dimensional (2D) and (b) three-dimensional (3D), variation nanosilica:silane (1) 1:0, (2) 1:0.5 and (3) 1:2.0.

(2D) (**Figure 7a(1–3)**) and three-dimensional (3D) (**Figure 7b(1–3)**) appearance. The analysis results using AFM shows that chitosan-nanosilica membrane without silane (chitosan-nanosilica:silane ratio = 1:0) shows that membrane topography tends to be regular and evenly distributed. It shows that the membrane still carries typical state of chitosan and nanosilica whose topography is not yet affected by other materials. It means between the two materials, chitosan and nanosilica, maximum interaction has not yet occurred. This result is in accordance with morphological analysis result with SEM which shows rough surface. In chitosan-nanosilica membranes with 50% silane addition to nanosilica, or the composition of chitosan-nanosilica = 1:0.5 indicates significant change in topography surface. The membrane surface forms irregular bumps (hills and valleys) caused by the certain amount of nanosilica composition that interacts well with chitosan on the membrane. On 100% silane addition or chitosan-nanosilica: of 1:2.0, bumps on membrane topography are no longer occurred, but the surface tends to be evenly distributed and more homogeneous, indicating prominent characteristics of each chitosan and nanosilica. The presence of nanosilica fillers interact with chitosan through hydrogen bonds as proven by FTIR analysis. The interaction will be maximized when the coupling agent is added in the form of silane at an optimum amount. In the presence of some silanes, the chitosan matrix will begin to be affected by the presence of silane-coupled nanosilica fillers to produce the formation of a certain number of hills and valleys at AFM analysis in line with capabilities of the existing filler and coupling agent. **Figure 7** shows a rough topography with fairly wide distribution of hills and valleys. The number of irregular areas is due to the presence of nanosilica hydrophobic fillers that interact with chitosan by the addition of an optimum amount silane as coupling agent. Rugged topography with many hills and valleys spread evenly over almost all surfaces that are found on the membrane with 50% silane addition. In this membrane, the chitosan and nanosilica matrices are no longer dominant, because the added nanosilica filler has interacted well with chitosan due to the GPTMS as coupling agent.

The interaction that occurs between the chitosan matrixes with nanosilica fillers in the presence of silane as coupling agent forms a unified and strong membrane. As a result, membrane topography is fused between two components, shown in more uniform color in **Figure 7**. The 3D images on the composition of chitosan-nanosilica:silane = 1:0.5 resulting from AFM analysis does not show any differences in the regions or parts of chitosan and nanosilica. The surfaces of both valleys and hills are joined in such a way that they show that the two materials interact very well. This phenomenon is supported by research [46] who makes a composite membrane from an organic matrix in the form of chitosan and alumina as an inorganic filler. The results of AFM analysis in [46] study shows that the chitosan/alumina composite membrane shows rough topography with membrane surface covered with granules show that chitosan has interacted strongly with alumina. Identical with [46] study, chitosan subtle areas are covered by nanosilica material so that the membrane topography in 3D shows the presence of hills and valleys that are evenly distributed to all surfaces. AFM analysis in [20] study also supports the facts and results of the AFM analysis in this study. As [20] research on chitosan-nanosilica supports mixed matrix membranes and the results of AFM analysis show that the incorporation of nanosilica and silane increases the membranes surface rippling which, as FTIR results shows, are attributed to strong bonds formed between nanosilica particles and chitosan matrix.

3.2.7 Thermal analysis using TGA

The thermal stability of the chitosan-nanosilica hybrid membrane is evaluated by thermo gravimetric analysis with TGA. The thermograms are shown in **Figure 8** and the change in mass percentage on each thermogram stage is presented in **Tables 2** and **3**. In TGA analysis, sample changes are marked by deviations from the horizontal line. As shown in **Figure 8** and **Table 2**, all membranes show three stages of weight-loss as area of change in mass percentage on TGA thermogram of membrane samples.

The first weight loss at around 35–150°C (**Table 2**) is attributed to the loss of water molecules present in hybrid membrane [10, 47] with mass reduction 12–14%. The difference in temperature range and the percentage of sample weight-loss is

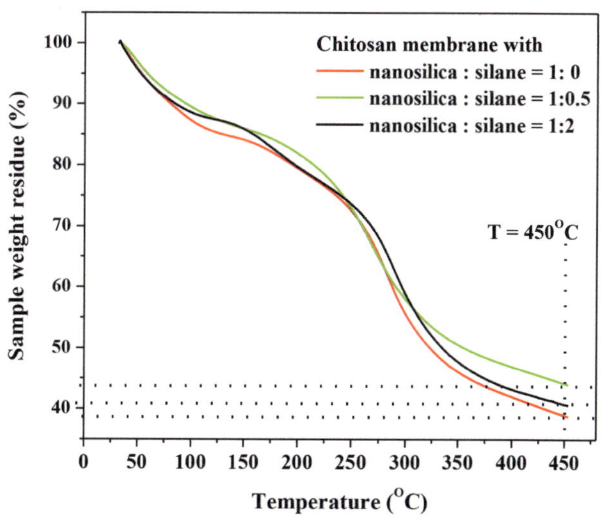

Figure 8.
TGA thermograms of chitosan-nanosilica membrane with silane addition.

Chitosan-nanosilica Membranes	Temperature range on an oblique curve (°C)			Mass reduction on an oblique curve (%)			Mass residue (%) at III
	I	II	III	I	II	III	
Nanosilica:silane = 1:0	35–120	120–220	220–430	12.72	9.96	36.02	41.58
Nanosilica:silane 1 = 1:0.5	40–150	150–330	330–450	13.72	34.06	7.79	44.09
Nanosilica:silane 1 = 1:2.0	35–140	140–335	335–450	14.35	36.16	8.78	37.93

Table 2.
The area of change (stages) in mass percentage reduction on TGA thermogram of membrane samples.

Chitosan-nanosilica Membranes	Mass residue (%) at 450°C	Temperature (°C) when sample is degraded and remaining 44%
Nanosilica:silane = 1:0	38.92	371.98
Nanosilica:silane 1 = 1:0.5	44.15	452.94
Nanosilica:silane 1 = 1:2.0	40.74	387.96

Table 3.
Chitosan-nanosilica:silane membrane thermal degradation.

not significant indicates that the water content in samples are not significantly different. Mass reduction in chitosan-nanosilica membrane samples with silane coupling agent is higher when compared to sample without silane. This shows that chitosan-nanosilica with silane is more hygroscopic.

The second weight-loss in **Table 2** appearing at around 150–335°C indicates the decomposition of chitosan polymer chain in hybrid membrane [10]. This chain decomposition is related to the loss of side groups as acetyl groups (as shown in **Figure 1**) in chitosan because the acetyl groups has a weak phi bond so it breaks easily first. The temperature range and the percentage of sample weight-loss is significantly different. The temperature range and the percentage of mass reduction in two samples with silane as coupling agent shows a large range. The second stage temperature range in the sample with silane shows a widened temperature region. This is due to the presence of the silane as coupling agent interact with the acetyl groups thus preventing the release of the acetyl groups. As a result, a large number of these groups release at higher temperatures.

The third weight-loss stage is observed near 335–450°C is due to unstable parts of the polymeric matrix whereas it is occurring due to complete decomposition of the backbone of polymeric matrix [48] and residual organic groups [8]. The temperature range and the percentage of sample weight-loss in this stage is significantly different. The temperature range in third stage in two samples with silane as coupling agent shows a large range while the percentage of mass reduction is a small range. The incorporation of silane has improved the interaction of nanosilica and chitosan by introducing more functional groups with hydrogen bond formation as **Figure 1** and FTIR results. Hydrogen bond between hydroxyl groups (–OH) in silane-coupled nanosilica interacts with the amine group ($-NH_2$) ether group (C–O–C) of chitosan that could form a good interface interaction [32]. This all makes it difficult for chitosan chain to degrade so that the degradation temperature range is greater and the remaining mass is less. In addition, chitosan membrane samples with silane-coupled nanosilica have higher component heterogeneity than chitosan-nanosilica without silane. High heterogeneity of polymer components causing a longer range of degradation and decomposition temperature. Therefore, it can be understood that **Table 3** shows that nanosilica:silane 1 = 1:0.5 have the highest thermal degradation. This result is in line with the results of FTIR and AFM analysis which finds that the composition nanosilica:silane 1 = 1:0.5 provides the best hydrogen bond interaction. This result is in accordance with [17] which state that the proper addition of modified nanosilica with silane enhanced the thermal performance by acting as superior insulator and mass transport barrier to the volatile products generated during decomposition.

4. Conclusions

The influence of silane as coupling agent on the chitosan-nanosilica membrane causes homogeneity and prevents agglomeration between nanosilica and chitosan. The higher silane compositions added, it decrease the water uptake, increases proton conductivity, decreases methanol permeability, and increases selectivity of the membrane. The best membrane performance is on the variation of nanosilica: silane 1:0.50 which has water uptake of 37.52%, tensile strength of 11.8 MPa, proton conductivity of 7.8988 x 10^{-4} S/cm at 25°C, methanol permeability of 5.8434 x 10^{-7} cm^2/s, and membrane selectivity of 13.5174 x 10^2 S s/cm^3. This membrane has high thermal stability of 452.94°C with mass residue 44%. Based on the results of membrane selectivity analysis, the best and most suitable chitosan membrane for electrolyte polymer applications is the chitosan membrane with the addition of nanosilica:silane = 1:0.5; 1:0.25; 1:1; 1:0; 1:1.5 and 1:2.0 (w/w). This is based on the order of membrane selectivity values from the highest to the lowest. The results of FT-IR, SEM and AFM analysis on membranes show that optimum silane addition provides the best interaction between chitosan matrix and silane-coupled nanosilica so that they have higher thermal resistance. When compared to Nafion 117 membrane, this membrane has lower proton conductivity value, but the advantages are it more environmentally friendly, has lower methanol permeability, higher temperatures stability and of course more economical in terms of cost. Further efforts are needed to increase the proton conductivity of this chitosan membrane for DMFC applications.

Acknowledgements

This research is supported by the Direktorat Riset dan Pengabdian Masyarakat (DRPM) under Kementerian Riset, Teknologi dan Perguruan Tinggi Republik Indonesia (Indonesian Ministry of Research, Technology and Higher Education/ Ristekdikti) through Pekerti Scheme Funding (Skim Penelitian Kerjasama Antar Perguruan Tinggi) in 2016. The authors are grateful to Chemistry Department of Institut Teknologi Sepuluh Nopember (ITS), Instrumentation and Analytical Science Laboratory of ITS and also Material Chemistry and Energy Laboratory of ITS for cooperation.

Author details

Ella Kusumastuti[1*], Fadila Mauliani[1], F. Widhi Mahatmanti[1], Jumaeri[1], Lukman Atmaja[2] and Nurul Widiastuti[2]

1 Faculty of Mathematics and Natural Sciences, Department of Chemistry, Universitas Negeri Semarang, Semarang, Indonesia

2 Faculty of Mathematics and Natural Sciences, Department of Chemistry, Institute Technology Sepuluh November, Surabaya, Indonesia

*Address all correspondence to: ella.kusuma@mail.unnes.ac.id

References

[1] Giorgi L, Leccese F. Fuel Cells: Technologies and Applications. Open Fuel Cells J. 2013;6:1-20.

[2] Fadzallah IA, Majid SR, Careem MA, Arof AK. A study on ionic interactions in chitosan – oxalic acid polymer electrolyte membranes. J Memb Sci. 2014;463(2014):65-72.

[3] Kim DJ, Jo MJ, Nam SY. A review of polymer – nanocomposite electrolyte membranes for fuel cell application. J Ind Eng Chem. 2015;21(2015):36-52.

[4] Prapainainar C, Kanjanapaisit S, Kongkachuichay P, Holmes SM, Prapainainar P. Surface Modification of Mordenite in Nafion Composite Membrane for Direct Ethanol Fuel Cell and Its Characterizations: Effect of Types of Silane Coupling Agent. Environ Chem Eng. 2016;4(3):2637-2646.

[5] Purwanto M, Atmaja L, Mohamed AM, Salleh MT, Jaafar J, Ismail AF, et al. Biopolymer-based electrolyte membranes from chitosan incorporated with montmorillonite- crosslinked GPTMS for direct methanol fuel cells. R Soc Chem (RSC Adv). 2016;6:2314-2322.

[6] Ma J, Sahai Y. Chitosan biopolymer for fuel cell applications. Carbohydr Polym. 2013;92(2):955-975.

[7] Vaghari H, Jafarizadeh-malmiri H, Berenjian A, Anarjan N. Recent advances in application of chitosan in fuel cells. Sustain Chem Process. 2013;1(16):1-12.

[8] Wang Y, Jiang Z, Li H, Yang D. Chitosan membranes filled by GPTMS-modified zeolite beta particles with low methanol permeability for DMFC. Chem Eng Process Process Intensif. 2010;49(3):278-285.

[9] Widhi Mahatmanti F, Nuryono, Narsito. Physical characteristics of chitosan based film modified with silica and polyethylene glycol. Indones J Chem. 2014;14(2):131-137.

[10] Atmaja L, Purwanto M, Taufiq M, Azuwa M, Jaafar J, Fauzi A, et al. GPTMS-Montmorillonite-filled biopolymer chitosan membrane with improved compatibility, physicochemical, and thermal stability properties. Malaysian J Fundam Appl Sci. 2019;15(4):492-497.

[11] Ying TP, Erfeida AM, Viet CX, Lan DNU. Effects of filler incorporation routes on mechanical properties of low density polyethylene/ natural rubber/silica (LDPE/NR/ SI) composites. Appl Mech Mater. 2014;679(Augustus):154-157.

[12] Lin H, Zhao C, Ma W, Shao K, Li H, Zhang Y, et al. Novel hybrid polymer electrolyte membranes prepared by a silane-cross-linking technique for direct methanol fuel cells. J Power Sources. 2010;195(3):762-768.

[13] Zou H, Wu S, Shen J. Polymer/ Silica Nanocomposites: Preparation, Characterization, Properties, and. Chem Rev. 2008;108(9):3893-3957.

[14] Kusumastuti E, Siniwi WT, Mahatmanti FW, Jumaeri J, Atmaja L, Widiastuti N. Modification of chitosan membranes with nanosilica particles as polymer electrolyte membranes. In: AIP Conference Proceedings. 2016. p. 1-9.

[15] Ifuku S, Yano H. Effect of a silane coupling agent on the mechanical properties of a microfibrillated cellulose composite. Int J Biol Macromol [Internet]. 2015;74:428-432. Available from: http://dx.doi.org/10.1016/j.ijbiomac.2014.12.029

[16] Lin H, Zhao C, Jiang Y, Ma W, Na H. Novel hybrid polymer

electrolyte membranes with high proton conductivity prepared by a silane-crosslinking technique for direct methanol fuel cells. J Power Sources [Internet]. 2011;196(4):1744-1749. Available from: http://dx.doi.org/10.1016/j.jpowsour.2010.10.003

[17] Nikje MMA, Tehrani ZM. Novel Hybrid Membranes Based on Chitosan and Organically-Modified Nano-SiO2. Des Monomers Polym. 2009;12(2009):315-322.

[18] Maharani DK, Kartini I, Aprilita NH. Nanosilica-Chitosan Composite Coating on Cotton Fabrics. In: AIP Conference Proceedings. 2010. p. 86-91.

[19] Connell LS, Romer F, Suarez M, Valliant EM, Zhang Z, Lee PD, et al. Chemical characterisation and fabrication of chitosan – silica hybrid scaffolds with 3-glycidoxypropyl trimethoxysilane. J Mater Chem B. 2014;2(2014):668-680.

[20] Asghari M, Sheikh M, Afsari M, Dehghani M. Molecular simulation and experimental investigation of temperature effect on chitosan-nanosilica supported mixed matrix membranes for dehydration of ethanol via pervaporation. J Mol Liq [Internet]. 2017;246:7-16. Available from: http://dx.doi.org/10.1016/j.molliq.2017.09.045

[21] Pandey RP, Shahi VK. Functionalized silica-chitosan hybrid membrane for dehydration of ethanol/water azeotrope: Effect of cross-linking on structure and performance. J Memb Sci [Internet]. 2013;444:116-126. Available from: http://dx.doi.org/10.1016/j.memsci.2013.04.065

[22] Vijayakumar V, Khastgir D. Hybrid composite membranes of chitosan/sulfonated polyaniline/silica as polymer electrolyte membrane for fuel cells. Carbohydr Polym [Internet]. 2018;179:152-163. Available from: http://dx.doi.org/10.1016/j.carbpol.2017.09.083

[23] Handayani S, Dewi EL. Influence of Silica/Sulfonated Polyether-Ether Ketone as Polymer Electrolyte Membrane for Hydrogen Fueled Proton Exchange Membrane Fuel Cells. Int J Sci Eng. 2011;2(2):27-30.

[24] Pandis C, Madeira S, Matos J, Kyritsis A, Mano JF, Ribelles JLG. Chitosan–silica hybrid porous membranes. Mater Sci Eng C. 2014;42(2014):553-561.

[25] Wang S, Shi L, Zhang S, Wang H, Cheng B, Zhuang X, et al. Proton-conducting amino acid-modified chitosan nano fi bers for nanocomposite proton exchange membranes. Eur Polym J. 2019;119(July):327-334.

[26] Yang CC, Lue SJ, Shih JY. A novel organic/inorganic polymer membrane based on poly(vinyl alcohol)/poly(2-acrylamido-2-methyl-1-propanesulfonic acid)/3-glycidyloxypropyl trimethoxysilane polymer electrolyte membrane for direct methanol fuel cells. J Power Sources [Internet]. 2011;196(10):4458-4467. Available from: http://dx.doi.org/10.1016/j.jpowsour.2011.01.051

[27] Mohanapriya S, Rambabu G, Bhat SD, Raj V. Pectin based nanocomposite membranes as green electrolytes for direct methanol fuel cells. Arab J Chem [Internet]. 2020;13(1):2024-2040. Available from: https://doi.org/10.1016/j.arabjc.2018.03.001

[28] Vanjeri VN, Goudar N, Kasai D, Masti SP, Chougale RB. Thermal and tensile properties study of 4-Hydroxycoumarin doped Polyvinyl alcohol/Chitosan blend films. Chem Data Collect [Internet]. 2019;23:100257. Available from: https://doi.org/10.1016/j.cdc.2019.100257

[29] Yang C, Jessie S, Shih J. A novel organic / inorganic polymer membrane based on poly (vinyl alcohol)/ poly (2-acrylamido-2-methyl-1-propanesulfonic acid) / 3- glycidyloxypropyl trimethoxysilane polymer electrolyte m ... A novel organic / inorganic polymer membrane based on poly. J Power Sources. 2011;196(10):4458-4467.

[30] Wu H, Zheng B, Zheng X, Wang J, Yuan W, Jiang Z. Surface-modified Y zeolite-filled chitosan membrane for direct methanol fuel cell. J Power Sources. 2007;173(2 SPEC. ISS.):842-852.

[31] Purwanto M, Atmaja L, Mohamed AM, Salleh MT, Jaafar J, Ismail AF, et al. Biopolymer-based electrolyte membranes from chitosan incorporated with montmorillonite-crosslinked GPTMS for direct methanol fuel cells. RSC Adv [Internet]. 2016;6(2314):2314-22. Available from: http://dx.doi.org/10.1039/C5RA22420A

[32] Arkles B. Silane Coupling Agents Connecting Across Boundaries. 3rd ed. Maddox A, Singh M, Zazyczny J, Matisons J, editors. Morrisville, PA: Gelest, Inc.; 2014.

[33] Hickner MA, Ghassemi H, Kim YS, Einsla BR, McGrath JE. Alternative polymer systems for proton exchange membranes (PEMs). Chem Rev. 2004;104(10):4587-4611.

[34] Zhu X, Hou X, Ma B, Xu H, Yang Y. Chitosan/gallnut tannins composite fiber with improved tensile, antibacterial and fluorescence properties. Carbohydr Polym [Internet]. 2019;226(July):115311. Available from: https://doi.org/10.1016/j.carbpol.2019.115311

[35] Suka IG, Simanjuntak W, Dewi EL. Pembuatan Membran Polimer Elektrolit Berbasis Polistiren Akrilonitril (SAN) untuk Aplikasi Direct Methanol Fuel Cell. J Natur Indones. 2012;13(1):1.

[36] Tripathi BP, Kumar M, Shahi VK. Highly stable proton conducting nanocomposite polymer electrolyte membrane (PEM) prepared by pore modifications: An extremely low methanol permeable PEM. J Memb Sci. 2009;327(1-2):145-154.

[37] Tripathi BP, Shahi VK. Organic-inorganic nanocomposite polymer electrolyte membranes for fuel cell applications. Prog Polym Sci [Internet]. 2011;36(7):945-979. Available from: http://dx.doi.org/10.1016/j.progpolymsci.2010.12.005

[38] Al-Sagheer F, Muslim S. Thermal and mechanical properties of chitosan/SiO2 hybrid composites. J Nanomater. 2010;2010.

[39] Wei B, Chang Q, Bao C, Dai L, Zhang G, Wu F. Surface modification of filter medium particles with silane coupling agent KH550. Colloids Surfaces A Physicochem Eng Asp [Internet]. 2013;434:276-280. Available from: http://dx.doi.org/10.1016/j.colsurfa.2013.05.069

[40] Oliveira AC De, Sabino RM, Souza PR, Muniz EC, Popat KC, Kipper MJ, et al. Chitosan/gellan gum ratio content into blends modulates the sca ff olding capacity of hydrogels on bone mesenchymal stem cells. Mater Sci Eng C [Internet]. 2020;106(August 2019):110258. Available from: https://doi.org/10.1016/j.msec.2019.110258

[41] Barzin J, Feng C, Khulbe KC, Matsuura T, Madaeni SS, Mirzadeh H. Characterization of polyethersulfone hemodialysis membrane by ultrafiltration and atomic force microscopy. J Memb Sci. 2004;237(1-2):77-85.

[42] Khulbe KC, Feng CY, Matsuura T. Synthetic Polymeric Membranes

Characterization by Atomic Force Microscopy. Springer LaboratoryManuals in Polymer Science; 2008.

[43] Ghaee A, Shariaty-Niassar M, Barzin J, Matsuura T. Effects of chitosan membrane morphology on copper ion adsorption. Chem Eng J [Internet]. 2010;165(1):46-55. Available from: http://dx.doi.org/10.1016/j.cej.2010.08.051

[44] Tamburaci S, Tihminlioglu F. Diatomite reinforced chitosan composite membrane as potential scaffold for guided bone regeneration. Mater Sci Eng C [Internet]. 2017;80:222-231. Available from: http://dx.doi.org/10.1016/j.msec.2017.05.069

[45] Arpornwichanop T, Polpanich D, Thiramanas R, Suteewong T, Tangboriboonrat P. Enhanced antibacterial activity of NR latex gloves with raspberry-like PMMA-N,N,N-trimethyl chitosan particles. Int J Biol Macromol [Internet]. 2015;81:151-158. Available from: http://dx.doi.org/10.1016/j.ijbiomac.2015.07.063

[46] Viswanathan N, Meenakshi S. Enriched fluoride sorption using alumina/chitosan composite. J Hazard Mater. 2010;178(1-3):226-232.

[47] Rana VK, Pandey AK, Singh RP. Enhancement of Thermal Stability and Phase Relaxation Behavior of Chitosan Dissolved in Aqueous l-Lactic Acid: Using "Silver Nanoparticles" as Nano Filler. Macromol Res. 2010;18(8):712-729.

[48] Dhawade P, Jagtap R. Comparative study of physical and thermal properties of chitosan-silica hybrid coatings prepared by sol-gel method. Der Chem Sin. 2012;3(3):589-601.

Chapter 10

Innovative Separation Technology Utilizing Marine Bioresources: Multifaceted Development of a Chitosan-Based System Leading to Environmentally-Friendly Processes

Keita Kashima, Tomoki Takahashi, Ryo-ichi Nakayama and Masanao Imai

Abstract

Chitosan, known as a most typical marine biological polymer, has a fruitful capability of biocompatible gel formation. Attempts of chitosan have been made to develop it from the multifaceted viewpoint of separation technology. The physico-chemical properties of chitosan containing a lot of hydroxyl groups and reactive amino groups help to build the characteristic polymer networks. The deacetylation degree of chitosan is found as the most influential factor to regulate properties of chitosan hydrogels. The antibacterial activity of the chitosan membrane is one of its notable abilities because of its practical application. The chitosan, its derivatives, and the complex formation with other substances has been used for applications in filtration and membrane separation processes. Adsorption processes based on chitosan have been also developed widely. Moreover, complex of chitosan gel helps to immobilize adsorbent particles. The chitosan membrane immobilizing Prussian-Blue for cesium ion removal from the aqueous phase is one of the leading cases. To elaborate the adsorption behavior on the chitosan immobilizing adsorbent, the isothermal equilibrium and mass transfer characteristics can be discussed. The adsorption process using chitosan-based membranes in combination with filtration in a flow process is advantageous compared with the batch process. More advanced studies of chitosan aerogel and chitosan nanofibers have been proceeded recently, especially for adapting to water purification and air filtration.

Keywords: chitosan, membrane, deacetylation degree, filtration, adsorption, water treatment, aerogel, nanofiber

1. Introduction

A separation process, which is often called the "downstream process", plays a key role of product manufacturing through chemical or biochemical reaction as well as a synthesis process called the “upstream process” [1]. To ensure quality and cost of final products, the separation process is important and has been developed in line with the social demands [2]. For chemical and biochemical industries, the separation process aims to purify objective substances, eliminate undesirable substances, and fractionate each component from their mixture. As environmental awareness around the world increases recently, new separation technologies, such as wastewater treatment [3], advanced desalination [4], air cleaning [5] *etc.*, are in great demand. In addition, materials used in such separation processes are not only expected to be efficient, low cost, easy operation, but also required to be environmentally-friendly.

Chitin and chitosan obtained from crustaceans possess sufficient environmental adaptability as well as an attractive potential to build various types of functional media, e.g., membranes [6–10], micro/nanoparticles [11, 12], and nanofibers [13–16]. Many studies have devoted to develop the novel media adapting separation processes using chitosan. The separation performance of such chitosan media should be strongly influenced from chemical properties characterized by deacetylation degree (*DD*) at amino groups in the chitosan molecular chain. Nevertheless, it was less mentioned that the *DD* could steer not only the structure of the prepared chitosan gels but also the characteristics as separation media.

This chapter describes the preparation and physicochemical properties of novel chitosan-based media and demonstrates the promising ability of chitosan with focus on principal studies for environmentally-friendly separation processes. The essential factors which regulate the performance of separation media prepared from chitosan, such as *DD*, molecular weight, and options of cross-linker, are explained. In particular, the notable impacts of *DD* on the mass transfer mediated by chitosan membrane, the mechanical property, and the antibacterial activity, are introduced based on our previous research [17–19]. Separation media prepared from chitosan are often combined with various adsorbents [20, 21], carbon nanotubes [22, 23], or other functional materials [24, 25]. In such case, the behavior of mass transfer into the chitosan hydrogel is complicated to quantitatively evaluate. The present chapter shows the determination of effective diffusion coefficient of cesium ions in chitosan membrane immobilizing Prussian Blue particles [20]. Furthermore, the chitosan aerogels with macro-porous structure is proposed for selective separation for anionic dye from aqueous phase. Chitosan nanofibers incorporated with polyethylene terephthalate (PET) non–woven are also covered to describe in an application of air filtration.

2. Physicochemical properties and gelling characteristics

Chitin and chitosan are known as secondary abundant polymers obtained from the external skeleton of crustaceans such as crab and shrimp [26]. Apart from this, chitin is also found in and produced from the exoskeleton of insects or the cell walls of fungi and yeast [6, 27, 28]; however, this contribution is much less than that from the marine resources. In recent years, chitin and chitosan have gained attention instead of raw materials of petroleum origin owing to the inevitable depletion of fossil fuels and the prevention of climate change [29]. This section looks at the physicochemical properties of chitosan and focuses on the gelling characteristics to build environmentally-friendly separation media.

2.1 Chemical composition and gelling ability for separation media

Figure 1 shows the chemical conversion between chitin and chitosan. The chemical composition of chitin can be described as a long-chain polymer,

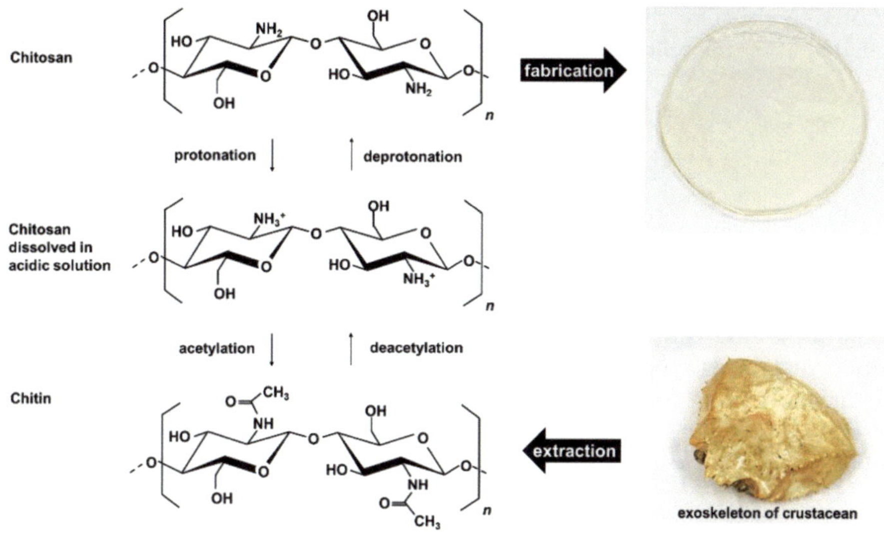

Figure 1.
Chemical conversion among chitin, chitosan, and furthermore functional separation media.

poly(β-(1 → 4)-*N*-Acetyl-D-glucosamine). Chitosan, poly(β-(1 → 4)-D-glucosamine), is obtained mainly by transforming partial deacetylation of chitin in an alkaline condition, such as using sodium hydroxide aqueous solution. It has been reported that chitosan and its oligosaccharides not only possess hydrophilicity, non-toxicity, biodegradability, and biocompatibility, but also possess antimicrobial activity, antioxidant properties, and an affinity for proteins [7, 26].

Chitosan is insoluble in water at neutral pH or in any organic solvent. Consequently, an acidic aqueous solution, such as acetate buffer solution, is usually employed to dissolve chitosan, whereby the acid dissociation constant of chitosan is found as pKa ≈ 6.5 [30]. Chitosan can be dissolved in acidic solutions by protonation of amino groups in glucosamine units.

Deprotonating a chitosan solution through an acid–base neutralization leads to formation of a water-insoluble gel structure without cross-linker due to intermolecular hydrogen bonding [8]. The salt (*e.g.* NaCl) coexisting with chitosan in an acidic solution acts as counter ions and disrupts intramolecular hydrogen bonding, and then the flexibility of chitosan molecular chains increases [31]. In addition, pH neutralization influences the formation of a polymeric network [9]. Therefore, the neutralization condition should be optimized. From the convenient gelling process, chitosan hydrogels have been developed widely as immobilizing matrices, with enzymes, carbon nanotubes, and electroconductive polymers as typical examples [10, 32, 33].

2.2 Deacetylation degree

Deacetylation degree (*DD*) is the most important factor to regulate physicochemical properties. The deacetylation degree of chitosan samples was determined using the colloidal titration method-based experimental conditions in previous works [34, 35]. We dissolved chitosan powder (0.5 g) in 5% acetic acid solution, and then increased the total weight of chitosan–acetic acid solution to 100.0 g by adding acetic acid. We mixed a 1 g sample of this chitosan–acetic acid solution to 30 ml of deionized water. The titrant was 0.0025 N potassium polyvinyl sulfate (PVS-K), and the indicator was 1% toluidine blue. The terminal point of titration was clearly indicated by the color changing from blue to claret. The deacetylation degree was calculated using the following equations.

$$X = f \times 0.0025 \times 10^{-3} \times v \times 161 \quad (1)$$

$$Y = 0.5 \times 10^{-2} - X \quad (2)$$

$$DD\ (\%) = \frac{X/161}{X/161 + Y/203} \times 100 \quad (3)$$

In these equations, X is the equivalent mass of glucosamine contained in a 1 g sample of the chitosan–acetic acid solution, v [ml] is the volume of 0.0025 N PVS-K solution, and f is its concentration factor. Y is the mass of acetyl glucosamine contained in a 1 g sample, calculated as the difference between the mass of the sample and the value of X. We evaluated the deacetylation degree of the chitosan samples as a molar fraction of glucosamine [36].

2.2.1 Control of the deacetylation degree

A chitosan membrane was prepared by the casting method in combination with *N*-acetylation reaction [17]. The deacetylation degree (*DD*) decreased linearly with increasing added amounts of acetic anhydride (**Figure 2**). Stoichiometric control of the deacetylation degree to the desired level was successfully performed. However, gelation reaction due to excess addition of acetic anhydride inhibited formation of the chitosan membrane.

The gelation behavior of chitosan, which has various degrees of acetylation (*DA*) of amino groups, was investigated to ensure preparation of the designed membrane structure [37]. The gelation behavior was evaluated by the gelation time and the quantity of syneresis, and useful information not only for preparing a membrane but also for preparing an immobilized carrier or a chemical reaction system was obtained in this work.

2.2.2 Water permeation mechanism of an N-acetylated chitosan membrane

A novel model of the water permeation mechanism in an *N*-acetyl-chitosan membrane with a cellular structure was proposed [18]. Although the entire membrane structure has a hydrophilic character, the cellular structure incorporates junction zones that practically prevent water permeation.

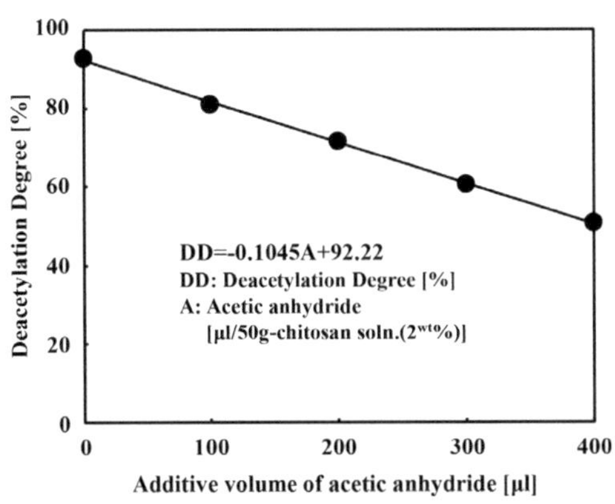

Figure 2.
Effect of acetic anhydride on N-*acetylation of chitosan. Acetic anhydride was added to 2 wt.% chitosan solution (50 g) [17] with permission from Elsevier.*

Chitosan membranes with a controlled degree of deacetylation (*DD*) were prepared using a casting method. Changes in the total water content and the pressure driven water flux of the membrane were observed with a change in *DD* (**Figure 3**). The membrane properties were analyzed and evaluated using water permeability measurements, scanning electron microscopy (SEM), X-ray diffraction (XRD), and differential scanning calorimetry (DSC). SEM observations indicated that the membrane structure was an individual cellular structure and that this cellular structure grew with decrease in *DD* (**Figure 4**). From XRD measurements, the intensity in the range from 10° to 20° were detected in the chitin (*DD* = 1.1%) and the chitosan membranes (71.3% < *DD* < 92.2%), which indicated that the crystal structure of the membrane was amorphous regardless of *DD*. The free water content (W_f), the freezable bound water (W_{fb}), and the bound water not able to freeze (W_b) were evaluated by DSC. The total water content and the sum of free water content ratios ($W_f + W_{fb}$) decreased with increasing *DD* whereas W_b gradually increased [18]. That suggests the membrane prepared from lower *DD* chitosan formed remarkable cellular structure. The free water was mainly contained inside of the cellular structure, and resulted in swelling the chitosan membrane (**Figure 5**).

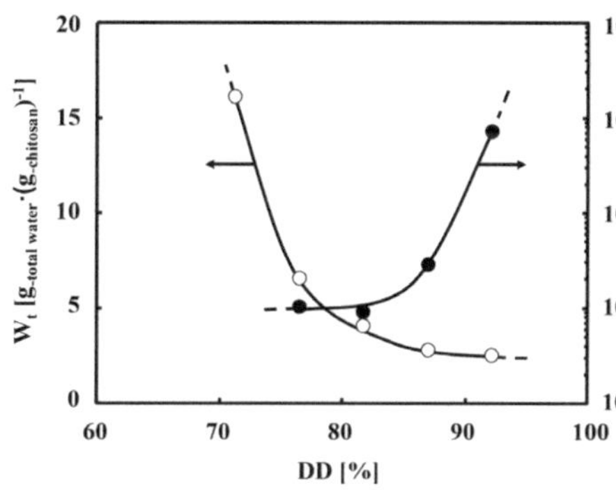

Figure 3.
Change in water flux (50 kPa, 298 K, μL = 0.901 mPa s) and total water content of membrane with regulated DD. *(●) water flux,* J_v *and (○) total water content,* W_t *[18] with permission from Elsevier.*

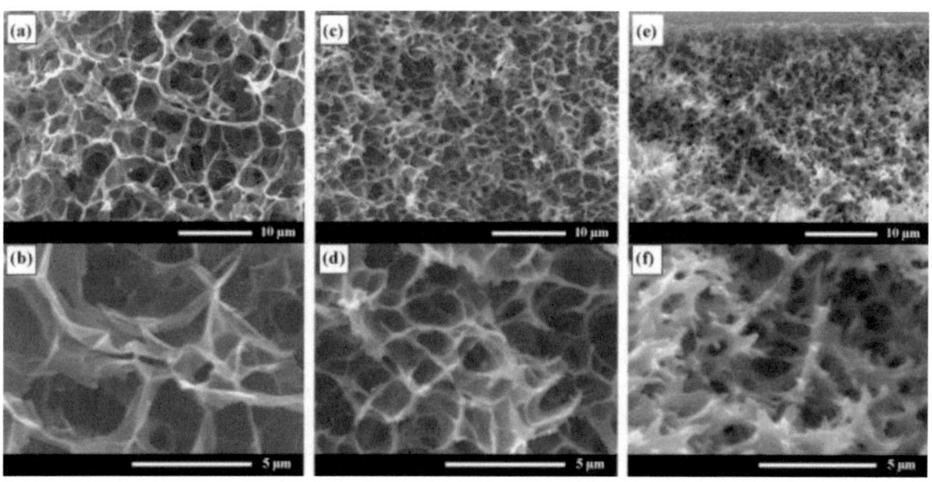

Figure 4.
SEM images observed at cross-section of the membranes prepared from different DD chitosan [18] with permission from Elsevier. (a) and (b) DD 71.3%; (c) and (d) DD 81.8%; (e) and (f) DD 92.2%. The full length of reference scaling measure in (a), (c) and (e) indicates 10 μm. The full length of reference scaling measure in (b), (d) and (f) indicated 5 μm.

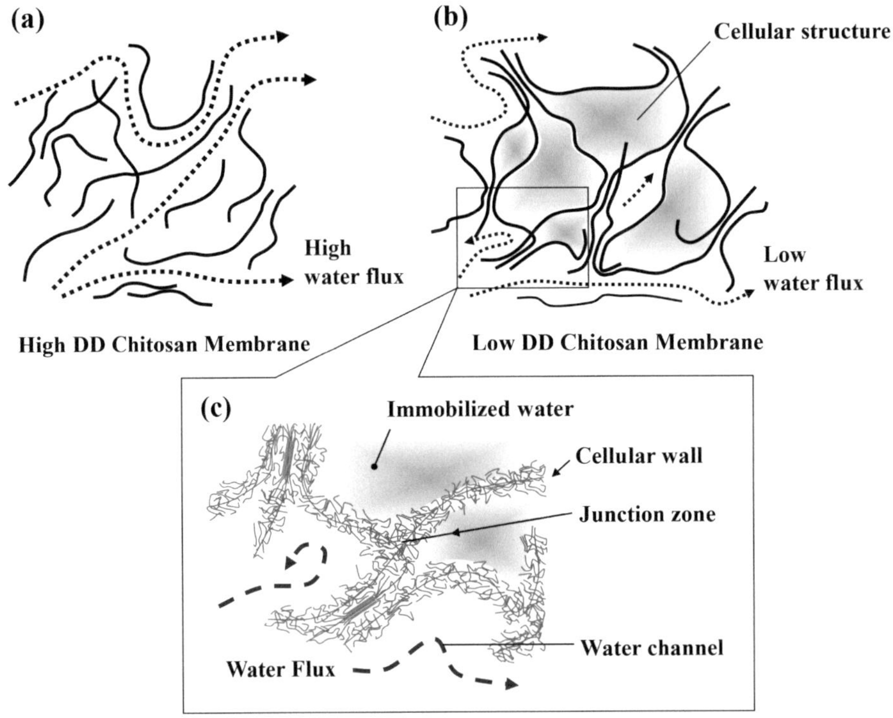

Figure 5.
Schematic illustration of the water permeation mechanism in an N*-acetyl-chitosan membrane with a cellular structure. (a) the structure of a high* DD *chitosan membrane, (b) the structure of a low* DD *chitosan membrane, and (c) the detailed image of a low DD chitosan membrane with its cellular structure and water channels. The cellular structure illustrated in (b) is composed of immobilized water, the cellular wall, and the junction zone. The structure prevents water flux [18] with permission from Elsevier.*

Pressure driven water flux was measured using the ultrafiltration apparatus; it was dependent on the operational pressure, membrane thickness, and the feed solution viscosity, and obeyed the Hagen-Poiseuille flow. At a higher *DD*, water permeation proceeded due to degradation of the cellular structure; the amount of water in permeation channels was greater than that for lower *DD* membranes even though the total water content in the membrane was less. The water flux of the chitosan membrane was determined by the water content constructing channels through the membrane and not by the total water content in the membrane.

2.2.3 Antibacterial activity

The antibacterial activity is also explained, because of its long-time practical application. The antibacterial activity of chitosan membranes was investigated by a conductimetric assay using a bactometer [19]. The growth of the gram-positive sample (*S. aureus*) was more strongly inhibited by chitosan than the gram-negative sample (*E. coli*). This inhibitory effect was recognized as a bactericidal effect. Antibacterial activity was also observed and was dependent on the shape and specific surface area of the powdered chitosan membrane. The influence of the *DD* of the chitosan on inhibiting the growth of *S. aureus* was investigated by two methods: incubation using a mannitol salt agar medium and a conductimetric assay (**Figure 6**). In both methods, chitosan with a higher *DD* successfully inhibited growth of *S. aureus*. Our findings regarding the dominant role of the *DD* of chitosan will be useful for designing lasting, hygienic, membrane-based processes.

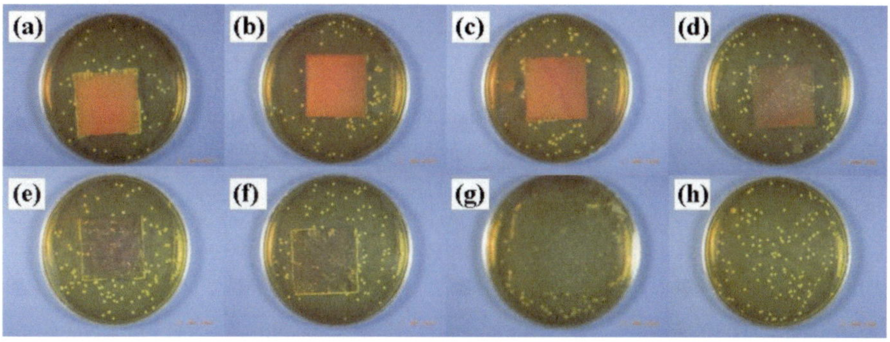

Figure 6
Photographical evidence of antibacterial activity of chitosan membrane immersed in mannitol salt agar culture involved with degree of deacetylation: (a) DD *92.2%; (b)* DD *90.1%; (c)* DD *88.0%; (d)* DD *83.9%; (e)* DD *79.7%; (f)* DD *75.5%; (g) PVC; (h) control [19] with permission from Elsevier.*

2.3 Molecular weight

The molecular weight of chitosan also plays a significant role in the properties of a prepared membrane. It was found that the tensile strength and elongation of membranes prepared from high molecular weight chitosan were higher than those prepared from low molecular weight chitosan; however, the permeability of membranes from high molecular weight chitosan is lower than those prepared from low molecular weight ones [38]. For convenient determination, the viscometric average molecular weight (M [g mol^{-1}]) can be calculated from the intrinsic viscosity (η [mL g^{-1}]) using the Mark-Houwink-Sakurada relationship as follows:

$$[\eta] = KM^{\alpha} \tag{4}$$

For a case, chitosan is dissolved in an acetate buffer composed of 0.2 M acetic acid +0.2 M sodium acetate at 25°C. The parameters in the above relationship were found as $K = 7.9 \times 10^{-2}$ [mL/g] and $\alpha = 0.796$ [−] [27, 39, 40].

2.4 Cross-linker

Various types of cross-linkers are often employed for the fabrication of suitable chitosan membranes adapted to separation processes. Glutaraldehyde (GA) is usually used as cross-linker, because it is extremely reactive in cross-linking chitosan polymer chains *via* the Schiff reaction between aldehyde groups and amino groups to form covalent imine bonds [41–43]; however, GA is also toxic. Genipin, which is produced from the hydrolysis of geniposide extracted from the fruits of *Gardenia jasminoides Ellis*, is attracted as a biocompatible and low biohazardous cross-linker [44, 45]. Tetraethyl orthosilicate (TEOS) is also used as a covalent cross-linker, which serves to cross-link with chitosan polymers at their hydroxyl groups, to immobilize adsorbent particles, and to hydrophobize membrane surfaces [22, 46–48]. Jóźwiak and coworkers widely investigated the effect of the cross-linker type occurring covalent bond or ionic bond on the chitosan hydrogel prepared for ionic dye adsorption. In case of covalent agents, it is suggested that epoxide functional groups prefer to attack hydroxyl groups of chitosan during cross-linking, and the free amine groups formed are responsible for anionic dye adsorption [49].

3. Filtration processes with sieving to molecules

Table 1 summarizes the recent studies of a filtration membrane consisting of chitosan or chitosan derivatives. In several membranes, the functional materials are additionally composed to improve the separation ability. The relevant studies are picked up and described below.

Membrane body	Additive functionalizing material	Cross-linker	Target substances	Expected application	Ref.	Year
chitosan	phosphotungstic acid	*NA*	proton, methanol	DMFCs [a]	[60]	2016
chitosan/PSF[b]	MOFs[c]	*NA*	NaCl, $MgCl_2$, $CaCl_2$, Na_2SO_4	nanofiltration	[50]	2017
chitosan	MWNT[d]	glycerin, PEGDE[e]	NaCl, $MgCl_2$ $MgSO_4$	nanofiltration	[51]	
carboxymethyl chitosan/PVDF [f]	*NA*	*NA*	humic acid	nanofiltration	[52]	
chitosan/PTFE [g]	*NA*	TEOS [h]	methanol, toluene	pervaporation	[46]	2018
chitosan/PVA [i]	*NA*	adipic acid	NaCl aerosol	air filtration	[56]	
phosphorylated chitosan/PAN [j]	*NA*	GA [k]	$MgCl_2$ Na_2SO_4 MB [l] MO [m], AYR [n]	nanofiltration	[41]	
chitosan	polyester nonwoven fabric	sulfuric acid	H_2O (vapor)	membrane drier	[61]	2019
chitosan/PVA [i]	motmorillonite	*NA*	Cr(VI)	nanofiltration	[53	
chitosan	MWNT[d]	*NA*	NaCl, $MgSO_4$	nanofiltration	[55]	2020
PVA [i]	microparticles of chitosan, phosphorylated chitosan, glycidol-modified chitosan, or sulphated chitosan	GA [k]	ethanol	pervaporation	[42]	
phosphorylated chitosan	graphene oxide	GA [k]	DB38 [o], PS [p], XO [q]	nanofiltration	[43]	
chitosan/PES [r]	MWNT[d]	*NA*	MG [s]	nanofiltration	[54]	
chitosan	*NA*	genipin	IPA [t]	pervaporation	[44]	
chitosan/PVA [i]	Cu-BTC [u]	*NA*	CO_2 (gas)	CO_2 separation	[57]	
carboxymethyl chitosan/ polyamidoamine	hydrotalcite	*NA*	CO_2 (gas)	CO_2 separation	[58]	
chitosan/ methoxy pectin	cutin	glycerol	H_2O (vapor)	packaging film	[62]	

Membrane body	Additive functionalizing material	Cross-linker	Target substances	Expected application	Ref.	Year
chitosan	*NA*	succinic acid	H_2O (vapor)	packaging film	[63]	2021

[a]DMFCs: direct methanol fuel cells
[b]PSF: polysulfone
[c]MOF: metal organic frameworks
[d]MWNT: multi-walled carbon nanotubes
[e]PEGDE: polyethyleneglycol diglycidyl ether
[f]PVDF: polyvinylidene fluoride
[g]PTFE: polytetrafluoroethylene
[h]TEOS: tetraethyl orthosilicate
[i]PVA: polyvinyl alcohol
[j]PAN: polyacrylonitrile
[k]GA: glutaraldehyde
[l]MB: Methylene Blue (dye)
[m]MO: Methyl Orange (dye)
[n]AYR: Alizarine Yellow (dye)
[o]DB38: Direct Black 38 (dye)
[p]PS: Ponceau S (dye)
[q]XO: Xylenol Orange (dye)
[r]PES: polyethersulfone
[s]MG: Malachite Green (dye)
[t]IPA: isopropanol
[u]Cu-BTC: copper-1,3,5-benzenetricarboxylic acid

Table 1.
Recent studies of a separation membrane consisting of chitosan.

3.1 Nanofiltration

As a common procedure, many types of chitosan-based membranes are prepared using the casting method. Chitosan membranes prepared by the casting method usually has a highly compacted gel structure since hydrogen bonding derived from a lot of hydroxyl groups. The dense membranes are beneficial for the nanofiltration process separating small substances, for instance, salts, organic acids, or organic dyes [41, 43, 50–54]. For improving the separation ability, combination of organic–inorganic polymeric hybrid membrane is an innovative approach. Metal–organic frameworks (MOFs) were incorporated into the chitosan polymeric matrix to obtain a positively charged membrane surface for cation removal [50]. Montmorillonite clay, which is dispersed uniformly in a porous matrix, enhances chromium removal [53]. Also, carbon nanotubes are combined with the chitosan membrane for improving solution permeability and salt rejection [51, 55].

3.2 Gas separation

The chitosan membranes with dense polymeric structures can also be used for gas separation processes. It is reported that a chitosan/polyvinyl alcohol (PVA)-blended membrane exhibited high air filtration with antibacterial activity [56]. Nowadays, CO_2 separation technology has caught attention due to the increase in concerns related to climate change because of greenhouse gases. Chitosan-based membranes for CO_2 separation membranes are developed to immobilize active sites, such as copper-1,3,5-benzenetricarboxylic acid (Cu-BTC) [57] and hydrotalcite [58].

3.3 Hydrophilicity-based process

Due to a high hydrophilicity derived from abundant hydroxyl groups in glucosamine units, the polymeric membranes consisting of chitosan are suitable for separation between water and organic solvents using pervaporation processes [42, 44, 46, 59]. From its hydrophilicity, the chitosan membrane has also been used as a separation membrane in direct methanol fuel cells (DMFCs) requiring blocking of methanol permeation as well as proton conductivity [60]. Regarding the hydrophilicity of chitosan, removal or blocking of water vapor was tested using a chitosan membrane for a part of a membrane drier apparatus [61] and packaging membrane [62, 63]. As a very recent issue, the interest in biodegradable films for packaging has recently been steadily increasing due to significant concerns on environmental pollution caused by nonbiodegradable packaging materials [64, 65].

4. Adsorption processes

Adsorption is frequently used in separation processes to recover worthy substances as well as water treatment due to its advantages, such as easy operation, high selectivity, and low operational costs [66, 67]. **Tables 2** and **3** summarize the recent studies of adsorption processes, which employed chitosan or chitosan derivatives as either main body of adsorbent or immobilizing media for other adsorbents (mainly fine particles).

As shown in **Figure 1**, chitosan has abundant functional groups. Thereby, chitosan and its derivatives also show adsorbent capabilities with various metal ions and organic substances depending on the pH and the concentration of ionic substances [13]. Hence, their adsorption properties are being widely researched [68]. A large portion of the study employed chitosan and its derivatives for adsorption processes devoted to remove heavy metal from polluted water [11, 13–16, 28, 69–71]. Considering organic compounds, chitosan can adsorb anionic dyes due to amine groups in a chitosan polymer chain [72–74]. It has been reported that nonionic compounds like naphthalene [22] are also adsorbed on chitosan. For improving the adsorption ability, fibrous membranes are fabricated to remove ionic substances or heavy metal ions [12–16, 71]. Fibrous membranes with the adsorption ability are beneficial due to not only large surface area but also high permeation flux. Chitosan gel have shown brilliant abilities as immobilizing media to various adsorbent substances not only to enhance absorption performance but also to enlarge coverage of adsorbates [20–21, 23, 25, 66, 73, 75–79].

Adsorption processes are liberally categorized into two types which are adsorption conducted in a batch process and adsorption in combination with filtration in a flow process.

4.1 Membrane-type absorbents adapting with filtration process

Chitosan have been focused as the media immobilizing various adsorbents since it is able to form stable gel structure through pH responsible [15, 20, 24, 71, 80]. From the advantages in both the adsorption process with selectivity and filtration with continuous operation, the membrane-type adsorbents have been recently developed to adopt the flow process. **Table 2** summarizes the studies of membrane-type adsorbents consisted of chitosan or chitosan derivatives.

4.1.1 Characterization of chitosan membrane immobilizing adsorbent in batch process

For a large portion of such studies, adsorption abilities are evaluated from isothermal adsorption in batch process to reveal adsorption mechanisms [24, 72]. As a typical case of an adsorption membrane immobilizing adsorbent particles, chitosan membrane incorporating Prussian Blue (PB) was developed for cesium

Membrane body	Membrane type	Additional adsorbent	Cross-linker	Target substances	Ref.	Year
chitosan/PVA[a]	porous membrane	CNTs[b]	TEOS [c]	naphthalene	[22]	2015
chitosan/polyester	fibrous membrane	*NA*	GA [d]	Cr(VI)	[14]	2016
chitosan/PET[e]	fibrous membrane	*NA*	GA [d]	Cu(II), Pb(II), Cd(II), Cr(VI)	[13]	2017
chitosan/cellulose	porous membrane	*NA*	GA [d]	Cu(II)	[69]	2018
chitosan	affinity membrane	multilayered molecularly imprinted	TEOS [c]	artemisinin	[47]	2019
chitosan/polyacrylonitrile	fibrous membrane	zirconium MOF [f]	*NA*	Pb(II), Cd(II), Cr(VI)	[15]	
chitosan	dense membrane immobilizing adsorbents	PB [g]	*NA*	Cs	[20]	
chitosan	*NA*	*NA*	glyoxal	RO16[h], MO[i]	[72]	
oxidized chitosan/PVA[a]	fibrous membrane	PHMG [j]	*NA*	Cu(II)	[16]	
chitosan/PVA[a]	dence membrane immobilizing adsorbents	ZIF-8 [k]	*NA*	MG [l]	[24]	2020
octyl-trimethoxysilane modified chitosan	dense membrane	*NA*	TEOS [c]	impurities from extracted artemisinin	[48]	
chitosan/PVA [a]	dence membrane immobilizing adsorbents	ZIF-8 [k]	*NA*	MG [l]	[80]	
chitosan/PVA[a] / PEI[m]	porous membrane	*NA*	*NA*	Cu(II)	[70]	
chitosan/PVDF[n]	fibrous membrane	ZIF-8 [k]	*NA*	BSA[o], Cr(VI)	[71]	
chitosan/polyamide6	fibrous membrane	*NA*	*NA*	Acid Fuchsin dye	[12]	

[a]*PVA: poly(vinyl alcohol)*
[b]*CNTs: carbon nanotubes*
[c]*TEOS: tetraethyl orthosilicate*
[d]*GA: glutaraldehyde*
[e]*PET: polyethylene terephthalate*
[f]*MOF: metal–organic frameworks*
[g]*PB: Prussian Blue (dye used as adsorbent)*
[h]*RO16: Reactive Orange 16 (dye)*
[i]*MO: Methyl Orange (dye)*
[j]*PHMG: poly-hexamethylene guanidine*
[k]*ZIF-8: zeolitic imidazole framework-8*
[l]*MG: Malachite Green (dye)*
[m]*PEI: polyethyleneimine*
[n]*PVDF: polyvinylidene fluoride*
[o]*BSA: bovine serum albumin*

Table 2.
Recent studies of adsorption membrane consisted of chitosan or its derivatives.

removal from the aqueous phase [20]. The maximum adsorption capacity can be evaluated from the equilibrium adsorption amount of absorbate on absorbent (q_e [mol g^{-1}]) and the equilibrium concentration of absorbate in aqueous phase

Matrix body	Additional adsorbent	Cross-linker	Type of adsorbent media	Adsorption process	Target substances	Ref.	Year
chitosan	*NA*	GA [a], EGDE [b]	spherical beads	through packed bed	Cu(II), Pb, Zn(II), *E. coli, S. aureus*	[28]	2016
chitosan/PAM [c]	EDTA [d]	MBA [e]	cylindrical tablets	batch adsorption	Cu(II), Pb(II), Cd(II)	[75]	2017
Rayon fibers coated with chitosan	*NA*	PB[f]	rayon fibers	batch adsorption	Cs	[76]	
chitosan	*NA*	CIT [g], TPP [h], SSA [i], OA[j], ECH [k], GA[a], TTE [l], EGDE [b]	spherical beads	batch adsorption	RB5 [m]	[49]	
chitosan	graphene nanopalates	GA [a]	spherical beads	batch adsorption	MO[n], AR1 [o]	[66]	2018
chitosan	PDMAEMA [p]	*NA*	magnetic spherical beads	batch adsorption	AG25 [q], RB19[r]	[73]	
chitosan	CA-CD [s]	EDC [t], NHS[u]	particles	through packed column	RB49 [v]	[77]	
chitosan	CB[8] [w]	*NA*	powder	batch adsorption	Pb, RO5 [x], AB25 [y], RY145 [z]	[84]	2019
chitosan	bentnite, cobalt oxide	*NA*	powder	batch adsorption	CR [A], Cr(VI)	[21]	
N-allylthiourea chitosan	*NA*	*NA*	powder	batch adsorption	AS(III)[B]	[74]	
chitosan	Fe_3O_4	*NA*	magnetic spherical beads	batch adsorption	AB [C]	[78]	
chitosan	CNTs[D]	*NA*	particles	batch adsorption	phenol	[23]	
chitosan	EDTA [d]-silane/mGO [E]	GA [a]	magnetic spherical beads	batch adsorption	Pb(II), Cd(II)	[25]	
oleoyl chitosan	*NA*	*NA*	nanoparticles	batch adsorption	Fe(II)	[11]	

DOI: http://dx.doi.org/10.5772/intechopen.95839

Matrix body	Additional adsorbent	Cross-linker	Type of adsorbent media	Adsorption process	Target substances	Ref.	Year
chitosan	DES [F] (choline chloride + urea or chline chloride + glycerol)	*NA*	spherical beads	batch adsorption	MG [G]	[79]	2020

[a]GA: glutaraldehyde
[b]EGDE: ethylene glycol diglycidylether
[c]PAM: polyacrylamide
[d]EDTA: ethylenediaminetetra-acetic acid
[e]MBA: N,N-methylenebis(acrylamide)
[f]PB: Prussian Blue (dye used as adsorbent)
[g]CIT: trisodium citrate
[h]TPP: tripolyphosphate
[i]SSA: sulfosuccinic acid
[j]OA: oxalic acid
[k]ECH: epichlorohydrin
[l]TTE: trimethylpropane triglycidyl ether
[m]RB5: Reactive Black 5 (dye)
[n]MO: Methyl Orange (dye)
[o]AR1: Acid Red 1 (dye)
[p]PDMAEMA: poly(2-(dimethylamino)ethyl meth&$$$;
[q]AG25: Acid Green 25 (dye)
[r]RB19: Reactive Blue (dye)
[s]CA-CD: citric acid modified β-cyclodextrin
[t]EDC: 3-(3-dimethylaminopropyl)-1-ethylcarbodiimide hydrochloride
[u]NHS: N-hydroxysuccinimide
[v]RB49: Reactive Blue (dye)
[w]CB[8]: cucurbit [8] uril
[x]RO5: Reactive Orange 5 (dye)
[y]AB25: Acidic Blue 25 (dye)
[z]RY145: Reactive Yellow 145 (dye)
[A]CR: Congo Red (dye)
[B]AS(III): Arsenazo III (dye)
[C]AB: Acid Blue (dye)
[D]CNTs: carbon nanotubes
[E]mGO: magnetic graphene oxide
[F]DES: deep eutectic solvents
[G]MG: Malachite Green (dye)

Table 3.
Recent studies of adsorption processes using chitosan or its derivatives (without membrane-type processes).

(C_e [mol g^{-1}]) using the Langmuir adsorption isotherm (Eq. 5). The Langmuir isotherm explains both monolayer and homogeneous adsorption.

$$q_e = \frac{Q_{max}\, K\, C_e}{1 + K\, C_e} \tag{5}$$

Where, Q_{max} and K are the maximum adsorption capacity [mol g^{-1}] and the equilibrium adsorption constant [L mol^{-1}], respectively. **Figure 7a** displays the effect of the mass fraction of immobilized PB in chitosan membrane (MF_{PB}) on the maximum adsorption capacity for the membrane and for the PB immobilized in membranes. Immobilization in the chitosan membrane achieved to improve cesium adsorption without inhibition of the adsorption ability of PB.

The diffusivity of adsorbate molecules in adsorbent media strongly influences the adsorption rate [81]. The effective diffusion coefficient (D_{eff} [m^2 s^{-1}]) of cesium ions in the chitosan membrane immobilizing PB was determined according to the mass transfer theory. **Figure 7b** depicts the effect of MF_{PB} on the D_{eff} of cesium ions in the initial period of isothermal adsorption. The obtained values of D_{eff} were lower than that of the diffusion coefficient in the bulk aqueous phase, which was previously reported as 2.17×10^{-9} m^2 s^{-1} [82]. The structure of the membrane consisting of a chitosan polymer chain and PB were observed to suppress the diffusion of cesium ions into the membrane. The diffusion of cesium ions was inhibited considerably by the immobilized PB, which became dominant in comparison to the mass transfer resistance by the chitosan polymer chain [20].

4.1.2 Adsorbed separation in flow process

The adsorbed separation in flowing process through the membrane immobilizing adsorbents is more efficiently than equilibrium adsorption in flask. Owing to the adsorption ability collaborated with molecular size screening ability, selective separation is exhibited higher than without adsorbent system. Moreover, adsorption capability of adsorbents is fully utilized in the adsorption in combination with filtration in a flow process whereas the batch adsorption is up to equilibrium by the decrease of solutes concentration in liquid phase [83].

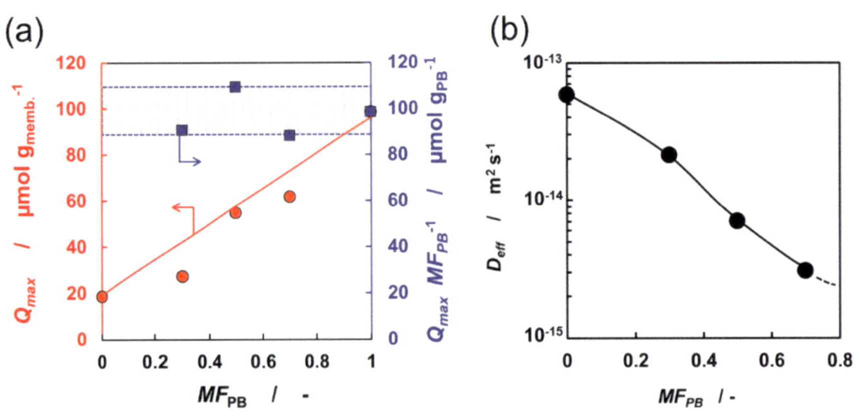

Figure 7.
Effect of the mass fraction of PB on (a) the maximum adsorption capacity for the chitosan membrane (left axis) and for the PB immobilized in the membrane (right axis). (b) the effective diffusion coefficient of cesium ions in the initial period of isothermal adsorption (25°C). Reprinted from Fujisaki, Kashima, Hagiri, Imai [20] with permission from WILEY-VCH.

Li and co-workers have prepared chitosan fibrous membrane and investigated the dynamic flow adsorption using filtration apparatus for removal of Cr (VI) from aqueous phase [14]. The study revealed the flow adsorption is advantageous than static batch adsorption for small concentration of feed Cr (VI). In addition, the flow rate of feed solution during the flow process directly influenced on the adsorption behavior. Lower flow rate increases the adsorbed amount of Cr (VI) nearly up to maximum adsorption capacity since the contact time between adsorbent and adsorbate increases.

Separation performance is strongly depended on the membrane structure as well as selection of absorbent. Khajavian and co-workers have demonstrated the removal of Malachite Green known as a cationic dye *via* flow process using chitosan/poly (vinyl alcohol) incorporating metal–organic frameworks (ZIF-8) [80]. The physical factors including membrane thickness and porous structure generated by pore generator (polyethylene glycol) inside membrane regulate water flux and dye rejection. The chitosan membrane adopting flow processes should be optimized for each system and target substances.

4.2 Other types of absorbent media composed of chitosan

Table 3 shows the recent studies of adsorption processes using chitosan or its derivatives. Here, the adsorption studies in membrane-type process, which were already showed in **Table 2** are eliminated.

A spherical bead is the basis of the adsorbent media. Chitosan beads can be obtained easily *via* droplet method which a small portion of chitosan dissolved in acid solution is dropped to alkaline solution for neutralization. Bead-type absorbents can be applied to packed bed adsorption operated in a flow process [28].

Adsorbents composed of chitosan and its derivatives incorporated with inorganic substances have been developed to enhance adsorption performance [21, 75, 78, 84]. In addition, it is a notable trend that the studies about the removal of organic dye from aqueous are clearly increasing.

As a very recent approach, Sadiq and coworkers developed chitosan beads incorporating with deep eutectic solvents (DES) which are prepared from choline chloride with urea or glycerol for removing organic dye [79]. The DES is paid attention to possess biocompatibility and low toxicity regardless of similar characteristics with ionic liquids [79, 85, 86].

4.3 Highly porous aerogels

Highly porous media from chitosan gel which have large surface area contributing high adsorption ability have been developed recently [87–91]. Such highly porous media are frequently called aerogel. Chitosan aerogels can be prepared *via* freeze-drying of chitosan aqueous solution and stabilization using cross-linker, such as glutaraldehyde (GA). Yi and coworkers reported chitosan aerogel silylated with methyl trimethoxysilane which has spring-like structure leading to high oil absorption [89]. Yu and coworkers revealed preparation of chitosan aerogel incorporating graphene oxide and montmorillonite without cross-linker exhibited efficiently Cr (VI) adsorption [88].

Chitosan aerogel prepared with GA has highly porous structure showed by a field emission scanning electron microscopy (FE-SEM) (**Figure 8a**). The chitosan aerogel prepared with GA adsorbed anionic dye (Methyl Orange: 327 Da) rapidly although cationic dye of similar molecular weight (Methylene Blue: 320 Da) is not removed from their aqueous solutions (**Figure 8b**). Selective removal of anionic substrates in high adsorption rate can be achieved by static attraction derived

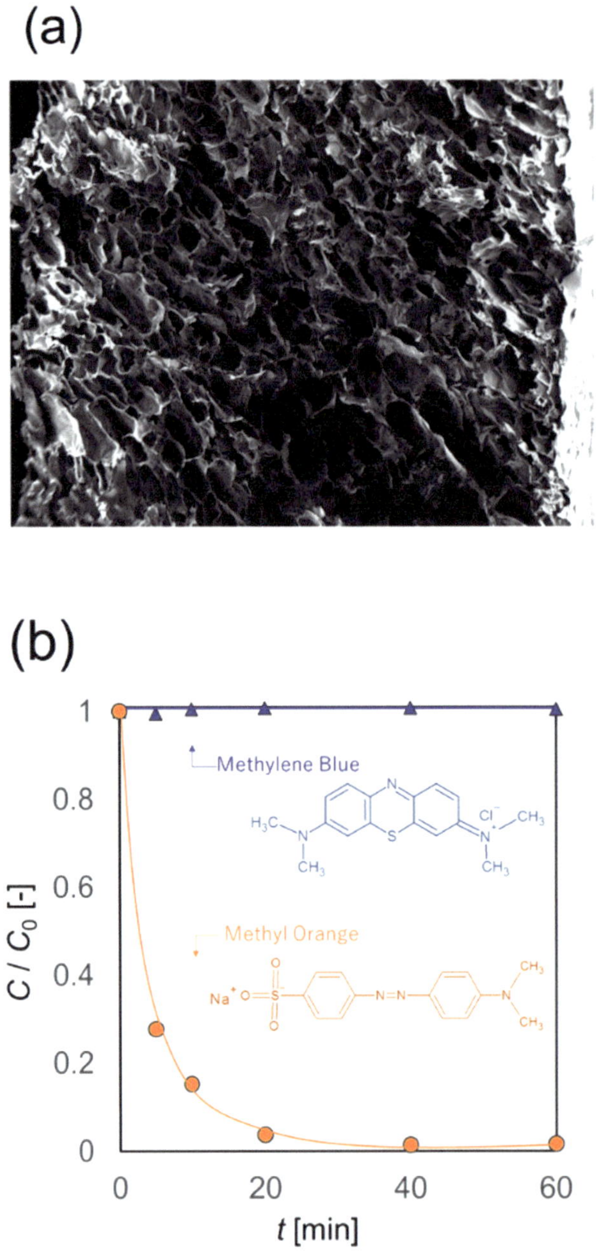

Figure 8.
Chitosan aerogel prepared with glutaraldehyde as a cross-linker. (a) Morphology observed by FE-SEM. (b) Isothermal adsorption of anionic dye (methyl Orange) and cationic dye (methylene blue) onto chitosan aerogel.

abundant amine and sufficient surface area. Moreover, chitosan aerogel can immobilize various functional particles hence it is expected to develop as new adsorbent media [88, 90].

5. Chitosan nanofiber

Recently, along with the expanding area of nanotechnologies, the area of nanofibers has been gaining interest [92]. Nanofibers with diameter in the range from several micrometers down to tens of nanometers have useful properties such as high specific surface area, porosity, as well as biocompatibility [93, 94]. Nanofibers made from chitosan have not yet been fully established as compared with other nanofibers [95–97]. Chitosan nanofibers have diameter around 10 nm and amino groups on its surface with positive charge [98]. Accordingly, chitosan nanofibers possess unique characteristics and advantages that other nanofiber does not have, that has been expected to be used in various sorts of industrial fields, for instance, filtrations, recovery of metal ions, adsorption of proteins, drug release, enzyme carriers, wound healing, cosmetics, and biosensors have been developed [99–106].

In particular, chitosan nanofiber can be expected as an alternative to air filter media (**Figure 9**). The particle collection performance across chitosan nanofiber media decreased with increase in the amount of chitosan nanofibers on the polyethylene terephthalate (PET) non–woven, as shown in **Figure 10**. In addition, it was

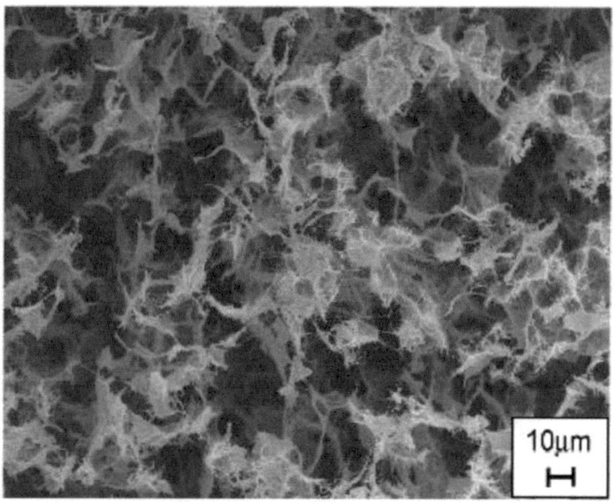

Figure 9.
SEM photograph of chitosan nanofibers filter media.

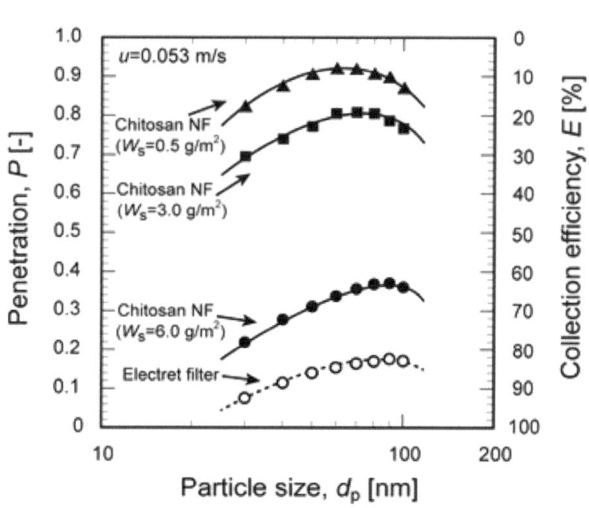

Figure 10.
Comparison of the fractional penetrations of various sorts of test filter media.

found that the particle collection performance of filter media with high weight of chitosan nanofibers (Weight of sheet (W_s) = 6.0 g/m^2) is compatible to commercial electret filter media.

6. Conclusion

This chapter described the promising development of utilizing chitosan and its derivatives with the focus on separation processes, and overviewed principal investigations. Reactive molecular chains of chitosan derived from hydroxyl and amino groups serve to form an attractive polymeric gel structure. In the filtration process, dense chitosan membranes separate small molecules of solvents, organic dyes, and toxic ions. The deacetylation degree is found to be the most notable factor for water permeation through membranes and its antibacterial ability. The complexity of chitosan-immobilizing systems with functional materials are found to be available in water purification processes. In recent years, adsorption combined with filtration is being developed for various types of water treatment. In this new type process, both the equilibrium adsorption capability and diffusivity in the absorbent matrix should be mentioned as describe in Section 4.1. Due to the easy operation, high selectivity, and low operational costs, adsorption process with chitosan gel is widely studied. As advanced separation media, highly porous chitosan aerogels and nanofibers, which possess large surface area that contribute to improved separation ability, have been developed recently.

Great demands of a novel biocompatible material and environmentally-friendly processes will continue to increase in future. Innovative separation technology utilizing chitosan obtained from bioresources is promising. The multifaceted development of chitosan-based system will lead to environmentally-friendly processes.

Acknowledgements

The authors sincerely appreciate the financial support from Mukai Science and Technology Foundation (Japan). Prussian Blue used in this study was kindly provided from Dainichiseika Color & Chemicals Mfg. Co., Ltd. (Japan). Authors are also grateful for laboratory students in National Institute of Technology, Oyama College and their great efforts on the studies presented in this chapter: Mr. Tomoyuki Fujisaki for the study on chitosan membrane immobilizing Prussian Blue particles and Mr. Haruki Koya for preparation of chitosan aerogel.

The authors sincerely thank Prof. Norikazu Namiki of Kogakuin University, who provide the experimental set up for evaluating particle collection performance of test filter media. Ms. Yuki Suto of laboratory student in Kogakuin University, who for the study on chitosan nanofibers filter media.

Author details

Keita Kashima[1,2]*, Tomoki Takahashi[3], Ryo-ichi Nakayama[2,4] and Masanao Imai[2]

1 Department of Materials Chemistry and Bioengineering, National Institute of Technology, Oyama College (Oyama KOSEN), Oyama, Tochigi, Japan

2 Course in Bioresource Utilization Sciences, Graduate School of Bioresource Sciences, Nihon University, Fujisawa, Kanagawa, Japan

3 Department of Liberal Arts and Basic Sciences, College of Industrial Technology, Nihon University, Narashino, Chiba, Japan

4 Department of Environmental Chemistry and Chemical Engineering, School of Advanced Engineering, Kogakukuin University, Hachioji, Tokyo, Japan

*Address all correspondence to: keitakashima@oyama-ct.ac.jp

References

[1] Quirino JP, Alejandro FM, Bissember AC. Towards cleaner downstream processing of biomass waste chemical products by liquid chromatography: A review and recommendations, Journal of Cleaner Production. 2532020:119937. DOI: 10.1016/j.jclepro.2019.119937

[2] Cuellar MC, Straathof AJJ. Downstream of the bioreactor: advancements in recovering fuels and commodity chemicals, Current Opinion in Biotechnology. 62;2020:189-195. DOI: 10.1016/j.copbio.2019.11.012

[3] Moreira VR, Lebron YAR, Santos LVS, Coutinho de Paula E, Amaral MCS. Arsenic contamination, effects and remediation techniques: A special look onto membrane separation processes, Process Safety and Environmental Protection. 148;2021:604-623. DOI: 10.1016/j.psep.2020.11.033

[4] Johnson DJ, Hilal N. Can graphene and graphene oxide materials revolutionise desalination processes?, Desalination. DOI: 10.1016/j.desal.2020.114852

[5] Chiaramonte de Castro BJ, Sartim R, Guerra VG, Aguiar ML. Hybrid air filters: A review of the main equipment configurations and results. Process Safety and Environmental Protection. 144;2020:193-207. DOI: 10.1016/j.psep.2020.07.025

[6] İlk S, Ramanauskaitė A, Bilican BK, Mulerčikas P, Çam D, Onses MS, Torun I, Kazlauskaitė S, Baublys V, Aydın Ö, Zang L-S, Kaya M. Usage of natural chitosan membrane obtained from insect corneal lenses as a drug carrier and its potential for point of care tests, Materials Science and Engineering: C. 112;2020:110897. DOI: 10.1016/j.msec.2020.110897

[7] Menezes JESA, dos Santos HS, Ferreira MKA, Magalhães FEA, da Silva DS, Bandeira PN, Saraiva GD, Pessoa ODL, Ricardo NMPS, Cruz BG, Teixeira AMR. Preparation, structural and spectroscopic characterization of chitosan membranes containing allantoin, Journal of Molecular Structure. 1199;2020:126968. DOI: 10.1016/j.molstruc.2019.126968

[8] Kashima K, Nomoto R, Imai M. Novel application of oceanic biopolymers -strategic regulation of polymer characteristics for membrane technology in separation engineering. In: Perveen FK, editor. Recent Advances in Biopolymers. Croatia: InTech; 2016. p. 189–220. DOI: 10.5772/61998

[9] Nomoto R, Imai M. Dominant role of acid–base neutralization process in forming chitosan membranes for regulating mechanical strength and mass transfer characteristics, Journal of Chitin and Chitosan Science. 2;2014: 197–204. DOI: 10.1166/jcc.2014.1055

[10] Sun J, Yendluri R, Liu K, Guo Y, Lvov Y, Yan X. Enzyme-immobilized clay nanotube–chitosan membranes with sustainable biocatalytic activities, Physical Chemistry Chemical Physics. 19;2017:562-567. DOI: 10.1039/C6CP07450B

[11] Jaber N, Aiedeh K. Sorption behavior and release kinetics of iron (II) ions by oleoyl chitosan polymeric nanoparticles, Journal of Drug Delivery Science and Technology. 54;2019:101354. DOI: 10.1016/j.jddst.2019.101354

[12] Jalalian N, Nabavi SR. Electrosprayed chitosan nanoparticles decorated on polyamide6 electrospun nanofibers as membrane for Acid Fuchsin dye filtration from water, Surfaces and Interfaces. 21;2020:100779. DOI: 10.1016/j.surfin.2020.100779

[13] Li L, Zhang J, Li Y, Yang C. Removal of Cr (VI) with a spiral wound chitosan

nanofiber membrane module via dead-end filtration, Journal of Membrane Science. 544;2017:333-341. DOI: 10.1016/j.memsci.2017.09.045

[14] Li L, Li Y, Yang C. Chemical filtration of Cr (VI) with electrospun chitosan nanofiber membranes, Carbohydrate Polymers, 140;2016:299-307. DOI: 10.1016/j.carbpol.2015.12.067

[15] Jamshidifard S, Koushkbaghi S, Hosseini S, Rezaei S, Karamipour A, Jafari rad A, Irani M. Incorporation of UiO-66-NH_2 MOF into the PAN/chitosan nanofibers for adsorption and membrane filtration of Pb(II), Cd(II) and Cr(VI) ions from aqueous solutions, Journal of Hazardous Materials, 368; 2019:10-20. DOI: 10.1016/j.jhazmat.2019.01.024

[16] Chen S, Li C, Hou T, Cai Y, Liang L, Chen L, Li M. Polyhexamethylene guanidine functionalized chitosan nanofiber membrane with superior adsorption and antibacterial performances, Reactive and Functional Polymers. 145;2019:104379. DOI: 10.1016/j.reactfunctpolym.2019.104379

[17] Takahashi T, Imai M, Suzuki I. Water permeability of chitosan membrane involved in deacetylation degree control, Biochemical Engineering Journal, 36;2007:43-48. DOI: 10.1016/j.bej.2006.06.014

[18] Takahashi T, Imai M, Suzuki I. Cellular structure in an *N*-acetyl-chitosan membrane regulate water permeability, Biochemical Engineering Journal. 42;2008:20-27. DOI: 10.1016/j.bej.2008.05.013

[19] Takahashi T, Imai M, Suzuki I, Sawai J. Growth inhibitory effect on bacteria of chitosan membrane regulated with deacetylation degree, Biochemical Engineering Journal. 40; 2008:485-491. 10.1016/j.bej.2008.02.009

[20] Fujisaki T, Kashima K, Hagiri M, Imai M. Isothermal Adsorption Behavior of Cesium Ions in a Novel Chitosan-Prussian Blue-Based Membrane, Chemical Engineering & Technology. 42;2019:910-917. DOI: 10.1002/ceat.201800603

[21] Abukhadra MR, Adlii A, Bakry BM. Green fabrication of bentonite/chitosan@cobalt oxide composite (BE/CH@Co) of enhanced adsorption and advanced oxidation removal of Congo red dye and Cr (VI) from water, International Journal of Biological Macromolecules. 126;2019:402-413. DOI: 10.1016/j.ijbiomac.2018.12.225

[22] Bibi S, Yasin T, Hassan S, Riaz M, Nawaz M. Chitosan/CNTs green nanocomposite membrane: Synthesis, swelling and polyaromatic hydrocarbons removal, Materials Science and Engineering: C. 46;2015: 359-365. DOI: 10.1016/j.msec.2014.10.057

[23] Alves DCS, Gonçalves JO, Coseglio BB, Burgo TAL, Dotto GL, Pinto LAA, Cadaval Jr. TRS. Adsorption of phenol onto chitosan hydrogel scaffold modified with carbon nanotubes, Journal of Environmental Chemical Engineering. 7;2019:103460. DOI: 10.1016/j.jece.2019.103460

[24] Khajavian M, Salehi E, Vatanpour V. Chitosan/polyvinyl alcohol thin membrane adsorbents modified with zeolitic imidazolate framework (ZIF-8) nanostructures: Batch adsorption and optimization. Separation and Purification Technology, 241;2020: 116759. DOI: 10.1016/j.seppur.2020.116759

[25] Shahbazi A, Marnani NN, Salahshoor Z. Synergistic and antagonistic effects in simultaneous adsorption of Pb(II) and Cd(II) from aqueous solutions onto chitosan functionalized EDTA-silane/mGO, Biocatalysis and Agricultural

Biotechnology. 22;2019:101398. DOI: 10.1016/j.bcab.2019.101398

[26] Liu C, Wang G, Sui W, An L, Si C. Preparation and characterization of chitosan by a novel deacetylation approach using glycerol as green reaction solvent, ACS Sustainable Chemistry and Engineering. 5;2017: 4690-4698. DOI: 10.1021/acssuschemeng.7b00050

[27] Younes I, Rinaudo M. Chitin and chitosan preparation from marine sources. structure, properties and applications, Marine Drugs. 13;2015: 1133-1174. DOI: 10.3390/md13031133

[28] Tayel AA, El-Tras WF, Elguindy NM. The potentiality of cross-linked fungal chitosan to control water contamination through bioactive filtration, International Journal of Biological Macromolecules. 88;2016:59-65. DOI: 10.1016/j.ijbiomac.2016.03.018

[29] Hülsey MJ. Shell biorefinery: A comprehensive introduction, Green Energy & Environment. 3;2018:318-327. DOI: 10.1016/j.gee.2018.07.007

[30] Claesson PM, Ninham BW. pH-dependent interactions between adsorbed chitosan layers, Langmuir. 8; 1992:1406-1412. DOI: 10.1021/la00041a027

[31] Tsai C-C, Morrow BH, Chen W, Payne GF, Shen J. Toward understanding the environmental control of hydrogel film properties: How salt modulates the flexibility of chitosan chains, Macromolecules. 50;2017:5946–5952. DOI: 10.1021/acs.macromol.7b01116

[32] Kim MY, Kim J. Chitosan microgels embedded with catalase nanozyme-loaded mesocellular silica foam for glucose-responsive drug delivery, ACS Biomaterials Science and Engineering 3; 2017:572–578. DOI: 10.1021/acsbiomaterials.6b00716

[33] Aswathy NR, Palai AK, Mohanty S, Nayak SK. Freestanding electrically conducting flexible membranes based on novel chitosan/PANI/rGO nanocomposites, Materials Letters. 259; 2020:126777. DOI: 10.1016/j.matlet.2019.126777

[34] Kina K, Tamura K, Ishibashi N, Use of 8-anilino-1-naphthalene sulfonate as a fluorescent indicator in colloid titration, Japan Analist. 23;1974:1082-1083 (in Japanese). DOI: 10.2116/bunsekikagaku.23.1082

[35] Hattoti T, Colloidal Titration of Chitosan, Chitin and Chitosan Research. 5;1999:14-18 (in Japanese).

[36] Takahashi T, Imai M, Suzuki I, High-potential molecular properties of chitosan and reaction conditions for removal of *p*-quinone from the aqueous phase, Biochemical Engineering Journal. 25; 2005:7-13. DOI: 10.1016/j.bej.2005.02.017

[37] Takahashi T, Yamada A, Shono A, Otake K, Imai M. Gelation behavior with acetylation of chitosan for membrane preparation, Desalination and Water Treatment, 17;2010:150–154. DOI: 10.5004/dwt.2010.1711

[38] Chen RH, Hwa H-D. Effect of molecular weight of chitosan with the same degree of deacetylation on the thermal, mechanical, and permeability properties of the prepared membrane, Carbohydrate Polymers. 29;1996:353-358. DOI: 10.1016/S0144-8617(96)00007-0

[39] Rinaudo M. Chitin and chitosan: Properties and applications, Progress in Polymer Science. 31;2006:603-632. DOI: 10.1016/j.progpolymsci.2006.06.001

[40] Brugnerotto J, Desbrières J, Roberts G, Rinaudo M. Characterization of chitosan by steric exclusion chromatography, Polymer. 42;2001: 09921-09927. DOI: 10.1016/S0032-3861(01)00557-2

[41] Song Y, Hu Q, Li T, Sun Y, Chen X, Fan J. Fabrication and characterization of phosphorylated chitosan nanofiltration membranes with tunable surface charges and improved selectivities, Chemical Engineering Journal. 352;2018:163-172. DOI: 10.1016/j.cej.2018.07.010

[42] Dudek G, Turczyn R, Konieczny K. Robust poly(vinyl alcohol) membranes containing chitosan/chitosan derivatives microparticles for pervaporative dehydration of ethanol, Separation and Purification Technology. 234;2020: 116094. DOI: 10.1016/j.seppur.2019.116094

[43] Song Y, Sun Y, Chen M, Huang P, Li T, Zhang X, Jiang K. Efficient removal and fouling-resistant of anionic dyes by nanofiltration membrane with phosphorylated chitosan modified graphene oxide nanosheets incorporated selective layer, Journal of Water Process Engineering. 34;2020:101086. DOI: 10.1016/j.jwpe.2019.101086

[44] Du JR, Hsu LH, Xiao ES, Guo X, Zhang Y, Feng X. Using genipin as a “green” crosslinker to fabricate chitosan membranes for pervaporative dehydration of isopropanol, Separation and Purification Technology. 244; 2020: 116843. DOI: 10.1016/j.seppur.2020.116843

[45] Ma H-F, Meng G, Cui B-K, Si J, Dai Y-C. Chitosan crosslinked with genipin as supporting matrix for biodegradation of synthetic dyes: Laccase immobilization and characterization. Chemical Engineering Research and Design. 132;2018:664-676. DOI: 10.1016/j.cherd.2018.02.008

[46] Moulik S, Vani B, Chandrasekhar SS, Sridhar S. Chitosan-polytetrafluoroethylene composite membranes for separation of methanol and toluene by pervaporation, Carbohydrate Polymers. 193;2018:28-38. DOI: 10.1016/j.carbpol.2018.03.069

[47] Zhang Y, Tan X, Liu X, Li C, Zeng S, Wang H, Zhang S. Fabrication of Multilayered Molecularly Imprinted Membrane for Selective Recognition and Separation of Artemisinin, ACS Sustainable Chemistry and Engineering. 7;2019:3127–3137. DOI: 10.1021/acssuschemeng.8b04908

[48] Cao Y, Zhang Y, Zhang Y, Wang L, Lv L, Ma X, Zeng S, Wang H. Biodegradable functional chitosan membrane for enhancement of artemisinin purification, Carbohydrate Polymers. 246;2020:116590. DOI: 10.1016/j.carbpol.2020.116590

[49] Jóźwiak T, Filipkowska U, Szymczyk P, Rodziewicz J, Mielcarek A. Effect of ionic and covalent crosslinking agents on properties of chitosan beads and sorption effectiveness of Reactive Black 5 dye, Reactive and Functional Polymers. 114;2017:58-74. DOI: 10.1016/j.reactfunctpolym.2017.03.007

[50] Ma X-H, Yang Z, Yao Z-K, Xu Z-L, Tang CY. A facile preparation of novel positively charged MOF/chitosan nanofiltration membranes, Journal of Membrane Science. 525;2017:269-276. DOI: 10.1016/j.memsci.2016.11.015

[51] Alshahrani AA, Al-Zoubi H, Nghiem LD, M in het Panhuis. Synthesis and characterisation of MWNT/chitosan and MWNT/chitosan-crosslinked buckypaper membranes for desalination, Desalination. 418;2017:60-70. DOI: 10.1016/j.desal.2017.05.031

[52] Ekambaram K, Doraisamy M. Fouling resistant PVDF/Carboxymethyl chitosan composite nanofiltration membranes for humic acid removal, Carbohydrate Polymers. 173;2017:431-440. DOI: 10.1016/j.carbpol.2017.06.017

[53] Sangeetha K, Angelin Vinodhini P, Sudha PN, Alsharani FA, Sukumaran A. Novel chitosan based thin sheet nanofiltration membrane for rejection of heavy metal chromium, International

Journal of Biological Macromolecules. 132;2019:939-953. 10.1016/j.ijbiomac.2019.03.244

[54] Mousavi SR, Asghari M, Mahmoodi NM. Chitosan-wrapped multiwalled carbon nanotube as filler within PEBA thin film nanocomposite (TFN) membrane to improve dye removal, Carbohydrate Polymers. 237;2020:116128. DOI: 10.1016/j.carbpol.2020.116128

[55] Alshahrani AA, Algamdi MS, Alsohaimi IH, Nghiem LD, Tu KL, Al-Rawajfehe AE, in het Panhuis M. The rejection of mono- and di-valent ions from aquatic environment by MWNT/chitosan buckypaper composite membranes: Influences of chitosan concentrations, Separation and Purification Technology. 234;2020:116088. DOI: 10.1016/j.seppur.2019.116088

[56] Wang Z, Yan F, Pei H, Li J, Cui Z, He B. Antibacterial and environmentally friendly chitosan/polyvinyl alcohol blend membranes for air filtration, Carbohydrate Polymers. 198;2018:241-248. DOI: 10.1016/j.carbpol.2018.06.090

[57] Jiamjirangkul P, Inprasit T, Intasanta V, Pangon A. Metal organic framework-integrated chitosan/poly (vinyl alcohol) (PVA) nanofibrous membrane hybrids from green process for selective CO_2 capture and filtration, Chemical Engineering Science. 221;2020:115650. DOI: 10.1016/j.ces.2020.115650

[58] Borgohain R, Mandal B. Thermally stable and moisture responsive carboxymethyl chitosan/dendrimer/hydrotalcite membrane for CO_2 separation, Journal of Membrane Science. 608;2020:118214. DOI: 10.1016/j.memsci.2020.118214

[59] Silvestre WP , Baldasso C, Tessaro IC. Potential of chitosan-based membranes for the separation of essential oil components by target-organophilic pervaporation, Carbohydrate Polymers, 247;2020: 116676. 10.1016/j.carbpol.2020.116676

[60] Wu Q, Wang H, Lu S, Xu X, Liang D, Xiang Y. Novel methanol-blocking proton exchange membrane achieved via self-anchoring phosphotungstic acid into chitosan membrane with submicro-pores, Journal of Membrane Science. 500;2016:203-210. DOI: 10.1016/j.memsci.2015.11.019

[61] Soontarapa K, Arnusan J. Dehydration of paddy rice in a chitosan membrane drier, Separation and Purification Technology. 209;2019:401-408. DOI: 10.1016/j.seppur.2018.07.048

[62] Ye Y, Zeng F, Zhang M, Zheng S, Li J, Fei P. Hydrophobic edible composite packaging membrane based on low-methoxyl pectin/chitosan: Effects of lotus leaf cutin, Food Packaging and Shelf Life. 26:2020:100592. DOI: 10.1016/j.fpsl.2020.100592

[63] Gabriele F, Donnadio A, Casciola M, Germani R, Spreti N. Ionic and covalent crosslinking in chitosan-succinic acid membranes: Effect on physicochemical properties, Carbohydrate Polymers. 251;2021:117106. DOI: 10.1016/j.carbpol.2020.117106

[64] Sun X, Wang Z, Kadouh H, Zhou K. The antimicrobial, mechanical, physical and structural properties of chitosan–gallic acid films, LWT - Food Science and Technology. 57;2014:83-89. DOI: 10.1016/j.lwt.2013.11.037

[65] Mujtaba M, Morsi RE, Kerch G, Elsabee MZ, Kaya M, Labidi J, Khawar KM. Current advancements in chitosan-based film production for food technology; A review, International Journal of Biological Macromolecules. 121;2019:889-904. DOI: 10.1016/j.ijbiomac.2018.10.109

[66] Zhang C, Chen Z, Guo W, Zhu C, Zou Y. Simple fabrication of Chitosan/Graphene nanoplates composite spheres for efficient adsorption of acid dyes from aqueous solution, International Journal of Biological Macromolecules. 112;2018:1048-1054. DOI: 10.1016/j.ijbiomac.2018.02.074

[67] Bulut Y, Aydın H. A kinetics and thermodynamics study of methylene blue adsorption on wheat shells, Desalination. 194;2006:259–267. DOI: 10.1016/j.desal.2005.10.032

[68] Upadhyay U, Sreedhar I, Singh SA, Patel CM, Anitha KL. Recent advances in heavy metal removal by chitosan based adsorbents, Carbohydrate Polymers. 251;2021:117000. DOI: 10.1016/j.carbpol.2020.117000

[69] Urbina L, Guaresti O, Requies J, Gabilondo N, Eceiza A, Corcuera MA, Retegi A. Design of reusable novel membranes based on bacterial cellulose and chitosan for the filtration of copper in wastewaters, Carbohydrate Polymers. 193;2018:362-372. DOI: 10.1016/j.carbpol.2018.04.007

[70] Sahebjamee N, Soltanieh M, Mousavi SM, Heydarinasab A. Preparation and characterization of porous chitosan–based membrane with enhanced copper ion adsorption performance, Reactive and Functional Polymers. 154;2020:104681. DOI: 10.1016/j.reactfunctpolym.2020.104681

[71] Pishnamazi M, Koushkbaghi S, Hosseini SS, Darabi M, Yousefi A, Irani M. Metal organic framework nanoparticles loaded- PVDF/chitosan nanofibrous ultrafiltration membranes for the removal of BSA protein and Cr (VI) ions, Journal of Molecular Liquids. 317;2020:113934. DOI: 10.1016/j.molliq.2020.113934

[72] Jawad AH, Norrahma SSA, Hameed BH, Ismail K. Chitosan-glyoxal film as a superior adsorbent for two structurally different reactive and acid dyes: Adsorption and mechanism study, International Journal of Biological Macromolecules. 135;2019:569-581. DOI: 10.1016/j.ijbiomac.2019.05.127

[73] Xu B, Zheng H, Zhou H, Wang Y, Luo K, Zhao C, Peng Y, Zheng X. Adsorptive removal of anionic dyes by chitosan-based magnetic microspheres with pH-responsive properties, Journal of Molecular Liquids. 256;2018:424-432. DOI: 10.1016/j.molliq.2018.02.061

[74] Bondock S, El-Zahhar AA, Alghamdi MM, Keshk SMAS. Synthesis and evaluation of N-allylthiourea-modified chitosan for adsorptive removal of arsenazo III dye from aqueous solutions, International Journal of Biological Macromolecules. 137;2019: 107-118. DOI: 10.1016/j.ijbiomac.2019.06.193

[75] Ma J, Zhou G, Chu L, Liu Y, Liu C, Luo S, Wei Y. Efficient removal of heavy metal ions with an EDTA functionalized chitosan/polyacrylamide double network hydrogel, ACS Sustainable Chemistry and Engineering. 5;2017:843–851. DOI: 10.1021/acssuschemeng.6b02181

[76] Dechojarassri D, Asaina S, Omote S, Nishida K, Furuike T, Tamura, H. Adsorption and desorption behaviors of cesium on rayon fibers coated with chitosan immobilized with Prussian blue, International Journal of Biological Macromolecules. 104;2017:1509-1516. DOI: 10.1016/j.ijbiomac.2017.03.056

[77] Zhao J, Zou Z, Ren R, Sui X, Mao Z, Xu H, Zhong Y, Zhang L, Wang B. Chitosan adsorbent reinforced with citric acid modified ⊠-cyclodextrin for highly efficient removal of dyes from reactive dyeing effluents, European Polymer Journal. 108;2018:212-218. DOI: 10.1016/j.eurpolymj.2018.08.044

[78] Xu P, Zheng M, Chen N, Wu Z, Xu N, Tang J, Teng Z. Uniform magnetic

chitosan microspheres with radially oriented channels by electrostatic droplets method for efficient removal of Acid Blue, Journal of the Taiwan Institute of Chemical Engineers. 104; 2019:210-218. DOI: 10.1016/j.jtice.2019.09.016

[79] Sadiq AC, Rahim NY, Suah FBM. Adsorption and desorption of malachite green by using chitosan-deep eutectic solvents beads, International Journal of Biological Macromolecules. 164;2020: 3965-3973. DOI: 10.1016/j.ijbiomac.2020.09.029

[80] Khajavian M, Salehi E, Vatanpour V. Nanofiltration of dye solution using chitosan/poly(vinyl alcohol)/ZIF-8 thin film composite adsorptive membranes with PVDF membrane beneath as support, Carbohydrate Polymers. 247; 2020:116693. DOI: 10.1016/j.carbpol.2020.116693

[81] Mihara Y, Sikder MT, Yamagishi H, Sasaki T, Kurasaki M, Itoh S, Tanaka S. Adsorption kinetic model of alginate gel beads synthesized micro particle-prussian blue to remove cesium ions from water, Journal of Water Process Engineering. 10;2016:9-19. DOI: 10.1016/j.jwpe.2016.01.001

[82] Sato H, Yui M, Yoshikawa H. Ionic diffusion coefficients of Cs^{+}, Pb^{2+}, Sm^{3+}, Ni^{2+}, $SeO^{2-}{}_4$ and $TcO^{-}{}_4$ in free water determined from conductivity measurements, Journal of Nuclear Science and Technology. 33;1996:950–955. DOI: 10.1080/18811248.1996.9732037

[83] Kashima K, Osawa K, Hagiri M, Imai M. High performance dye removal by embedded activated carbon in calcium alginate membrane combination with low-molecular-weight polyethylene glycol, In: Proceedings of The 5th Asian Conference on Innovative Energy and Environmental Chemical Engineering (ASCON-IEEChE 2016); 13-16 November 2016; Yokohama (Japan); 2016. p. 70-73.

[84] Li Z, Li L, Hu D, Gao C, Xiong J, Jiang H, Li W. Efficient removal of heavy metal ions and organic dyes with cucurbit [8] uril-functionalized chitosan, Journal of Colloid and Interface Science. 539;2019:400-413. DOI: 10.1016/j.jcis.2018.12.078

[85] Ozdemir N, Zengin H, Yavuz A. Electropolymerization of pyrrole from a deep eutectic solvent for supercapacitor applications, Materials Chemistry and Physics. 256;2020:123645. DOI: 10.1016/j.matchemphys.2020.123645

[86] Liu C, Mei G, Yu M, Cheng Q, Yang S. New applications of deep eutectic solvents for separation of quartz and magnetite, Chemical Physics Letters. 762;2021:138152. DOI: 10.1016/j.cplett.2020.138152

[87] Yang H, Sheikhi A, van de Ven TGM. Reusable green aerogels from cross-linked hairy nanocrystalline cellulose and modified chitosan for dye removal, Langmuir. 32;2016:11771–11779. DOI: 10.1021/acs.langmuir.6b03084

[88] Yu P, Wang H-Q, Bao R-Y, Liu Z, Yang W, Xie B-H, Yang M-B. Self-assembled sponge-like chitosan/reduced graphene oxide/montmorillonite composite hydrogels without cross-linking of chitosan for effective Cr(VI) sorption, ACS Sustainable Chemistry and Engineering. 5;2017:1557–1566. DOI: 10.1021/acssuschemeng.6b02254

[89] Yi L, Yang J, Fang X, Xia Y, Zhao L, Wu H, Guo S. Facile fabrication of wood-inspired aerogel from chitosan for efficient removal of oil from water, Journal of Hazardous Materials. 385; 2020:121507. DOI: 10.1016/j.jhazmat.2019.121507

[90] de Luna MS, Ascione C, Santillo C, Verdolotti L, Lavorgna M, Buonocore GG, Castaldo R, Filippone G, Xia H, Ambrosio L. Optimization of dye adsorption capacity and mechanical

strength of chitosan aerogels through crosslinking strategy and graphene oxide addition, Carbohydrate Polymers. 211;2019:195-203. DOI: 10.1016/j.carbpol.2019.02.002

[91] Keshipour S, Mirmasoudi SS. Cross-linked chitosan aerogel modified with Au: Synthesis, characterization and catalytic application, Carbohydrate Polymers. 196;2018:494-500. DOI: 10.1016/j.carbpol.2018.05.068

[92] Kalantari K, Afifi AM, Jahangirian H, Webster TJ. Biomedical application of chitosan electrospum nanofibers as a green polymer-review, Carbohydrate Polymers. 207 ;2019: 588-600. DOI: 10.1016/j.carbpol.2018.12.011

[93] Yoshimoto H, Shin YM, Terai H, Vacanti JP. A biodegradable nanofiber scaffold by electrospinning and its potential for bone tissue engineering, Biomaterials. 24;2003:2077-2082. DOI: 10.1016/S0142-9612(02)00635-X

[94] Min BM, Lee G, Kim SH, Nam YS, Lee TS, Park WH, Electrospinning of silk fibroin nanofibers and its effect on the adhesion and spreading of normal human keratinocytes and fibroblasts in vitro, Biomaterials. 25;2004:1289-1297. DOI: 10.1016/j.biomaterials.2003.08.045

[95] Sciffman JD, Schauer CL. Cross-linking chitosan nanofibers, Biomacromolecules. 8;2007:594-601. DOI: 10.1021/bm060804s

[96] Sciffman JD, Schauer CL. One-step electrospinning of cross-linked chitosan fibers, Biomacromelecules. 8;2007:2665-2667. DOI: 10.1021/bm7006983

[97] Shalumon KT, Binulal NS, Selvamurugan N, Nair SV, Menon D, Furuike T, Electrospinning of carboxymethyl chitin/poly (vinyl alcohol) nanofibrous scaffolds for tissue engineering applications, Carbohydrate Polymers. 77:2009:863-869. DOI: 10.1016/j.carbpol.2009.03.009

[98] Jayakumar R, Prabaharan M, Nair SV, Tamura H. Novel chitin and chitosan nanofibers in biomedical applications, Biotechnology Advances. 28;2010:142-150. DOI: 10.1016/j.biotechadv.2009.11.001

[99] Almasi H, Jafarzadeh P, Mehryar L. Fabrication of novel nanohybrids by impregnation of CuO nanoparticles into bacterial cellulose and chitosan nanofibers: Characterization, antimicrobial and release properties, Carbohydrate Polymers. 186;2018:273-281. DOI: 10.1016/j.carbpol.2018.01.067

[100] Zhang Y, Huang X, Duan B, Wu L, Li S. Yuan X. Preparation of electrospun chitosan/poly (vinyl alcohol) membranes, Colloid and Polymer Science. 285;2007:855-863. DOI: 10.1007/s00396-006-1630-4

[101] Amirabad LM, Jonoobi M, Mousavi NS, Oksman K, Kaboorani A, Yousefi H, Improved antifungal activity and stability of chitosan nanofibers using cellulose nanocrystal on banknote papers, Carbohydrate Polymers. 189; 2018:229-237. DOI: 10.1016/j.carbpol.2018.02.041

[102] Kawasaki T, Hirabayashi T, Masuyama K, Fujita S, Kitano H. Complex film of chitosan and carboxymethyl cellulose nanofibers, Colloids and Surfaces B: Biointerfaces. 139;2016:95-99. DOI: 10.1016/j.colsurfb.2015.11.056

[103] Aliabadi M, Irani M., Ismaeili J, Najafzadeh S. Design and evaluation of chitosan/hydroxyapatite composite nanofiber membrane for the removal of heavy metal ions from aqueous solution, Journal of the Taiwan Institute of Chemical Engineers. 45;2014:518-526. DOI: 10.1016/j.jtice.2013.04.016

[104] Tao F, Cheng Y, Shi X, Zheng H, Du Y, Xiang W, Deng H. Applications of chitin and chitosan nanofibers in bone regenerative engineering, Carbohydrate

Polymers. 230;2020:115658. DOI: 10.1016/j.carbpol.2019.115658

[105] Li Q-H, Dong M, Li R, Cui Y-Q, Xie G-X, Wang X-X, Long Y-Z. Enhancement of Cr(VI) removal efficiency via adsorption/photocatalysis synergy using electrospun chitosan/g-C3N4/TiO2 nanofibers, Carbohydrate Polymers. 253;2021:117200. DOI: 10.1016/j.carbpol.2020.117200

[106] Show PL, Ooi CW, Lee XJ, Yang C-L, Liu B-L, Chang Y-K. Batch and dynamic adsorption of lysozyme from chicken egg white on dye-affinity nanofiber membranes modified by ethylene diamine and chitosan, International Journal of Biological Macromolecules. 162;2020:1711-1724. DOI: 10.1016/j.ijbiomac.2020.08.065

Section 7

Industrial Application

Chapter 11

The Application of Chitosan-Based Compounds against Metallic Corrosion

Brahim El Ibrahimi, Lei Guo, Jéssica Verger Nardeli and Rachid Oukhrib

Abstract

Biopolymers-based compounds were used by different manners for metal protection toward corrosion phenomena, namely via inhibiting additive and coating strategies. In the last decade, the application of these compounds or their chemically modified forms as effective replacements for toxic inorganic and organic inhibitors attracts more attention. Additionally to their intrinsic chemical stability, biodegradability, eco-friendly, low cost and renewability, biopolymers set were shown the remarkable effect to control the dissolution of several metallic materials in various corrosive environments. Among a large variety of available biopolymers, chitosan and its functionalized form, as well as its nanoparticle composites, have been reported and widely used as good anti-corrosion compounds for different metal/medium systems. In this context, the current chapter aims to shed more light on this subject.

Keywords: corrosion, chitosan, chitin, inhibitor, metal, solution, coating, composite

1. Introduction

Metal corrosion is defined as a spontaneous deterioration of metallic materials caused by the adjacent environments (e.g. during the acidic cleaning process) through electrochemical and/or chemical processes. Such a phenomenon is inevitable due to its undesirable outcomes on technological and industrial applications, which leads to huge loss of natural resources, human lives and economic [1]. In this regard, researchers were compelled to perform several scientific investigations to extend the working life of these materials and to overcome the devastating impact of corrosion. Among available technical solutions, the addition of corrosion inhibitors into the aggressive environment seems to be an attractive and economic technic to effectively control corrosion [2]. Diverse organic and inorganic substances have been employed as anti-corrosion compounds for many metal-environment systems [3]. Another efficient strategy to extend the life of metallic-based materials and protect them against corrosion is the application of organic coatings [4–7].

In the present chapter, we interesting in the organic inhibitor category, which its protection ability is related to its adsorption onto the metal surface via electrostatic attraction or/and chemical bond formation, leading to the formation of a protective layer resulting in corrosion mitigation [8]. It is well known that the adsorption process of these compounds occurs via their electron-donator sites like heteroatoms (N, O, S and P) and multiple bonds (π-bonds) and aromatic rings as well. Even they are efficient, recently, the exploitation of organic substances as corrosion retarders has been limited by several strict environmental rules, and such a trend aims to limit their unsafe menace to ecology and health [9].

At this time, the use of biopolymers as promising replacement of toxic corrosion inhibitors is considered a new trend and novel strategy to omit metallic corrosion. One of the key advantages of these bio-macromolecules is their increased attachment sites to the metallic substrate, rising to good film-formation and adhesion as compared to small molecule inhibitors [10]. This capability can be further boosted by the insertion of additional adsorption sites, i.e. functional groups, within the biopolymers backbone [11]. On the other hand, these biopolymers are biodegradable, biocompatible, cheap and non-toxic. Besides, they are readily available and renewable sources of materials [12]. All these characteristics have made them ideal candidates to mitigate ecologically metallic corrosion. In this regards, a large variety of natural polymers are reported to act as anti-corrosion agents to secure the metal against dissolution such as alginate, sodium chitosan, pectin, carboxymethyl and hydroxylethyl cellulose [13, 14].

Among available biopolymers, chitosan (**Figure 1(a)**) was especially exhibited a noticeable ability to control corrosion. It is characterized by the existence of oxygen (of alcohol and ether functional groups) and nitrogen (of amine group) atoms within its backbone chain. These sites are known to act as the effective centers of adsorption to metallic substrates. Chitosan can be obtained by the deacetylation of chitin (**Figure 1(b)**), a natural polysaccharide and the main structural component of crustacean exoskeletons, and is soluble in acid media as compared to chitin, which is a highly insoluble and a non-reactive biopolymer [15]. Furthermore, chitosan exhibits a polycationic character and is non–toxic and biodegradable [16, 17].

Recently, the application of chitosan-based compound as ecofriendly corrosion inhibitor was extended to the use of its functionalized form instead of pure one. Such tendency aims to decorate chitosan backbone with particular functional motifs, generally, through the chemical modification of the amino groups. Furthermore, the enhancement of chitosan capability for protection purposes has been also attained via its combination with other chemical materials to prepare nanoparticle composites, which are served to act as the effective coatings to mitigate corrosion. In this context, it has been reported that the combination of nano-scaled organic and inorganic fillers can successfully improve mechanical, adhesion and barrier qualities of polymer coatings [18]. Among the used additives in the matrix of chitosan-based coatings, there are zinc oxide, graphene oxide and hydroxyapatite nanoparticles.

On this basis, we aimed in the present chapter to shed more light on the merits to employ different chitosan forms as sustainable compounds for corrosion controlling of metallic materials in different aggressive environments.

(a) (b)

Figure 1.
Molecular structure of (a) chitosan and (b) chitin bio-macromolecules.

2. Application of pure chitosan form as corrosion inhibitor

Chitosan is a naturally occurring polymer that meets the requirements to be classified as a green corrosion inhibitor, which is a low-cost alternative of widely used inhibitors in industrial applications. The solubility of inhibitor into the target corrosive media is one of among key prerequisites that judges its utilization. For instance, such property has limited the use of chitin biopolymer as a corrosion inhibitor. In this regard, temperature, degree of deacetylation, solution's pH and molecular weight are the main factors affecting the solubilization of chitosan in the aqueous media. For instance, at the higher temperatures, with higher molecular weight (>29.2 kDa) and lower deacetylation degree, low water-solubility of pure chitosan was observed [19, 20].

As an inhibiting additive, pure chitosan has been reported to act as an effective retarder of corrosion in different aggressive environments, namely saline and acidic solutions, as well as natural ones such as seawater. Up to now, pure chitosan compound is widely applied for iron and its alloys, like mild steel and carbon steel. This particular attention owing to the fact that these metallic materials are extensively used in numerous industrial applications in which their corrosion is more intense. **Table 1** collects the obtained inhibition efficiency (IE) for pure chitosan for some metallic materials in different corrosive environments. From tabulated data, it is clear that pure chitosan can act as a potent ecofriendly corrosion inhibitor even in the most aggressive environments. This is attributed to the formation of a protective layer upon the metal surface, which prevents it attack by the aggressive species present in the solution.

Metallic material	Aggressive medium	IE(%) at [chitosan]	Ref.
mild steel	0.1 M HCl	93% at 1.8 mM	[21]
carbon steel	1.0 M HCl	93% at 5000 ppm	[22]
mild steel	3.65% NaCl	90% at 1.2 wt%	[23]
copper	1.0 M HCl	87% at 0.1 mg L^{-1}	[24]
copper	Synthetic seawater +20 ppm Na_2S	89% at 800 ppm	[25]
316 austenitic	0.1 M HCl	71% at 11 mM	[26]
mild steel	0.1 M HCl	69% at 4 μM	[27]

Table 1.
Some works on the use of pure chitosan form as corrosion inhibitor.

As mentioned above, the molecular weight of chitosan biopolymer can affect its solubility, consequently, the attained prevention efficiency. In this context, lower inhibition efficiency has been obtained for mild steel in seawater employing chitosan with higher molecular weights [28]. Furthermore, the role of exposure time to the corrosive solution on the ability of pure chitosan to reduce metallic dissolution was also evaluated. In the CO_2-saturated saline environment, the extension of immersion time has implied an improvement in the inhibitive action of chitosan [29]. In another study, the opposite behavior is outlined from which the reduction of the inhibition efficiency is attributed to the destruction of the dense adsorbed film on the metal surface at longer exposure times [30]. Concerning the influence of temperature on the inhibition process of pure chitosan, there is no commune agreement, which a favorable effect is observed by some researchers, whereas the opposite one is reported by other ones [26].

To improve the inhibition property of pure chitosan form for some metal/solution systems, the synergistic corrosion inhibiting effect was applied. In this enhancement strategy, additional compounds such as cations and anions species are added into the corrosive solution with chitosan. As result, remarkable enhancement of protection capabilities of chitosan is pointed out. For instance, the combination of pure chitosan (200 ppm) with 5 ppm of KI was led to a significant improvement of the inhibition efficiency for mild steel in acidic solution, which 90% prevention percentage was achieved instead of 74% in the case of chitosan alone [31]. In this regard, a similar tendency is noted for another steel variety, i.e. St37 steel, in concentrated sulfuric acid solution in which 92% inhibition efficiency is attained [32]. On the other hand, it was found that the adsorption mechanism of chitosan onto metal surface depends on the adopted circumstances. Chitosan can adsorb on the metal surface either via physisorption or chemisorption modes [22, 27].

3. Functionalized chitosan forms as anti-corrosion agents

The current trend in the use of chitosan-based compounds as corrosion inhibitors is its functionalization, afterward its application. This novel approach aims to increase the solubilization of these bio-compounds in almost corrosive media and to enhance their adsorption and adhesion abilities to the metallic surface. In this respect, further polar functional groups are attached to the chitosan molecular skeleton. The chemical modifications of chitosan biopolymer are often performed at amine group, which is an active site. As result, various chitosan-based derivatives with different structural compositions have been synthesized and then used to retard or suppressed metal corrosion in different aggressive environments. Even the simplest chitosan derivative, i.e. carboxymethyl chitosan (**Figure 2(a)**), an improved inhibition efficiency is attained as compared to the pure chitosan form, which is increased from 23 to 38% for steel in wastewater liquids [33].

Depending on the molecular structure, functionalized chitosan-based inhibiting additives could be classified into several categories, namely, chitosan Schiff bases, chitosan surfactants, triazole modified chitosan, chitosan polymeric salts, PEG cross-linked chitosan, carboxymethyl hydroxypropyl chitosan, chitosan thiocarbohydrazide, acid grafted chitosan, acetyl thiourea chitosan, polymer and biomaterial grafted chitosan. Here, we limit to present the inhibition activity of the three first functionalized chitosan sets.

During the last decade, Schiff bases class compounds have been attracted exceptional attention to be applied in the field of corrosion inhibition owing to the presence of imine linkage, i.e. –CH=N–. They are reported to act as potent anti-corrosion compounds for different metallic materials, especially in acidic solutions [34]. In this respect, the synthesis of chitosan Schiff bases derivatives via condensation reaction and/or under microwave irradiations are conducted. It was found that the introduction of Schiff bases functional group into the chitosan skeleton leads to a significant enhancement in the inhibition property and film adhesion of polymer on the metal surface. Generally, the achieved prevention efficiencies using those chitosan-based derivatives were higher than 80%, which outlined that chitosan Schiff base could be an appropriate candidate to employ as effective anti-corrosion agents [35]. Recently, three chitosan Schiff bases derivatives (CSB-1, –2 and – 3, **Figure 2(b)**) have been synthesized under microwave irradiations and tested as corrosion inhibitors for mild steel in acidic solution. According to the obtained experimental data, these modified chitosan compounds were exhibited significant tendencies to reduce metallic corrosion even at a lower concentration, which the supreme prevention efficiencies of 91, 87 and 85% (at 50 ppm) were attained for CSB-3, –2 and – 1, respectively [36]. Another chitosan-modified Schiff base, namely, the salicylaldeyde-chitosan Schiff base (**Figure 2(c)**), has been reported to act as a good inhibitor (IE(%) = 95.4% at

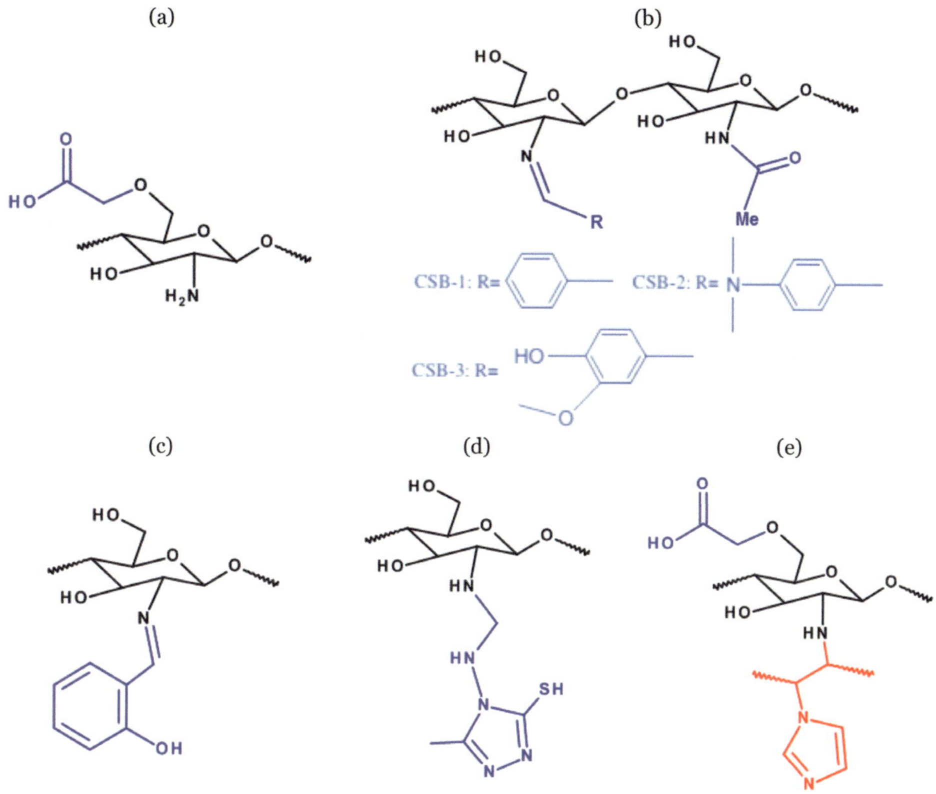

Figure 2.
Molecular structure of some chitosan derivatives used as corrosion inhibitors.

150 ppm) for J55 steel-variety in 3.5% NaCl solution saturated with carbon dioxide at elevated temperature. The merits of developed functionalized chitosan including its eco-friendly aspect, safe, simple and cheap synthesize of used Schiff base, as well as the improvement of inhibitor solubility compared to unmodified chitosan. All these listed advantages make salicylaldeyde-chitosan Schiff base as a good anti-corrosion agent for the oil and gas industries [37].

Surfactant is a surface-active agent that characterizes by the presence of hydrophilic and hydrophobic groups per molecule. These chemical compounds are largely served as effective corrosion inhibitors in the petrochemical industry dues to their affinity to be oriented at the metal/solution interface. In 2012, over 26% was the demand for surfactants as anti-corrosion components only for the petrochemical industry, as well as this request grew by 4.1% per year [38]. To combine the attractive anti-corrosion property of surfactant set with chitosan biopolymer, several chitosan-surfactants macromolecules are synthesized and then evaluated as potential retarders of corrosion. In this regard, the introduction of hydrophobic moiety into the chitosan skeleton has been led to an increase of its hydrophobic property to become surface-active polymers, which in result an enhancement of the prevention capability of chitosan. For instance, a sequence of seven modified hydrophobically chitosan surfactants were produced and their anti-corrosion property is measured for carbon steel in acid medium. As compared to the pure chitosan, good inhibition efficiencies between 93 and 74% (at 250 ppm) are achieved for those surfactants functionalized chitosan derivatives [39]. It was found that carboxymethyl chitosan thio-derivative provides the highest protection. This finding is related to its high surface activity and the presence of more active adsorption centers within its molecular skeleton as well.

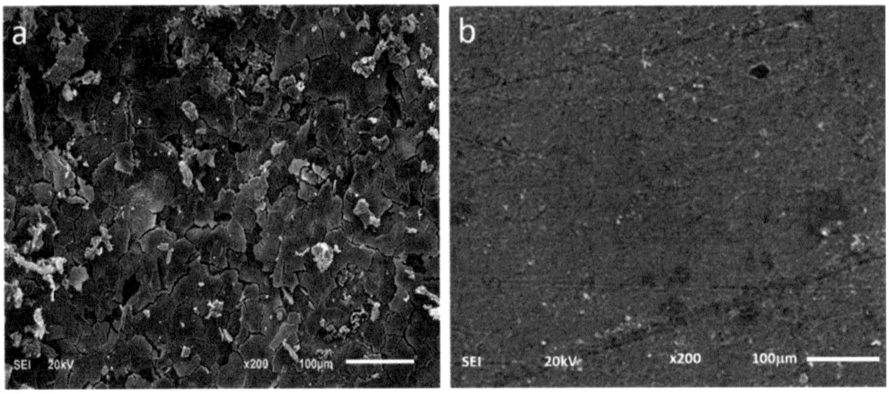

Figure 3.
SEM images of carbon steel surface (a) without and (b) with the addition of developed triazole-modified chitosan at 200 ppm [42].

A wide range of organic heterocyclic molecules has been employed to face against metallic corrosion. In this context, azole-based compounds have shown an excellent capacity to act as good anti-corrosion compounds for several metallic materials in different corrosive environments, especially in acidic ones. The latter molecule set includes N-azole, thiazole and oxazole cyclic molecules with different architectures [40]. The chemical incorporation of azole moieties or their derivatives into the chitosan backbone has shown excellent results in terms of inhibition efficiency. Recently, a novel triazole modified chitosan (**Figure 2(d)**) has been reported to act as an efficient retarder of carbon steel corrosion, which a maximum inhibition efficiency of 97% is reached using just 200 ppm of developed chitosan derivative [41]. The benefic effect of this triazole-modified chitosan biomacromolecule against corrosion can be revealed from the reported scanning electron microscopy (SEM) images as depicted in **Figure 3**. It is clear from **Figure 3(a)** that the morphology of carbon steel surface is more rough and damaged in the absence of modified chitosan inhibitor. Nevertheless, in its presence (**Figure 3(b)**) the morphology of steel surface become smoother, which supports the protection capacity of the developed chitosan derivative. In this work, it was found that the synthesized compound could block cathodic sites at the metal surface via the physical and chemical adsorption process.

In addition to the amine group, i.e. $-NH_2$, the functionalization of chitosan can be also carried out on both extra-functional groups including –OH group. This approach to amplify the inhibiting effect of chitosan has been attracting interest. We can list the example of poly (N-vinyl-imidazole) grafted carboxymethyl chitosan (**Figure 2(e)**), which is a polymer grafted chitosan. The newly synthesized chitosan derivative has exhibited interesting corrosion protection for steel metal in acid solution [43].

4. Using chitosan-nanoparticle composite form for metal protection

Chitosan composites have mainly been applied as inhibiting coatings and are used for corrosion protection purposes in different media. Some works have reported the preparation of composite coatings with chitosan to obtain protective systems to metal substrates [42, 44, 45]. To improve the anti-corrosion properties of the polymeric matrix, it is necessary to invest in improving the mechanical and adhesion properties through the incorporation of inorganic and organic fillers. It is reported that Nano-scaled fillers imply better barrier properties in the polymer coatings compared to the micron-size additives [15]. In general, a schematic for composite formation is shown in **Figure 4**.

Biopolymer chitosan-based nanocomposite coatings have been investigated for protection against copper corrosion [46]. Coatings are composed of chitosan matrix with 2-mercaptobenzothiazole and silica nanoparticles. Overall, the combination of the organic corrosion inhibitor and inorganic nanoparticles enhanced the protection efficiency of chitosan coatings, which is an important advance toward developing sustainable corrosion protection coatings for different metals.

Several researchers have reported the viability of chitosan composites films e.g. chitosan/ZnO nanoparticle composite for protection against corrosion for steel [47, 48] and bio-corrosion inhibition for S150 carbon steel [49]. All of these studies showed that the quality of the chitosan film was improved due to the addition of ZnO nanoparticles. Another study [50] evaluated and compared corrosion protection of carbon steel using two different systems of chitosan e.g. oleic acid-modified chitosan-graphene oxide composite coating and pure chitosan coating. In this case, it was observed that the corrosion protection of oleic acid-modified chitosan-graphene oxide composite coating improved by 100 folds when compared with pure chitosan coating. Thus, oleic acid-modified chitosan-graphene oxide composite is more effective in corrosion protection of carbon steel.

Another composite coating [51] consisting of graphene oxide-chitosan-silver on Cu-Ni Alloy with enhanced anti-corrosive and antibacterial properties show graphene oxide retards the diffusion of corrosive ions to the substrate and minimizes the electron transport between the electrolyte and metal, while chitosan prevents the galvanic coupling of graphene oxide with the metal surface [51].

The system chitosan/hydroxyapatite nanoparticle composites revealed that they could inhibit corrosion in steel, however, it was found that the combination of chitosan/hydroxyapatite nanoparticle with other species provides more effective protection against corrosion [52–54]. Different composites such as chitosan/hydroxyapatite-Mg [55], chitosan/hydroxyapatite-Si [56], chitosan/hydroxyapatite-multiwalled carbon nanotube [57], chitosan/hydroxyapatite-$CaSiO_3$ [58], and chitosan/hydroxyapatite-cellulose acetate [59] were synthesized and tested as corrosion protective layers. The composites chitosan/hydroxyapatite-Mg, chitosan/hydroxyapatite-Si, chitosan/hydroxyapatite-$CaSiO_3$ and chitosan/hydroxyapatite-cellulose acetate demonstrated that the insertion of the third component exhibit a representative improvement in the corrosion protection of the chitosan/hydroxyapatite nanoparticle composite, except for composite chitosan/hydroxyapatite-multiwalled carbon nanotube. Consequently, it could be expected that the presence of carbon nanotubes in any non-conductive polymer coating provides lower protection against corrosion.

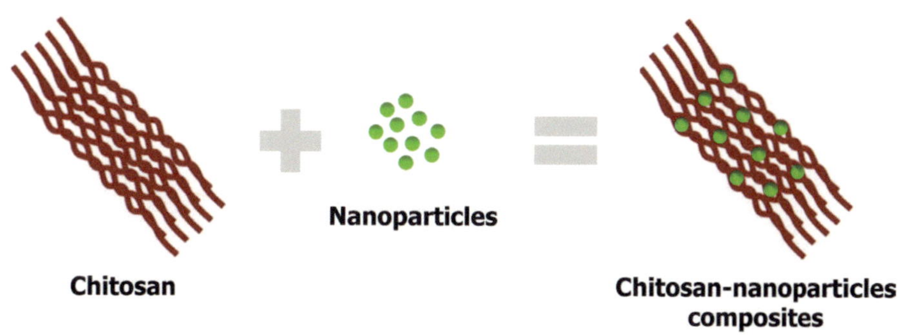

Figure 4.
Schematic presentation of chitosan-nanoparticles composites formation process.

As shown previously, the use of chitosan-nanoparticle composites led to improvements in the corrosion protection of different surfaces, i.e. copper and steel. Bahari *et al.* [46] concluded that addition of nanoparticles contributes to the reduction in swelling of chitosan coatings and crosslinked chitosan coatings are superior to the non-crosslinked ones vis-a-vis in mitigation of corrosion of copper surface. When, John *et al.* [47] concluded that mitigation of corrosion of mild steel by nanostructured chitosan/ZnO nanoparticle films was obtained based on chemical stability, oxidation control of coatings. Therefore, the process of corrosion control depends on the structure of the coating (polymeric matrix, crosslinking, adhesion, among other parameters).

5. Conclusion

Efficient inhibitors and organic coatings are able to extend the life of some metal surfaces. Chitosan is a component with a high potential for protection against corrosion of metals when exposed to corrosive media. In the first stage, the pure chitosan form is used to omit the corrosion of numerous metallic-based materials. Nevertheless, the lower water-solubility of this biopolymer was limited to its application and its utilization to a broad aggressive media range such as near-neutral solutions. Although, such weakness can overcome and the prevention performance of chitosan can be improved through different functionalization demarches. Furthermore, chitosan can be combined with the other materials to develop new chitosan-nanoparticles composites that can apply as coatings.

Author details

Brahim El Ibrahimi[1], Lei Guo[2*], Jéssica Verger Nardeli[3] and Rachid Oukhrib[1]

1 Team of Physical Chemistry and Environment, Faculty of Science, Ibn Zohr University, Agadir, Morocco

2 School of Material and Chemical Engineering, Tongren University, Tongren, China

3 Universidade Estadual Paulista-UNESP, Instituto de Química, Araraquara, São Paulo, Brazil

*Address all correspondence to: cqglei@163.com

References

[1] D.Q. Huong, T. Duong, and P.C. Nam, *Effect of the Structure and Temperature on Corrosion Inhibition of Thiourea Derivatives in 1.0 M HCl Solution*, ACS Omega 4 (2019), p. 14478-14489.

[2] A. Jmiai, B. El Ibrahimi, *A. Tara*, S. El Issami, O. Jbara, and L. Bazzi, *Alginate biopolymer as green corrosion inhibitor for copper in 1 M hydrochloric acid: Experimental and theoretical approaches*, J. Mol. Struct. 1157 (2018), pp. 408-417.

[3] B. EL Ibrahimi, L. Bazzi, and S. EL Issam, *The role of pH in corrosion inhibition of tin using the proline amino acid: theoretical and experimental investigations*, RSC Adv. 10 (2020), p. 29696.

[4] J.V. Nardeli, C.S. Fugivara, M. Taryba, M.F. Montemor, and A.V. Benedetti, *Biobased self-healing polyurethane coating with Zn micro-flakes for corrosion protection of AA7475*, Chemical Engineering Journal 404 (2021), p. 126478.

[5] J.V. Nardeli, C.S. Fugivara, M. Taryba, M.F. Montemor, and A. Benedetti, *Self-healing ability based on hydrogen bonds in organic coatings for corrosion protection of AA1200*, Corrosion Science 177 (2020), p. 108984.

[6] J.V. Nardeli, C.S. Fugivara, M. Taryba, M.F. Montemor, S.J.L. Ribeiro, and A.V. Benedetti, *Novel healing coatings based on natural-derived polyurethane modified with tannins for corrosion protection of AA2024-T3*, Corrosion Science 162 (2020), p. 108213.

[7] J.V. Nardeli, C.S. Fugivara, M. Taryba, E.R.P. Pinto, M.F. Montemor, and A.V. Benedetti, *Tannin: a natural corrosion inhibitor for bare and coated aluminum alloys*, Progress in Organic Coatings 135 (2019), pp. 368-381.

[8] B. El Ibrahimi, A. Jmiai, A. Somoue, R. Oukhrib, M. Chadili, S. El Issami, and L. Bazzi, *Cysteine Duality Effect on the Corrosion Inhibition and Acceleration of 3003 Aluminium Alloy in a 2% NaCl Solution*, Port. Electrochimica Acta 36(6) (2018), pp. 403-422.

[9] B. El Ibrahimi, A. Jmiai, K. El Mouaden, A. Baddouh, S. El Issami, L. Bazzi, and M. Hilali, *Effect of solution's pH and molecular structure of three linear α-amino acids on the corrosion of tin in salt solution: A combined experimental and theoretical approach*, J. Mol. Struct. 1196 (2019), pp. 105-118.

[10] R. Oukhrib, B. El Ibrahimi, H. Abou Oualid, Y. Abdellaoui, S. El Issami, L. Bazzi, M. Hilali, and H. Bourzi, *In silico investigations of alginate biopolymer on the Fe (110), Cu (111), Al (111) and Sn (001) surfaces in acidic media: Quantum chemical and molecular mechanic calculations*, Journal of molecular liquids 312 (2020), p. 113479.

[11] Jiyaul Haque, Vandana Srivastava, Dheeraj S. Chauhan, Hassane Lgaz, and Mumtaz A. Quraishi, *Microwave-Induced Synthesis of Chitosan SchiffBases and Their Application as Novel and Green Corrosion Inhibitors: Experimental and Theoretical Approach*, ACS Omega 3 (2018), p. 5654-5668.

[12] D. Elieh-Ali-Komi, and M.R. Hamblin, *Chitin and Chitosan: Production and Application of Versatile Biomedical Nanomaterials*, Int. J. Adv. Res. 4 (2016), pp. 411-427.

[13] S.A. Umoren, M.M. Solomon, A. Madhankumar, and I.B. Obot, *Exploration of natural polymers for use as green corrosion inhibitors for AZ31 magnesium alloy in saline environment*, Carbohydrate Polymers 230 (2020), p. 115466.

[14] S.A. Umoren, and U.M. Eduok, *Application of carbohydrate polymers as*

corrosion inhibitors for metal substrates in different media: A review, Carbohydrate Polymers 140 (2016), pp. 314-341.

[15] H. Ashassi-Sorkhabi, and A. Kazempour, *Chitosan, its derivatives and composites with superior potentials for the corrosion protection of steel alloys: A comprehensive review*, Carbohydrate Polymers 237 (2020), p. 116110.

[16] M.R. Dedloff, C.S. Effler, A.M. Holban, and M.C. Gesta!, *Use of biopolymers in mucosally-administered vaccinations for respiratory disease*, Materials 12 (2019), p. 2445.

[17] L.M. Anaya-Esparza, J.M. Ruvalcaba-Gómez, C.I. Maytorena-Verdugo, N. González-Silva, R. Romero-Toledo, S. Aguilera-Aguirre, A. Pérez-Larios, and E. Montalvo-González, *Chitosan-TiO2: A Versatile Hybrid Composite*, Materials 13 (2020), p. 811.

[18] J. Fang, K. Xu, L. Zhu, Z. Zhou, and H. Tang, *A study on mechanism of corrosion protection of polyaniline coating and its failure*, Corrosion Science 49 (2007), pp. 4232-4242.

[19] S.-H. Chang, H.-T.V. Lin, G.-J. Wu, and G.J. Tsai, *pH Effects on solubility, zeta potential, and correlation between antibacterial activity and molecular weight of chitosan*, Carbohydrate Polymers 134 (2015), pp. 74-81.

[20] M. Rinaudc, G. Pavlov, and J. Desbrieres, *Solubilization of chitosan in strong acid medium*, International Journal of Polymer Analysis and Characterization 5 (1999), pp. 267-276.

[21] T. Rabizadeh, and S.K. Asl, *Chitosan as a green inhibitor for mild steel corrosion: Thermodynamic and electrochemical evaluations*, Materials and Corrosion 70 (2019), pp. 738-748.

[22] A.E. Okoronkwo, S.J. Olusegun, and O.O. Oluwasina, *The inhibitive action of chitosan extracted from Archachatina marginata shells on the corrosion of plain carbon steel in acid media*, Anti-Corrosion Methods and Materials 62 (2015), pp. 13-18.

[23] O.S.I. Fayomi, I.G. Akande, O.O. Oluwole, and D. Daramola, *Effect of watersoluble chitosan on the electrochemical corrosion behaviour of mild steel*, Chemical Data Collections 17 (2018), pp. 321-326.

[24] A. Jmiai, B. El Ibrahimi, *A. Tara*, R. Oukhrib, S. El Issami, O. Jbara, L. Bazzi, and M. Hilali, *Chitosan as an eco-friendly inhibitor for copper corrosion in acidic medium: protocol and characterization*, Cellulose 24 (2017), pp. 3843-3867.

[25] K. El Mouaden, B. El Ibrahimi, R. Oukhrib, L. Bazzi, B. Hammouti, O. Jbara, *A. Tara*, D.S. Chauhan, and M.A. Quraishi, *Chitosan polymer as a green corrosion inhibitor for copper in sulfide-containing synthetic seawater*, International Journal of Biological Macromolecules 119 (2018), pp. 1311-1323.

[26] A. Eddib, Y.A. Albrimi, A.A. Addi, J. Douch, R.M. Souto, and M. Hamdani, *Inhibitory action of non toxic compounds on the corrosion behaviour of 316 austenitic stainless steel in hydrochloric acid solution: Comparison of chitosan and cyclodextrin*, International Journal of Electrochemical Science 7 (2012), pp. 6599-6610.

[27] S.A. Umoren, M.J. Banera, T. Alonso-Garcia, C.A. Gervasi, and M.V. Mirífico, *Inhibition of mild steel corrosion in HCl solution using chitosan*, Cellulose 20 (2013), pp. 2529-2545.

[28] X. Yang, L. Shao, S. Zhang, W. Jiao, Y. Li, and B. Hou, *Corrosion inhibition properties of water soluble chitosan and its degradation products for mild steel in seawater*, Journal of Chinese Society for Corrosion and Protection 28 (2009), pp. 325-330.

[29] I.O. Arukalam, C.O. Alaohuru, C.O. Ugbo, K.N. Jideofor, P.N. Ehirim, and I.C. Madufor, *Effect of xanthan gum on the corrosion protection of aluminium in HCl medium*, International Journal of Advanced Research and Technology 3 (2014), pp. 5-16.

[30] S. Yang, Y. Wen, P. Yi, K. Xiao, and C. Dong, *Effects of chitosan inhibitor on the electrochemical corrosion behavior of 2205 duplex stainless steel*, International Journal of Minerals, Metallurgy, and Materials 24 (2017), pp. 1260-1266.

[31] N.K. Gupta, P.G. Joshi, V. Srivastava, and M.A. Quraishi, *Chitosan: A macromolecule as green corrosion inhibitor for mild steel in sulfamic acid useful for sugar industry*, International Journal of Biological Macromolecules 106 (2018), pp. 704-711.

[32] M.M. Solomon, H. Gerengi, T. Kaya, E. Kaya, and S.A. Umoren, *Synergistic inhibition of St37 steel corrosion in 15% H_2SO_4 solution by chitosan and iodide ion additives*, Cellulose 24 (2017), pp. 931-950.

[33] H. Sun, H. Wang, H. Wang, and Q. Yan, *Enhanced removal of heavy metals from electroplating wastewater through electrocoagulation using carboxymethyl chitosan as corrosion inhibitor for steel anode*, Environmental Science: Water Research & Technology 4 (2018), pp. 1105-1113.

[34] K. Benbouguerra, S. Chafaa, N. Chafai, M. Mehri, O. Moumeni, and A. Hellal, *Synthesis, spectroscopic characterization and a comparative study of the corrosion inhibitive efficiency of ana-aminophosphonate and Schiff base derivatives: Experimental and theoretical investigations*, J. Mol. Struct. 1157 (2018), pp. 165-176.

[35] N.L. Chen, P.P. Kong, H.X. Feng, Y.Y. Wang, and D.Z. Bai, *Corrosion mitigation of chitosan Schiff base for Q235 steel in 1.0 M HCl*, Journal of Bio-and Tribo-Corrosion 5 (2019), p. 27.

[36] J. Haque, V. Srivastava, D.S. Chauhan, H. Lgaz, and M.A. Quraishi, *Microwaveinduced synthesis of chitosan schiffbases and their application as novel and green corrosion inhibitors: Experimental and theoretical approach*, ACS Omega 5 (2018), pp. 5654-5668.

[37] K.R. Ansari, D.S. Chauhan, M.A. Quraishi, M.A.J. Mazumder, and A. Singh, *Chitosan Schiff base: an environmentally benign biological macromolecule as a new corrosion inhibitor for oil & gas industries*, International Journal of Biological Macromolecules 144 (2020), pp. 305-315.

[38] O. Kaczerewska, R. Leiva-Garcia, R. Akid, B. Brycki, I. Kowalczyk, and T. Pospieszny, *Effectiveness ofO-bridged cationic gemini surfactants as corrosion inhibitors for stainless steel in 3 M HCl: Experimental and theoretical studies*, Journal of Molecular Liquids 249 (2018), pp. 1113-1124.

[39] A.M. Alsabagh, M.Z. Elsabee, Y.M. Moustafa, A. Elfky, and R.E. Morsi, *Corrosion inhibition efficiency of some hydrophobically modified chitosan surfactants in relation to their surface active properties*, Egyptian Journal of Petroleum 23 (2014), pp. 349-359.

[40] B. El Ibrahimi, and L. Guo, *Azole-based compounds as corrosion inhibitors for metallic materials*, in *Azoles - Synthesis, Properties, Applications and Perspectives*, 2020.

[41] D.S. Chauhan, M.A. Quraishi, A.A. Sorour, S.K. Saha, and P. Banerjee, *Triazole-modified chitosan: a biomacromolecule as a new environmentally benign corrosion inhibitor for carbon steel in a hydrochloric acid solution*, RSC Advances 9 (2019), pp. 14990-15003.

[42] Q. Bao, D. Zhang, and Y. Wan, *2-Mercaptobenzothiazole doped chitosan/11-alkanethiolate acid composite*

coating: dual function for copper protection, Applied Surface Science 257 (2011), pp. 10529-10534.

[43] U. Eduok, E. Ohaeri, and J. Szpunar, *Electrochemical and surface analyses of X70 steel corrosion in simulated acid pickling medium: Effect of poly (N-vinyl imidazole) grafted carboxymethyl chitosan additive*, Electrochimica Acta 278 (2018), pp. 302-312.

[44] X. Pang, and I. Zhitomirsky, *Electrodeposition of hydroxyapatite–silver–chitosan nanocomposite coatings*, Surface and Coatings Technology 202 (2008), pp. 3815-3821.

[45] J. Carneiro, J. Tedim, and M.G.S. Ferreira, *Chitosan as a smart coating for corrosion protection of aluminum alloy 2024: A review*, Progress in Organic Coatings 89 (2015), pp. 348-356.

[46] H.S. Bahari, F. Ye, E.A.T. Carrillo, C. Leliopoulos, H. Savaloni, and J. Dutta, *Chitosan nanocomposite coatings with enhanced corrosion inhibition effects for copper*, International Journal of Biological Macromolecules 162 (2020), pp. 1566-1577.

[47] S. John, A. Joseph, A.J. Jose, and B. Narayana, *Enhancement of corrosion protection of mild steel by chitosan/ZnO nanoparticle composite membranes*, Progress in Organic Coatings 84 (2015), pp. 28-34.

[48] Z.F. Lin, *P. Wang*, D. Zhang, and Y. Wang, *A ZnO/chitosan composite film: Fabrication and anticorrosion characterization*, Advanced Materials Research 152 (2011), pp. 1199-1202.

[49] P.A. Rasheed, K.A. Jabbar, K. Rasool, R.P. Pandey, M.H. Sliem, M. Helal, A. Samara, A.M. Abdullah, and K.A. Mahmoud, *Controlling the biocorrosion of sulfate-reducing bacteria (SRB) on carbon steel using ZnO/chitosan nanocomposite as an eco-friendly biocide*, Corrosion Science 148 (2019), pp. 397-406.

[50] E.M. Fayyad, K.K. Sadasivuni, D. Ponnamma, and M.A.A. Al-Maadeed, *Oleic acid-grafted chitosan/graphene oxide composite coating for corrosion protection of carbon steel*, Carbohydrate Polymers 151 (2016), pp. 871-878.

[51] G. Jena, B. Anandkumar, S.C. Vanithakumari, R.P. George, J. Philip, and G. Amarendra, *Graphene oxide-chitosan-silver composite coating on Cu-Ni alloy with enhanced anticorrosive and antibacterial properties suitable for marine applications*, Progress in Organic Coatings 139 (2020), p. 105444.

[52] A.S. Hammood, M.A.S. Mahdi, L. Thair, and H. Haddad, *Evaluating the effect of hydroxyapatite-chitosan coating on the corrosion behavior of 2205 duplex stainless steel for biomedical applications*, Materials Research Express 6 (2019), p. 85411.

[53] X. Pang, and I. Zhitomirsky, *Electrophoretic deposition of composite hydroxyapatite-chitosan coatings*, Materials Characterization 58 (2007), pp. 339-348.

[54] I. Zhitomirsky, and X. Pang, *Fabrication of chitosan-hydroxyapatite coatings for biomedical applications*, ECS Transactions 3 (2007), pp. 15-22.

[55] S. Sutha, N.R. Dhineshbabu, M. Prabhu, and V. Rajendran, *Mg-doped hydroxyapatite/chitosan composite coated 316l stainless steel implants for biomedical applications*, Journal of Nanoscience and Nanotechnology 15 (2015), pp. 4178-4187.

[56] S. Sutha, K. Kavitha, G. Karunakaran, and V. Rajendran, *In-vitro bioactivity, biocorrosion and antibacterial activity of silicon integrated hydroxyapatite/chitosan composite coating on 316 L stainless steel implants*, Materials Science and Engineering: C 33 (2013), pp. 4046-4054.

[57] F. Batmanghelich, and M. Ghorbani, *Effect of pH and carbon*

nanotube content on the corrosion behavior of electrophoretically deposited chitosan–hydroxyapatite–carbon nanotube composite coatings, Ceramics International 39 (2013), pp. 5393-5402.

[58] X. Pang, T. Casagrande, and I. Zhitomirsky, *Electrophoretic deposition of hydroxyapatite–*$CaSiO_3$*–chitosan composite coatings*, Journal of Colloid and Interface Science 330 (2009), pp. 323-329.

[59] Z. Zhong, J. Qin, and J. Ma, *Cellulose acetate/hydroxyapatite/chitosan coatings for improved corrosion resistance and bioactivity*, Materials Science and Engineering: C 49 (2015), pp. 251-255.

Chapter 12

Modified Chitosan Forms for Cr (VI) Removal

Şerife Parlayici and Erol Pehlivan

Abstract

The forms of utility in the wastewater treatment of chitosan-based adsorbents acquired from natural substances have attracted numerous attentions in recent years. The use of chitosan modified adsorbents for removal of the chromium has aroused great interest. When chitosan-based modified adsorbents are considered, they have got large amount of an amino (-NH_2) and hydroxyl (-OH) functional groups. Such adsorbents display that they have high activity and therefore they may be extensively utilized in wastewater treatment plants to cast off chromium. In this chapter, the utility outcomes of chitosan-based substances will be explained after applying different parameters to remove Cr (VI) from the aqueous surrounding with the information obtained the use of batch adsorption systems. Application of various chitosan-based adsorbents for Cr (VI) removal application will be demonstrated in a detailed way and they will be discussed within the textual content.

Keywords: adsorption, chitosan, Cr (VI), modification

1. Introduction

There is an increasing demand for new materials of natural origin to remove metal ions which include chromium from industrial liquid waste. Many researchers have evaluated and implemented the possibility of the usage of renewable-based materials to get this element. Cellulose, starch, pectin or crusted shell waste such as chitin and chitosan (Cts), which are polysaccharides as a class of natural macromolecules. They're exceedingly bioactive substances in the removal of water pollutants and are commonly derived from marine crustaceans and agricultural raw substances or by-products [1, 2].

There are thousands of different chemicals, physical and biological processes occurring between the solid adsorbent surface and the adsorbate. In the adsorption process, the active interactions occurring at the adjacent intermediate surfaces between a solid adsorbent and certain toxic metal solution and the changes in its concentration between both phases stand out. Biopolymers can be thought of as a polymer that we are used to (such as proteins, lignin, polysaccharides, chitin, chitosan, cellulose and hemicellulose) and originate from living organisms in nature. These polymers are often found in sources of carbon origin and are mostly derived from carbohydrates. Origins of some biopolymers are shown in **Figure 1**.

There have been numerous articles on chitin and Cts in recent years and it is more directed towards applications of these polymers and their modification. Some biological activities exhibited through such polymers, specially Cts and its derivatives which include biocompatibility, biodegradability, low toxicity, mucousness

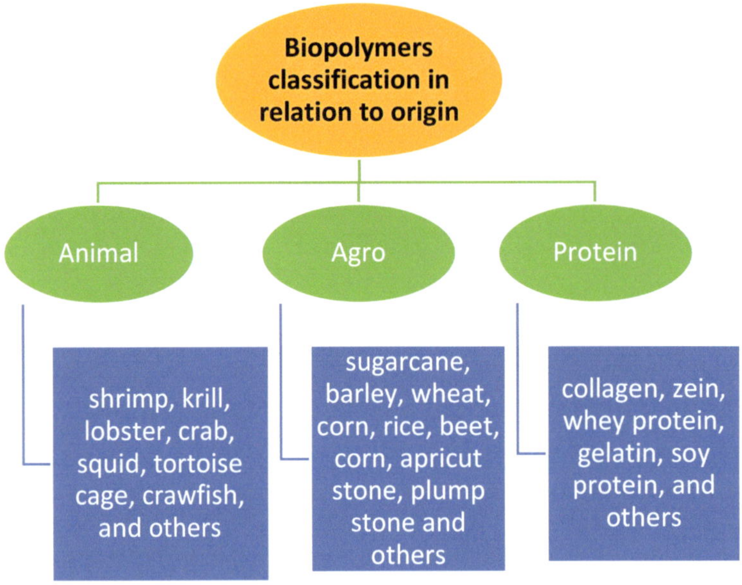

Figure 1.
Biopolymer classification: Animal, agro, and protein origins.

and antimicrobial activity have attracted interest in medicine, pharmacy, biomedical and health-related applications [1, 3, 4]. Numerous Cts-based materials were produced such as nano and microparticles, gels, sponges, films and membranes [4]. They have created an efficient field of application for drug delivery, wound healing and tissue regeneration applications.

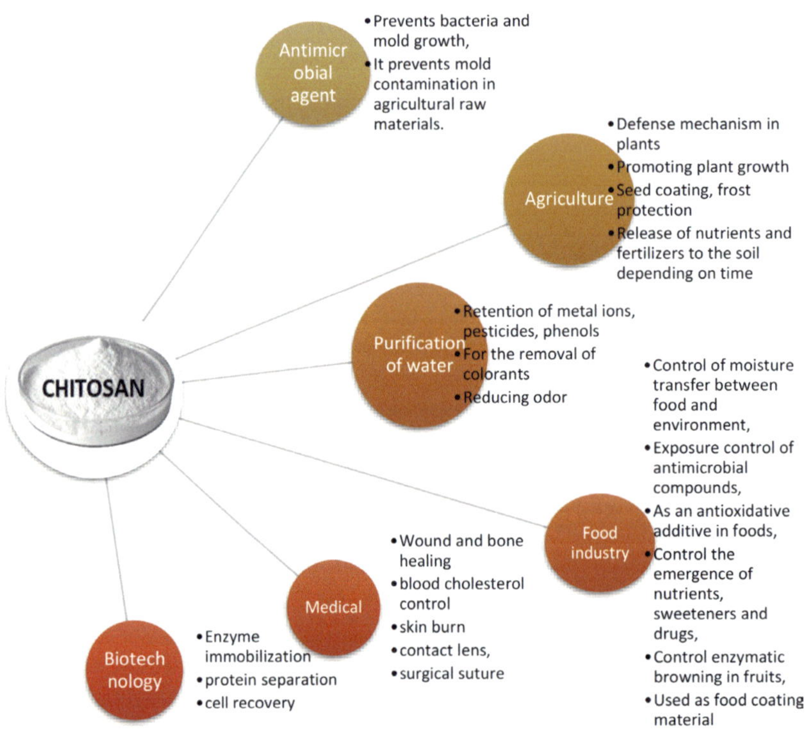

Figure 2.
Field of application of Cts.

The biggest advantage of chitin and Cts is that they are a renewable resource and an environmentally friendly natural biopolymer. With these features, it has been used in many different sectors in recent years. Cts attracts more attention compared to chitin because it is commercially available and can be used in many forms. Application areas of Cts can be listed as pharmacy (controlled drug release), medical (bandage), wastewater treatment, biotechnology, cosmetics, food, textile and agriculture (**Figure 2**).

2. Chitosan

2.1 Sources of chitin and chitosan

Chitin is usually found in nature in seaweeds, protozoa (flagellates, amoebic ciliates, etc.), selenterals, mollusks, arthropods, bacteria, fungi, insects and some plants. The richest sources of chitin are; crab, shrimp, lobster and crayfish shells. Chitin biopolymers of animal origin are mainly found in animals. Chitin is the most considerable polysaccharide on the planet. Due to its strong hydrogen bonds and high crystal property of its cohesive structure forces, chitin is an insoluble substance in normal solvents, including water and organic matter [5]. Cts is abundant in marine or terrestrial animals such as crustaceans, insects, mollusks and fungi [6] without a backbone, such as various insects and marine diatoms. Commercial production of Cts by alkaline hydrolysis by extracting chitin from shrimp shell waste has been carried out by a group scientist [3, 6].

2.2 Chitosan production

Cts is obtained from chitin. Rouget became aware of Cts in 1859 while conducting experimental activities on deacetylated forms of chitin [7]. Cts is obtained from waste shrimp and crab shells from the seafood industries by chemical methods in industrial processes around the world [4]. Basic structure properties of Cts are related to its molecular weight and degree of deacetylation. The crystallinity of Cts is important in classifying its particle size, moisture and ash content, which is based on its surface morphology properties. The solubility, antibacterial, polycationic character, biocompatibility and bio adhesiveness of Cts are the basis that define the Cts structure and application areas [8, 9].

Chemical modifications of Cts have been extensively described in recent studies. Cts contains the reactive amino (-NH_2) and hydroxyl (-OH) groups in its structure and can very easily be converted into new modified forms. The presence of amino groups in the Cts matrix (deacetylation degree) in macromolecular chains indicates that Cts has a polycationic property in acidic aqueous solutions. The protonation of amino groups in the polymer structure seriously affects the structure of macromolecules in solutions, and it is known that the structure behavior may be managed through changing the pH or ionic strength of the solution it is in.

Cts production is carried out by two basic methods, chemically and biologically. Classically, the chemical method is the method in which physical and chemical methods are used together. Chemical method; the isolation of the chitin is carried out by removing other substances in the shell. The procedures for this can be grouped under four main steps:

1. Preparation of shells.

2. Demineralization (removal of minerals).

3. Deproteinization (removal of proteins).

4. Decoloration (removal of pigments). These stages are given in **Figure 3**.

Biological methods are processes that can be examined under several headings. Cell cultures of various organisms such as Mucor rouxii, Phycomyces blakesleeanus are used in the microbial synthesis of Cts. The product obtained is developed by adding Aspergillus niger to the culture medium. Thus, this production mechanism also leads to the deacetylation of the Cts. Cts is obtained at the end of the 96-hour incubation period [10]. As a biological method, the use of proteolytic enzymes such as pepsin, trypsin, protease, proteinase and papain in protein removal can be counted [11]. Today, many other polymers used in industry are produced synthetically. Synthetic polymers have limited biocompatibility and biodegradability compared to Cts, a natural polymer.

2.3 Resolution of chitosan

Chitin is insoluble in dilute acids and many organic solvents. Because it is a semi-crystalline polymer and has a large amount of intramolecular and intermolecular hydrogen bonds in its structure. The effect of changes in the N-acetyl-D-glucosamine and D-glucosamine groups on solubility is shown in **Figure 4**.

The solubility concerning chitosan in a range of solvents and its polycationic properties are essential for the advent of chitosan derivatives and the use of them in many different applications. Many different factors are taken into account to control the solubility of Cts in aqueous solutions. These are the result of applying certain parameters such as temperature, alkali concentration, type and

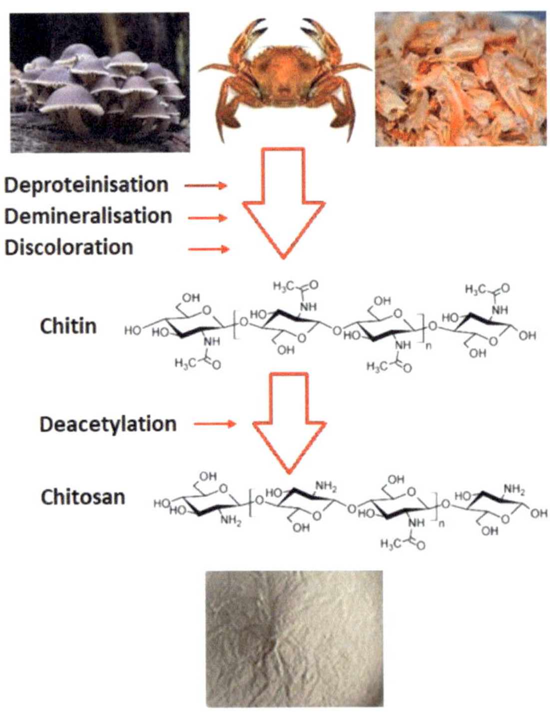

Figure 3.
General steps for chitin and Cts production.

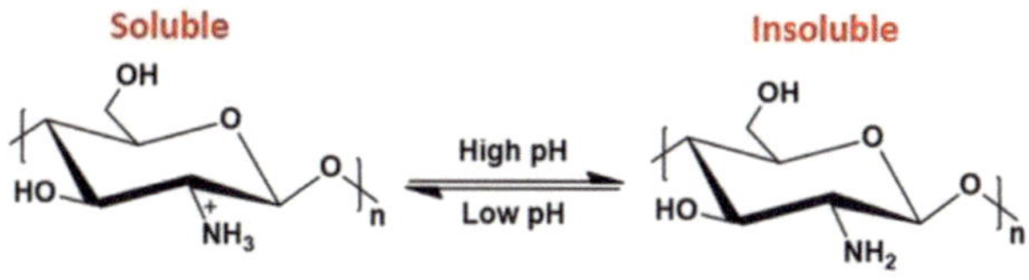

Figure 4.
Nature of Cts below its pKa.

concentration of acid, pH, deacetylation time, previous treatments applied to chitin isolation and chitin particle size, etc. Cts is regarded as a strong base as it has amino groups with a pKa value equal to 6.3. Cts dissolves easily in dilute acid solutions below pH 6.0. It is a reality that the amino groups in the chitosan structure have significantly changed its charged state and its properties. H^+ ions present in low pH solutions protonate the NH_2 groups of Cts and causes the formation of $(NH_3)^+$. Thus, Cts has a polycationic biopolymer structure. This situation turns Cts into a very easily soluble polyelectrolyte in acidic medium. The most functional amine groups of Cts are protonated in low pH buffer solution. On the other hand, Cts amines are transformed to deprotonated form as the pH rises above 6 and the polymer loses its charge and becomes insoluble in basic solutions [12, 13]. The soluble–insoluble transition forms at its pKa value between pH 6.0 and 6.5. The soluble and insoluble situation of Cts in aqueous media occurs pKa valued around pH between 6 and 6.5. pKa is considered to be significantly dependent on the degree of N-acetylation, hence Cts solubility varies depending on the deacetylation method applied [14]. Acetic formic, formic acid, lactic acids and organic acids are the most commonly used solvents. They can easily dissolve Cts as it is easily converted into quaternary nitrogen salts in aqueous solution at low pH values [15]. Cts, on the other hand, can be easily dissolved in some solutions (formic acid, acetic acid, etc.) at pH <6 due to its cationic structure. Considering the best solvents, it has been found that it is the formic acid from which solutions are obtained in aqueous systems containing 0.2–100% formic acid. 1% acetic acid solution around pH 4.0 has been applied as the most utilized solvent in the Cts applications [15]. Likewise, Cts is very easily soluble in hydrochloric acid and dilute nitric acid solutions, however insoluble in sulfuric and phosphoric acids. These acids are not shown since they can break the chitosan polymeric chains promoting depolymerization. Cts is very difficult to break up in organic solvents, for example, dimethylformamide and dimethyl sulfoxide. Its solubility in the acidified polyol is generally very excellent [15, 16].

2.4 Cross-linkers used for chitosan cross-linking

Cts is soluble in both organic and mineral acids. This is a limiting parameter for industrial applications in wastewater treatments. By using different chemical reagents in the cross-linking process, the structure of Cts is strengthened mechanically and chemically (**Figure 5**). The chemical modification of Cts can be done for two purposes:

a. To prevent the dissolution of the polymer when metal adsorption or metal desorption is performed in acidic solutions,

b. To increase the metal adsorption properties (increase in adsorption capacity or increase in adsorption selectivity).

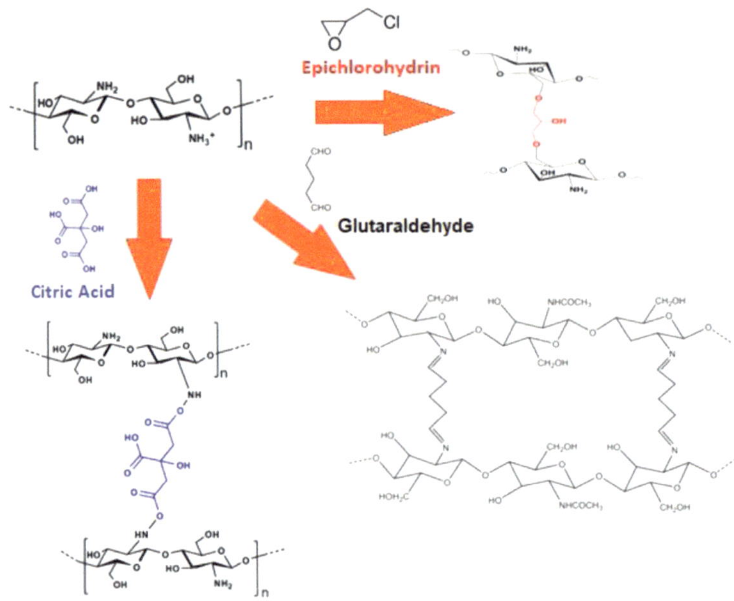

Figure 5.
Schematic representations of the cross-linking reaction of Cts with cross-linking agents.

The modifications carried out in the structure of Cts are aimed to further improve its biological and chemical properties and to change its solubility in the solvent or aquatic environment to be used. The behavior of Cts chains in diluted acids has been studied and at higher ionic strength or lower pH values, amino groups will undergo a higher protonation. In these cases, the macromolecules have been found to behave more like stiff rods [17, 18]. Cross-linking process can be applied by the reaction of Cts with different crosslinking agents such as glutaraldehyde, epichlorohydrin, polyethyleneimine etc. (**Figure 5**). Today, there are many activating agents and newly developed methods. Glutaraldehyde is widely used crosslinking agent because it is cheap and very effective crosslinker. Glutaraldehyde, a linear 5-carbon dialdehyde, is transparent and colorless [19]. It is an oily liquid with a pungent odor that can be dissolved in water, alcohol and organic solvents in all proportions. Chemical stability may be enhanced through chemical procedures which include crosslinking with glutaraldehyde for utility in a chemically acidic environment (**Figure 6**). By means of developing new bonds between Cts chains, the polymer is proof against dissolution even in strong solutions such as hydrochloric acid solution. In addition, the reality that Cts includes amine groups is an essential factor in being a good adsorbent. It is possible to observe an increase in the adsorption

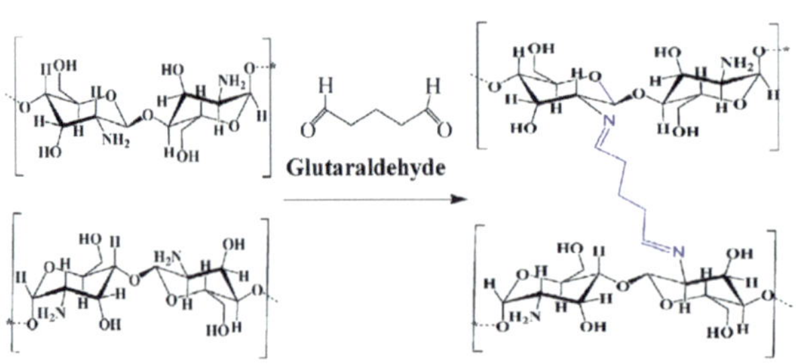

Figure 6.
Schematic representation of the cross-linking reaction of Cts with glutaraldehyde [20].

capacity by increasing the amine groups. The most important advantage of glutaraldehyde is the presence of suitable amino groups that perform the binding on Cts surfaces [21, 22]. Adding cross-links to the structure of Cts gives it a solid three-dimensional structure. To form cross-linked Cts, chemicals with a wide variety of groups are used to form the cross-link structure. Condensation reaction resulting from the reaction between an aldehyde function and a primary amine group from the Cts chain. It quickly joins the matrix structure of Cts.

There may be the opportunity to apply mono-functional reagents (epichlorohydrin) by opening the ether group for grafting to an amine function through the Schiff base reaction, which is likewise capable of react, at the same time the chloride group then interacts with different functional groups or other amines. Tri-polyphosphate is also selected as a possible cross-linking agent, which may be used to prepare Cts gel beads by its coagulation and neutralization impact.

3. Adsorption of Cr (VI) using chitosan-based materials

3.1 Toxicity of chromium

The toxicity of chromium compounds varies according to the pH, temperature and oxidation step. Cr (VI) ions are much more toxic than chromium (III) (Cr (III)) in terms of toxicity [23, 24]. EPA's threshold limit value is 10 times lower than Cr (III). When Cr (VI) solution mixes with seawater, it prevents some aquatic plants from photosynthesis, reduces reproduction in fish and can cause fish death. Cr (VI) causes burns in human body, in case of contact, causing irritation, wounds and ulcers on the skin and respiratory tract [25]. Sensitivity to Cr (III) and Cr (VI) causes allergic reactions, redness in the eyes and nose, itching and rash. Taking Cr (VI) with the digestive system can cause ulcers, necrosis and death in the stomach and intestines. The recommended Cr (VI) limit in drinking water is 0.05 ppm [26, 27]. One of the most important effects of Cr (VI), which is a very oxidizing substance, is that it causes lung cancer [28, 29].

3.2 Removal of hexavalent chromium

Chromium pollution is especially caused by chrome plating, automotive, leather and paint industry wastes [18]. Traditional refining methods used for chromium refining are not highly efficient, and these techniques require large amounts of chemicals and energy. Because they are costly, their use is impractical. Adsorption has superior properties in these aspects. In chromium treatment, the ability to use many different sources as adsorbents, such as plants, animal materials and various microorganisms, are easy to obtain, they can be produced by cheap and simple methods, regeneration ease and high removal efficiency are the features that make the adsorption approach preferred [23, 30].

Chemical precipitation, microfiltration, ultrafiltration, flotation, reduction, dialysis, membrane technologies, chelating, ion exchange, evaporation, solvent extraction, reverse osmosis, and adsorption can be listed among the methods used for dechroming of industrial wastewater [20]. In the selection of these methods, the acidic or basic character of the wastewater, the target envisaged for removal and recovery, the type and concentration of the chromium compound in the waste, the cost, chemical and energy consumption, the management of the waste generated by treatment and the removal efficiency should be taken into consideration.

The pH of the aqueous chromium influences the surface charge of the modified adsorbent, the degree of ionization and the adsorbate species. Depending on the current pH, the Cr (VI) waste solution can be found in dichromate ($Cr_2O_7^{2-}$), chromate (CrO_4^{2-}) and hydrogen chromate ($HCrO_4^-$) forms. According to the pH of the Cr (VI) in aqueous solution varying ion types are given (**Figure 7**). $HCrO_4^-$ and $Cr_2O_7^{2-}$ are the dominant species on the experimental concentration and these are the dominant components at a low pH. When the pH is excessive, CrO_4^{2-} is the dominant component of Cr (VI), so the initial solution pH is an important parameter for the adsorption of chromium. Electrostatic attraction forces between positively charged adsorbent surface groups and anionic Cr (VI) types occurs. This is particularly more common because Cr (VI) are present as oxyanions ($HCrO_4^-$, $Cr_2O_7^{2-}$ and CrO_4^{2-}) in solution phase. Positively charged adsorbent surface groups are formed as a result of the formation of both protonation or quaternization of the groups on the adsorbent matrix.

The equilibrium among the various Cr (VI) species may be represented by using Eqs (1), (2) and (3).

$$H_2CrO_4^- \rightleftharpoons H^+ + HCrO_4^- \quad (1)$$

$$HCrO_4^- \rightleftharpoons H^+ + CrO_4^{2-} \quad (2)$$

$$2HCrO_4^- \rightleftharpoons Cr_2O_7^{2-} + H_2O \quad (3)$$

A few other parameters affecting the adsorbent capacity; solution pH, equilibration time, temperature of solution, coexisting ions and adsorbent dosage concentration.

3.3 Application of various chitosan form for Cr (VI) removal

Cts modifications can be by physical methods or by chemical methods in which crosslinking or gridding of functional groups is performed. Cts can form perfect gels in the membrane, bead, capsule or different forms thanks to its ability to dissolve in organic acids and to combine with polyionic compounds. Four different methods are used to create Cts gels. These are solvent evaporation method, neutralization method, cross-linking method and ionotropic gelation method. Among these methods, the neutralization method is the main method used in the preparation of spherical Cts beads of different sizes and pore properties. It is carried out by

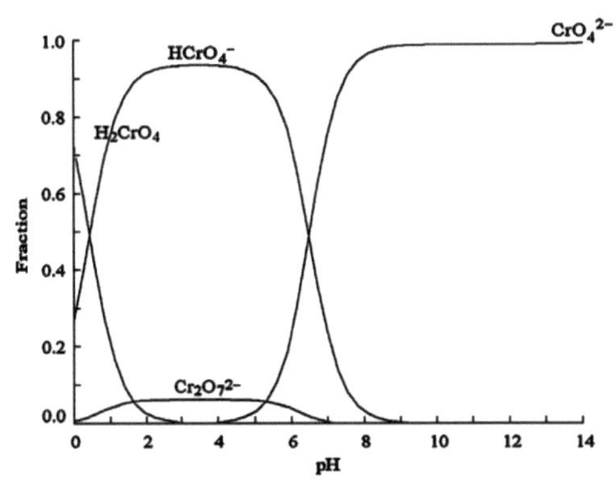

Figure 7.
Chromium speciation as a function of pH.

dropping Cts solution into NaOH and ethanol solution drop by drop. Since Cts is not soluble at high pH values, the drops become solid due to polymer precipitation.

The chelating property of Cts enables it to retain various metal ions. Cts is a polymer that has a wide range of uses in the removal of heavy metals from wastewater systems, in the food industry, in the encapsulation of drugs that need to be released slowly or overtime, in cosmetics, agriculture, pulp and paper industry to have all these properties. Cts, a poly N acetyl glucosamine, contains cationic amino and polar hydroxyl groups, which are chemically reactive groups attached to the glucosamine chain it carries. These groups support immobilization methods such as adsorption and covalent bonding. Amino capability qualifies Cts for a lot of chemical reactions inclusive of quaternization, grafting, alkylation and reaction with carbonyl compounds. The presence of the hydroxyl group lets in Cts to form hydrogen bonding together with polar atoms, grafting, crosslinking, and some chemical reactions inclusive of O-acetylation [31].

Recently, more composite materials containing modified Cts used in chromium removal have been reported. Eliodório et al. investigated the synthesis and characterization of Cts functionalized with three different ionic liquids and their application in Cr (VI) waste treatment [28]. Amino groups in Cts are significantly effective in the adsorption mechanism, especially when protonated at acidic pHs. Besides, intermolecular interaction is also effective. This phenomenon will increase the cationic ability on the Cts surface, facilitating the adsorption of negative groups via electrostatic interaction and improving the elimination of Cr (VI) (**Figure 8**). At acidic pHs, there has been a strong protonation of the amine groups that gave the Cts surface a positive character. Thus, the electrostatic attraction force between this positively charged surface and negatively charged anion ($HCrO_4^-$) will boom and this could motive the increase of adsorption of Cr (VI).

Parlayıcı and Pehlivan inspect the adsorption of chromium ions on Cts doped with multiwalled carbon nanotubes (Cht-MWCNT) [32]. Under optimum

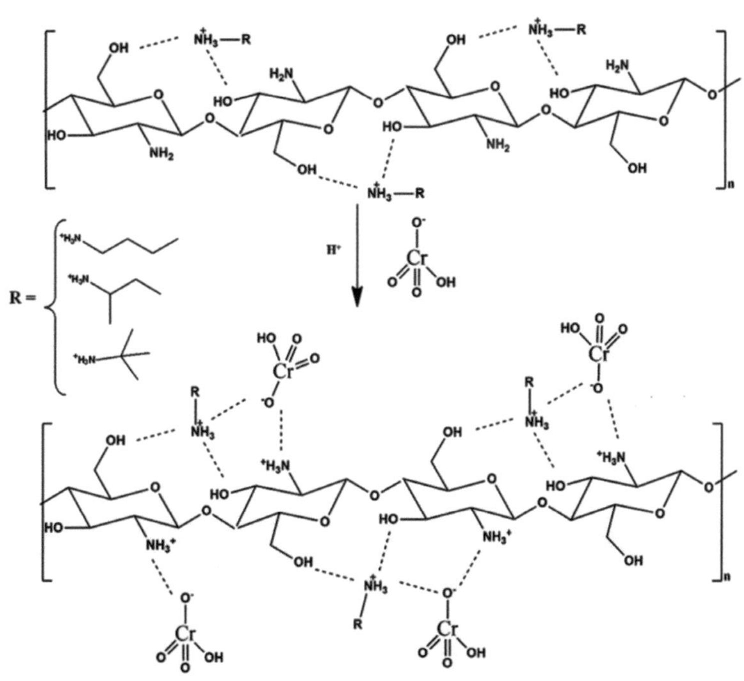

Figure 8.
Adsorption mechanism of chromium hexavalent [28].

situations, maximum adsorption capacity of Cr (VI) determined by using Langmuir model have been improved 26.14 mg/g. When carbon nanotubes (CNT) is examined, carbon atoms are allotropes of carbon that have an aromatic surface while in a sp^2-type hybridization rolled together like a tubular system (1D system) [33]. The structural properties of the CNT's surface permit because of an intense together with solids having organic molecules through non-covalent forces, for example, hydrogen bonding, π–π stacking, electrostatic forces, van der Waals forces and hydrophobic interactions. Moreover, the CNT structure allows the incorporation of one or more chemicals functional groups and they can build selectivity and stability of the subsequent framework. The chemical functional groups may be anchored to CNT surface through functionalization or purification processes [34]. This carbon allotrope structure is an interesting nanostructure that harbors remarkable electronic and mechanical properties that are directly dependent on chirality and diameter. Its excellent properties combined with unusual morphology make it extremely attractive for some many practical applications in wastewater systems; for example, with the improvement of solid composite materials for ideal adsorption, high impact ability, selectivity and productivity [35].

Cts-humic acid-graphene oxide (Cts-HA-GO) composite was produced (**Figure 9**) and the removal of toxic Cr (VI) ion from aqueous solutions was

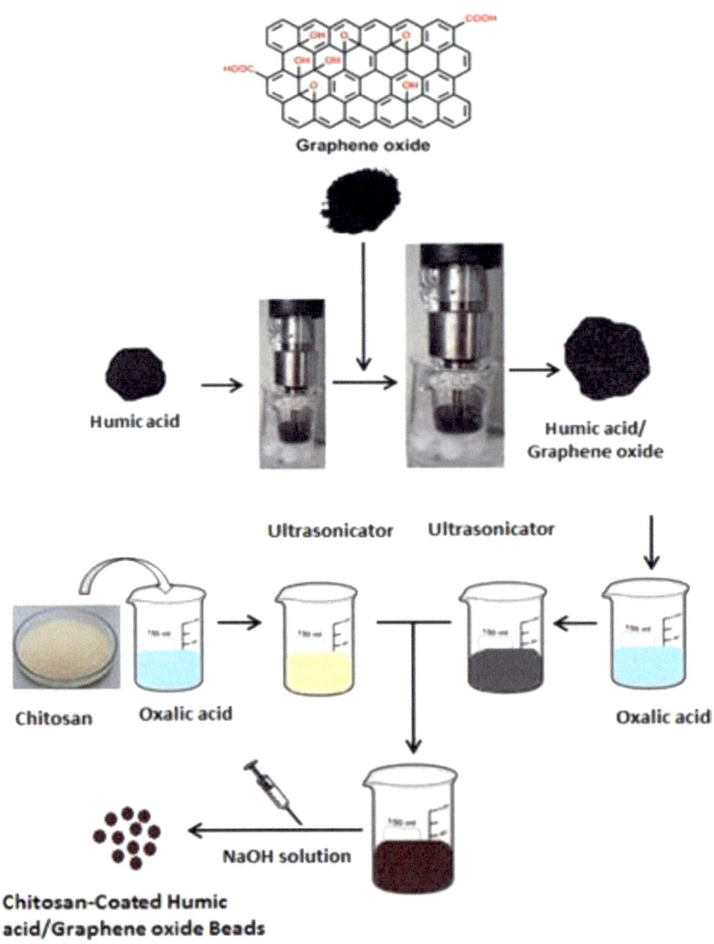

Figure 9.
Preparation of Cts coated humic acid/graphene oxide composite beads.

studied by batch equilibration technique [27]. The percentage adsorption was 85.28% by using 200 ppm Cr (VI) and 2 g/L composite. Maximum adsorption of Cr (VI) ion has occurred at pH 2 value as 83.64 mg g^{-1}.

Graphene oxide (GO)-based adsorbents are very useful substances because of the presence of functional groups on the surface of imparted for the duration of the oxidation of graphite. The presence of those groups is taken into considered best adsorbent for the use of graphene oxide for metal ion chelation because they show off hydrophilic properties due to hydrogen bond. Some other exciting magnificence adsorbents is GO-based adsorbent, mainly due to the plentiful –OH, –COOH and –C=O functional groups on their surface imparted throughout oxidation of graphite. The presence of such groups makes GO an ideal adsorbent for metal ion chelation and outstanding hydrophilic properties owing to hydrogen bonding [36].

Humic acids (HA) are a principal component of humic substances, which are the significant constituents of soil, peat and coal. It is a perplexing combination of various acids containing carboxyl and phenolate groups so that the mixture acts functionally as a dibasic acid or, occasionally, as a tribasic acid. There are functional groups in the HA structure. These consist of various types of carboxylic (COOH), phenolic hydroxyl and carbonyl (C=O) structures. HAs have an extraordinary effect on metal adsorption as they have carboxyl and phenolic -OH groups that interact with various metal ions [37]. The carboxyl groups, which render the polymers negatively charged at neutral pH values, are in particular effective in complexing metal ions in aqueous solution.

Cts consists of a crystal phase and a non-crystalline (amorphous) phase [38]. Crystalline Cts molecules are in the form of layers and are tightly held by intra-layer hydrogen bonds (**Figure 10**). Fenton reaction was used to increase the Cr (VI) adsorption capacity of Cts [24]. With the Fenton modification, Cts both efficiently adsorbed Cr (VI) and converted it to less toxic Cr (III) over a wide pH range as a result of layer formation defined by a sandwich model. The highly reactive HO• destroyed the hydrogen bond in the Cts structure, and Fe^{3+} ions were chelated with the Cts biopolymer. After Fenton modification, $HCrO_4^{4-}$entered the gap in the Cts structure and was adsorbed in the newly formed adsorption sites. The adsorption process of Cr (VI) using the Fenton modified Cts, adsorption isotherms of Freundlich were investigated and the adsorption capacity exceeded 120 mg/g.

Cellulose and lignin-based adsorbents are the most basic biopolymer widely used for toxic metal removal from wastewater. Cellulose exhibits natural qualities such as strength, reward, biodegradability, non-toxicity, and mechanical stability. However, cellulose is an odorless water-soluble linear polymer covalently bonded in its structure by the monomeric units of b-D-anhydroglucopyranose C1-C4 b-glycosidic bonds [39]. Due to its water solubility, it has some drawbacks in terms of being used as an

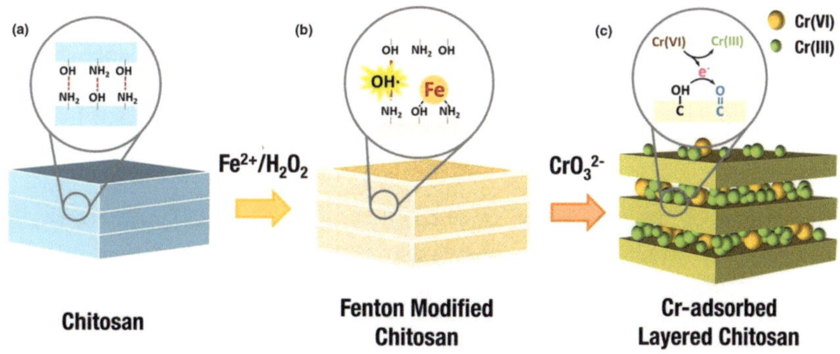

Figure 10.
The adsorption of Cr (VI) onto the Fenton modified Cts by a sandwich model [24].

adsorbent in the raw structure; thus, it is made more suitable for adsorption applications by applying various functionalization techniques. Abundance, low cost and availability of various functions groups in agro-based byproducts (hydroxyl, carboxyl, carbonyl etc.) like shells, barks, straws, stem and seeds, they attract the attention of researchers to explore potential applications in the removal process of toxic metals from wastewater. These substances have the small surface area and their low stage of adsorption performance in wastewater treatment limits their implementation among their ordinary state. Valorization of agricultural waste in adsorption processes is an environmentally friendly approach for wastewater treatment studies. Parlayıcı and Pehlivan reported the preparation of glutaraldehyde crosslinked Cts coated Rosehip (*Rosa canina*) seed shell (Cts/RS) capsules to evaluate the adsorption of Cr (VI) ions from aqueous solution [40]. The data were fitted to the Langmuir adsorption isotherm. The maximum capacity of Cts/RS was determined as 34.13 mg/g.

Black sesame (*Sesamum indicum* L.) seed pulp (BSSP) and Cts (Cts)-coated black sesame seed pulp bead (Cts-BSSP) composite were produced (**Figure 11**) and tested to remove hexavalent chromium (Cr (VI)) during the adsorption process [23]. BSSP is an agricultural waste that has no economic value. After reaching equilibrium time, the Cr (VI) removal capacity was calculated as 31.44 mg/g for Cts-BSSP and 18.32 mg/g for BSSP. The results indicated the feasibility of using Cts-BSSP as adsorbents for the removal of Cr (VI) from aqueous medium.

Let us talk about the synergy in adsorption and reduction during removal of Cr (VI) from aqueous media by natural adsorbents. Anionic Cr (VI) is adsorbed onto protonated solid adsorbent surfaces preferentially by electrostatic bonding or anion exchange. When the strong redox nature of Cr (VI) is examined, it causes the adsorbent surface to oxidize while it is reduced to Cr (III). If the aqueous medium has acidic conditions and the presence of electrons is large, this further catalyzes the reduction. Heteroatoms such as; O, N and S in the adsorbent structure carry these electrons, making Cr (VI) easier to reduce. In the case of Cr (VI) adsorption, the chelation mechanism is very touchy to pH in the attachment of the metal by the adsorbent and it is generally found that no adsorption happens at low pH. Therefore, a simple change in the pH of the solution medium can reverse the adsorption reaction completely.

$$Cr_2O_7^{2-} + 14H^+ + 6e^- \rightleftharpoons 2Cr^{3+} + 7H_2O \; E_0 = 1.33\,V \tag{4}$$

$$Cr_2O_7^{2-} + 4H_2O + 3e^- \rightleftharpoons Cr(OH)_3 + 5OH^- \; E_0 = -0.13\,V \tag{5}$$

It has been determined that there are two feasible reduction mechanisms in the removal of Cr (VI) from aqueous solutions (Reaction 1 and Reaction 2). In mechanism I indicates that Cr (VI) will be reduced to Cr (III) in acidic conditions at some stage in adsorption. Mechanism II is an indirect reduction. Adsorbed Cr (VI) species take electrons from an adjoining location of the adsorbent mass to facilitate the reduction. Besides, reducing agents such as ferrous iron, zerovalent iron, and iron composites have also been used [29].

The physicochemical properties of Cts can be improved by making a stronger structure, especially by making a composite by chemical modification or the inclusion of some strengthening agents. New adsorbents can also be formed by integrating into the Cts matrix with composite materials that allow continuous capture of toxic materials [41]. Clay mineral has a long enchanting history of metal-binding capacity while being used independently or combined with other natural polymers. Clay-biopolymer composites show excellent potential and high efficiency for the removal of chromium from aqueous solution. Parlayıcı prepared Cts coated perlite

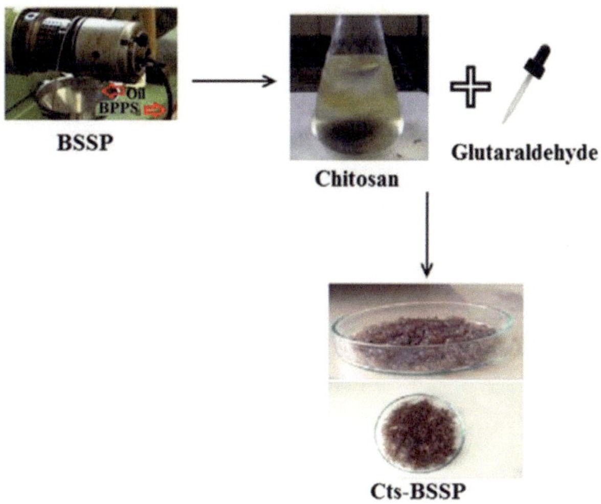

Figure 11.
Preparation of Cts-coated black sesame seed pulp beads (Cts-BSSP) [23].

composite beads (**Figures 12** and **13**) and investigated systematically the process parameters influencing the adsorptions of Cr (VI) ions [30].

Jiang et al. synthesized magnetic $Fe_3O_4@SiO_2$–Cts (MFSC) biopolymers successfully by a simple cross-link method assisted and evaluated them for Cr (VI) removal [21]. The maximum adsorption capacities reported were 336.7 mg g^{-1}. Chitosan was successfully coated with inert substrate perlite and prepared as spherical beads. In this modified form of chitosan, it exhibits two types of accessory mechanisms to remove Cr (VI) from the aqueous environment, one of which is electrostatic interaction and the other is a procedure involving reduction (**Figure 14**).

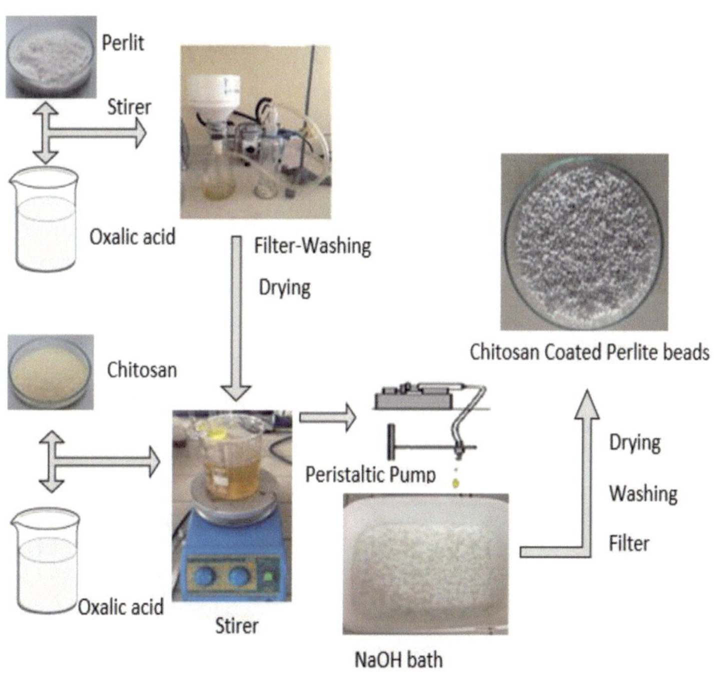

Figure 12.
Preparation of Cts coated perlite beads.

Figure 13.
Photograph images of Cts coated perlite beads (a) wet beads (b) dry beads.

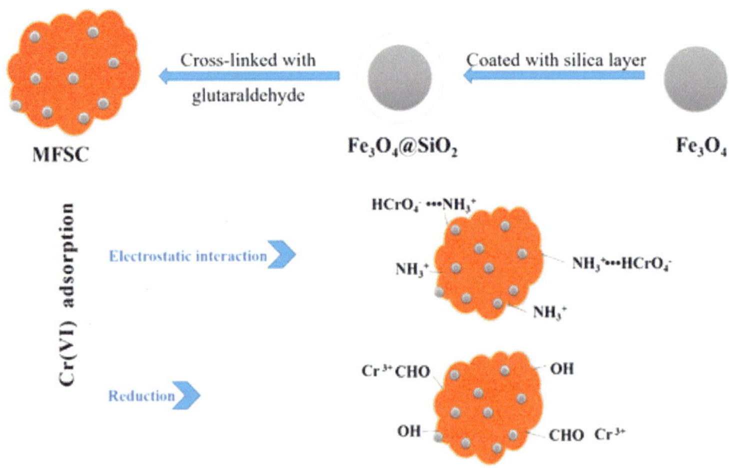

Figure 14.
Adsorption mechanisms involved in adsorption of hexavalent chromium onto $Fe_3O_4@SiO_2$–Cts (MFSC) biopolymers [21].

4. Conclusion

The main purpose of this article was to review some Cts based modified composite adsorbents for the adsorptive removal of Cr (VI). Most adsorbents used for longer in its pristine form, but with few changes carried out. Superior properties of adsorbents can be achieved by functionalizing and converting them into composite forms. The layers are formed by using different levels of polymers. Some of these layers are made up by different adsorbents, eg. Humic acid/graphene oxide, MWCNT/Cts, nanoparticle/Cts and Cts/perlite. Graphene oxide, nanoparticles and silica are known to have high surface areas, while biomass such as black sesame seed pulp and chitosan exhibit low surface area, but contain abundant functional groups. Cross-linking or modification is very effective in functionalization of adsorbents.

Features and development history together chromium toxicity, sources, removal mechanism from the aqueous medium, latest trends in composite adsorbent synthesis and preparation were discussed at length. The interpretation of the Cr (VI) removal mechanism summarized in the text. It is possible to reach higher capacities to determine the appropriate conditions in studies with composite adsorbents. The adsorption of Cr (VI) alone electrostatically is not sufficient. If Cr (VI) is considered to be suitable for strong oxidation, it can be removed from the aqueous environment by finding sensitive electrons from the groups in Cts-based adsorbents to transform into Cr (III) form. Heteroatoms such as O and N responsible for donating electrons are found in almost every Cts-based adsorbent structure. The removal mechanism of adsorbents is also affected by surface chemistry and physical properties in the Cr (VI) elimination process. Their high surface area properties, usability, porosity and cost effectiveness play an important role in increasing the capacity of the Cts-based adsorbent for Cr (VI) removal in wastewater. The modified adsorbents together with Cts for Cr (VI) removal have been shown to possess high capacities and they can be economically used in wastewater treatment units. By using Cts modified adsorbent, a faster and more efficient process will be provided to remove Cr (VI) for water treatment applications.

Conflict of interest

The authors declare no conflict of interest.

Notes

The authors declare no competing financial interest.

Abbreviations

Cts	Chitosan
MWCNT	Multiwalled Carbon Nanotubes
CNT	Carbon Nanotubes
GO	Graphene Oxide
HA	Humic Acids
RS	Rosehip Seed Shell
BSSP	Black sesame (*Sesamum indicum* L.) seed pulp
MFSC	Magnetic Fe_3O_4@SiO_2–Cts

Author details

Şerife Parlayıcı and Erol Pehlivan*
Department of Chemical Engineering, Faculty of Engineering and Natural Sciences, Konya Technical University, Konya, Turkey

*Address all correspondence to: erolpehlivan@gmail.com

References

[1] Wan-Ngah WS, Teong LC, Hanafiah MA. Adsorption of dyes and heavy metal ions by chitosan composites: a review. Carbohydr Polym,. 2011:83:1446-1456. DOI: 10.1016/j.carbpol.2010.11.004

[2] Ramesh HP, Tharanathan RN. Carbohydrates the renewable raw materials of high biotechnological value. Crit Rev Biotechnol, 2003:23(2):149-173. DOI: 10.1080/713609312

[3] Islam SZ, Khan M, Nowsad Alam KM. Production of chitin and chitosan from shrimp shell wastes. J Bangladesh Agril Univ. 2016:14:253-259. DOI: 10.3329/jbau.v14i2.32701

[4] Ravi Kumar MN. A review of chitin and chitosan applications. React Funct Polym. 2000:46: 1-27. DOI: 10.1016/S1381-5148(00)00038-9

[5] Roy JC, Salaün F, Giraud S, Ferri A, Chen G, Guan J. Solubility of chitin: solvents, solution behaviors and their related mechanisms. Solubility of polysaccharides. 2017:10. DOI: 10.5772/intechopen.71385

[6] Campana-Filho SP, Britto DD, Curti E, Abreu FR, Cardoso MB, Battisti MV, Sim PC, Goy RC, Signini R, Lavall RL. Extraction, structures and properties of alpha-and beta-chitin. Química Nova. 2007:30(3):644-650. DOI: 10.1590/S0100-40422007000300026

[7] Rouget C. Des substances amylacees dans le tissue des animux, specialement les articules (Chitine). Compt Rend. 1859:48:792.

[8] Rinaudo M. Chitin and chitosan: properties and applications. Prog Polym Sci. 2006:31:603-632. DOI: 10.1016/j.progpolymsci.2006.06.001

[9] Dotto GL, Souza VC, Pinto LA. Drying of chitosan in a spouted bed: The influences of temperature and equipment geometry in powder quality. LWT -. Food Sci Technol (Campinas). 2011:44:1786-92. DOI: 10.1016/j.lwt.2011.03.019

[10] Özdemir ZÖ. Functional properties and uses of chitin, chitosan. Chemistry and Industry. 2014:44(323):104-117. DOI: 10.1111/j.1365-2621.1982.tb10131.x

[11] Mehmood MA, Latif M, Hafeez FY. Heterologous expression and characterization of an antifungal chitinase Chi39 from Bacillus thuringiensis serovar konkukian. Pakistan Journal of Life and Social Sciences. (2012):10(2):116-122.

[12] Yi H, Wu LQ, Bentley WE, Ghodssi R, Rubloff GW, Culver JN, Payne GF. Biofabrication with chitosan. Biomacromolecules. 2005:6(6):2881-2894. DOI: 10.1021/bm050410l

[13] Prashanth KVH, Tharanathan RN. Chitin/chitosan: modifications and their unlimited application potential—an overview. Trends Food Sci Technol. 2007:18(3):117-131. DOI: 10.1016/j.tifs.2006.10.022

[14] Cho YW, Jang J, Park CR, Ko SW. Preparation and solubility in acid and water of partially deacetylated chitins. Biomacromolecules. 2000:1(4):609-614. DOI: 10.1021/bm000036j

[15] Kim KM, Son JH, Kim SK, Weller CL, Hanna MA. Properties of chitosan films as a function of pH and solvent type. Journal of Food Sci. 2006:71(3):119-124. DOI: 10.1111/j.1365-2621.2006.tb15624.x

[16] Pillai CKS, Paul W, Sharma CP. Chitin and chitosan polymers: Chemistry, solubility and fiber formation. Progress in Polymer Science. 2009:34(7):641-678. DOI: 10.1016/j.progpolymsci.2009.04.001

[17] Pa JH, Yu TL. Light scattering study of chitosan in Acetic Acid Aqueous Solutions. Macromol. Chem. Phys. 2001:202:985-991. [CrossRef] DOI: 10.1002/1521-3935(20010401)202:7<985::AID-MACP985>3.0.CO;2-2

[18] Piegat A, Goszczyńska A, Idzik T, Niemczyk A. The importance of reaction conditions on the chemical structure of N, O-acylated chitosan derivatives. Molecules. 2019:24(17): 3047. DOI: 10.3390/molecules24173047

[19] Nashchekina YA, Lukonina OA, Darvish DM, Nashchekin AV, Elokhovskii VY, Yudin VE, Mikhailova NA. Biological and Rheological Properties of Collagen Cross-Linked with Glutaraldehyde. Technical Physics. 2020:65(9):1535-1540. DOI: 10.1134/S1063784220090224

[20] Islam N, Dmour I, Taha MO. Degradability of chitosan micro/nanoparticles for pulmonary drug delivery. Heliyon. 2019:5(5):e01684. DOI: 10.1016/j.heliyon.2019.e01684

[21] Jiang Y, Cai W, Tu W, Zhu M. Facile cross-link method to synthesize magnetic Fe_3O_4@ SiO_2–chitosan with high adsorption capacity toward hexavalent chromium. Journal of Chemical & Engineering Data. (2018) 64(1):226-233. DOI: 10.1021/acs.jced.8b00738

[22] Reddy N, Reddy R, Jiang Q. Crosslinking biopolymers for biomedical applications. Trends in biotechnology. 2015:33(6):362-369. DOI: 10.1016/j.tibtech.2015.03.008

[23] Parlayıcı Ş, Tuna K, Özdemir E, Pehlivan E. Chitosan-coated black sesame (Sesamum indicum L.) seed pulp as a novel candidate adsorbent for Cr (VI) elimination. Water Science and Technology. 2019:79(4):688-698. DOI: 10.2166/wst.2019.085

[24] Jia N, Yun L, Huang J, Chen H, Shen C, Wen Y. A sandwich model of Cr (VI) adsorption and detoxification by Fenton modified chitosan. Water Environment Research. 2020:DOI: 10.1002/wer.1397

[25] Sharma N, Sodhi KK, Kumar M, Singh DK. Heavy metal pollution: Insights into chromium eco-toxicity and recent advancement in its remediation. Environmental Nanotechnology, Monitoring & Management. 2021:15:100388. DOI: 10.1016/j.enmm.2020.100388

[26] Almeida JC, Cardoso CE, Tavares DS, Freitas R, Trindade T, Vale C, Pereira E. Chromium removal from contaminated waters using nanomaterials–a review. TrAC Trends in Analytical Chemistry. 2019:118:277-291. DOI: 10.1016/j.trac.2019.05.005

[27] Parlayıcı Ş, Avcı A, Pehlivan E. Fabrication of novel chitosan-humic acid-graphene oxide composite to improve adsorption properties for Cr (VI). Arabian Journal of Geosciences,. 2019:12(19):1-13. DOI: 10.1007/s12517-019-4828-8

[28] Eliodório KP, Pereira GJ, de Araújo Morandim-Giannetti A. Functionalized chitosan with butylammonium ionic liquids for removal of Cr (VI) from aqueous solution. Journal of Applied Polymer Science,. 2021:138(9):49912. DOI: 10.1002/app.49912

[29] Yang R, Wang Y, Li M, Hong Y. A new carbon/ferrous sulfide/iron composite prepared by an in situ carbonization reduction method from hemp (*Cannabis sativa* L.) stems and its Cr(VI) removal ability. ACS Sustainable Chem. Eng. 2014:2:1270-1279. DOI: 10.1021/sc500092z.

[30] Parlayıcı Ş. The development of new natural and synthetic composite adsorbents to remove some heavy

metal ions. Doctoral Thesis, Selcuk University. 2016.

[31] Croisier F, Christine J. Chitosan-based biomaterials for tissue engineering. European Polymer Journal. 2013:49 (4):780-792. DOI: 10.1016/j.eurpolymj.2012.12.009

[32] Parlayıcı Ş, Pehlivan E. Removal of Cr (VI) from aqueous solution using chitosan doped with carbon nanotubes. Materials Today: Proceedings. 2019:18:1978-1985. DOI: 10.1016/j.matpr.2019.06.689

[33] Iijima S, Ichihashi T. Single-shell carbon nanotubes of 1-nm diameter. Nature. 1993:363: 603-605. DOI: 10.1038/364737d0

[34] Chatterjee S, Chatterjee T, Lim SR, Woo SH. Effect of the addition mode of carbon nanotubes for the production of chitosan hydrogel core–shell beads on adsorption of Congo red from aqueous solution. Bioresour Technol. 2011:102:4402-4409. DOI: 10.1016/j.biortech.2010.12.117

[35] Geim AK. Graphene: status and prospects. Science. 2009:324:1530-1534. DOI: 10.1126/science.1158877

[36] Chen JH, Xing HT, Guo HX, Weng W, Hu SR, Li S, Huang YH, Xue S, Bo Zhen S. Investigation on the adsorption properties of Cr(VI) ions on a novel graphene oxide (GO) based composite adsorbent. J. Mater. Chem. A. 2014:2:12561-12570. DOI: 10.1039/c4ta02004a.

[37] Pehlivan E, Arslan G. Uptake of Metal Ions on Humic Acids. Energy Sources, Part A,. 2006:28:1099-1112. DOI: 10.1080/009083190910451

[38] Mogilevskaya EL, Akopova TA, Zelenetskii AN, Ozerin AN. The crystal structure of chitin and chitosan. Polymer Science Series A. (2006) 48(2):116-123. DOI: 10.1134/S0965545X06020039

[39] Heinze T. Cellulose: Structure and Properties. Adv. Polym. Sci., 2015:271: 1-52. DOI: 10.1007/12.

[40] Parlayıcı S, Pehlivan E. Chitosan based a new bio-composite adsorbent for the removal of Cr (VI) from aqueous solution. Annals of Ecology and Environmental Science. 2018:2(4):30-35.

[41] Ferraro V, Cruz IB, Jorge RF, Malcata FX, Pintado ME, Castro PM. Valorisation of natural extracts from marine source focused on marine by-products: A review. Food Research International. 2010:43 (9):2221-2233. DOI: 10.1016/j.foodres.2010.07.034

Chapter 13

Chitosan for Using Food Protection

Sadik Büyükyörük

Abstract

Chitosan is a collective name used for a group of compounds having various molecular weights, which are produced from chitin by partially or fully de-acetylating and is prepared of β 1,4-linked glucosamine, and it is in deacetylated form of chitin acquired from fungi and/or crustaceans. Due its hydrophilic, cationic and biodegradable nature, chitosan has been cared for a biomaterial, medical, pharmaceutical, drug efficiency, textile, agricultural, food additive for preserving, wastewater clarification, plant pesticide agents and in wound healing. As a compound obtained using various methods, the most prominent features of chitosan are attributable to its antimicrobial and antioxidant properties. Among all the antibacterial compounds from crustaceans, chitosan and its derivatives have been widely used for providing the safety of the foods (especially marine based foods) and shelf life extension. This study presents information about antibacterial activity of chitosan, its mode of action against microorganisms, factors affecting its antimicrobial property and its application in food industry and for public health.

Keywords: Antimicrobial activity, chitosan, food safety, mode of action, public health

1. Introduction

The discovery of chitosan dates back to 1811 when Professor Henri Braconnot, director of the botanical garden in Nancy, France, isolated what he called "fungine" from fungal cell walls. About 30 years before the isolation of cellulose, in 1823, Odier conducted a study on insects and found that the same structure was present in insects as well as plants. Odier later named the fungine "chitin" a word derived from Greek that means membrane or envelope. The concept of chitin became more understandable when Lassaigne showed the presence of nitrogen in the structure of chitin in 1843. The term "chitosan" emerged following a discovery by Rouget in 1859. When heating chitin in a concentrated potassium hydroxide solution Rouget observed that the chitin became soluble with the chemical and heat treatment. Ledderhose described in 1878 that chitin consists of glucosamine. Hoppe-Seyler adapted the term chitosan from chitin in 1894. At the beginning of the 20th century, many studies on chitosan from sources of chitin were conducted. Rammelberg proved that chitosan was found in crab shells and fungi through his work in 1930. In addition, chitin was hydrolyzed in many ways and found to be a glucosamine polysaccharide. Studies on the formation of chitin and chitosan in mushrooms were performed with x-ray analyses in the 1950s. The first book on chitosan was published in 1951, 140 years after Braconnot's first observations. In the early

1960s, studies were conducted on the ability of chitosan to bind red blood cells. In the same year, chitosan was also considered as a hemostatic agent. In the next 30 years, chitosan was used in treatment plants to provide asepsis water. In the last 20 years, research on chitosan has intensified due to its many important properties [1]. Today, chitosan has many industrial applications and after cellulose, it is the most common polysaccharide chitin in the world. As one of the most important derivatives of chitin, chitosan is a polycationic biopolymer obtained by partial or complete deacetylation (removal of an acetyl functional group from an organic compound) of chitin in an alkaline environment [2]. The only difference between cellulose and chitosan biopolymer is the presence of the acetyl ($-NH_2$) functional group instead of the hydroxyl (-OH) functional group in the cellulose structure. This difference ensures that the chain structure of the chitosan biopolymer is polycationic. Many superior properties of chitosan arise from this polycationic structure. In addition to this advantage, the presence of both –OH and $–NH_2$ groups in the chain structure of chitosan and the fact that these groups can be modified in different ways is a situation that highlights its uses [3]. Chitosan, which can be obtained in large quantities from many natural sources containing chitin, such as the exoskeleton of mushrooms, crayfish, shrimp, and crabs, is more advantageous than other biopolymers including chitin in terms of non-toxicity to organisms, easy biodegradability, and biocompatibility. For these reasons, chitosan is a natural, safe, cheap, raw material biopolymer used in many industrial areas such as food, medicine, pharmaceuticals, cosmetics, agriculture, wastewater treatment, and textiles. Besides having antiviral, antibacterial, and antifungal properties, chitosan is also an effective agent in controlling and reducing the spread of diseases by promoting the defense system of plants. In addition, chitosan is being used for improvement in agriculture because it chelates metal ions in the environment (water, soil, etc.) and prevents the uptake of toxic metals in plants [4].

Chitosan is a natural and biodegradable biopolymer used in different industrial applications as an agent for flocculation and chelating, permeability control, and as an antimicrobial, among other processes. Predominantly produced today by the deacetylation of chitin on an industrial scale, chitosan is found in the exoskeleton of crustaceans and insects, and the cell walls of many fungi and some algae. Although the main source of chitin is crab, shrimp, crayfish, and shrimp residues, the importance of insect chitosan depends on the role insects play as a sustainable protein source. Insects are seen as an alternative to traditionally consumed proteins derived predominantly from traditional livestock (mainly cows, chickens, and pigs) and fish. In addition, using the insect as a protein source produces two by-products of interest to the industry, lipids that can be used as biofuels (30–40% total dry weight) as well as a residual material made of chitin with some bioactive properties from which chitosan can be produced [5].

2. Antimicrobial activity

Chitin and chitosan have interesting physicochemical, biological, and mechanical properties. One such property of chitosan is related to its antimicrobial activity. There are several studies demonstrating the antimicrobial and antifungal properties of chitosan and many derivatives [6–11]. Recently, the effect of the physical form of chitosan on its antibacterial activity against pathogenic bacteria was studied. Researchers examined chitosan coating as an inhibitor of *Listeria monocytogenes* on vacuum-packed pork fillets and fresh cheese. The antibacterial effect is reported to be generally rapid, eliminating bacteria within a few hours. As for the physical properties of chitosan, these are mainly governed by two factors: deacetylation degree (DD) and molecular weight (MW). Natural origin, as well as variability in chemical composition, can affect the properties of chitosan and have an impact on its industrial uses. Some studies have revealed that DD correlates with the antimicrobial activity of chitosan. The effect of chitosan as an antimicrobial in the agriculture and food industry has been studied. According to these studies, the antimicrobial activity of chitosan depends on several external and internal factors as well as a number of environmental factors. The type of microorganism, physiological state, pH, temperature, ionic strength, metal ions, ethylenediaminetetraacetic acid (EDTA), the presence of organic matter, MW, DD, solvent, and concentration are all influencing factors [12].

Chitosan is a commercial biopolymer produced predominantly from crab and shrimp residues. The physicochemical properties of chitosan affect the functional properties that differ according to crustacean type and preparation methods. Chitosan has been studied to compare the functionality of commercial products obtained from crustacean and insect chitosan as antimicrobials. The results indicated differences between commercial insect chitosan and crustacean chitosan with regard to their antimicrobial capacity. Generally speaking, crustacean chitosan with a pH of 5,0 during a 49-hour incubation period displayed a greater antimicrobial capacity than insect chitosan at the same pH. This behavior was seen mostly in Salmonella cases where crustacean chitosan resulted in more than 4 logarithmic decreases, whereas insect chitosan was only bacteriostatic resulting in about a 1 logarithmic decrease. The similar behavior was noticed for *Escherichia coli*, despite the smaller differences in antimicrobial influence in Salmonella cases. As noted, some studies have pointed out potential differences between the functions and physical properties of chitosan in different species of crustaceans. This may be even more pronounced among chitosan obtained from various sources such as crustaceans and insects [6].

Antimicrobial activity can be adversely affected by pH, and as such pH plays an important role in the antimicrobial capacity of chitosan. Low pH chitosan appears to have more antimicrobial activity than high pH chitosan [13]. A study was conducted to determine the effect of two different concentrations of chitosan at pH 6,5 and 5,5 on different pathogenic microorganisms, including *Salmonella* Typhimurium, *E. coli*, and *L. monocytogenes*. The author concluded that chitosan with a pH of 6.5 had a rather weak effect on pathogenic microorganisms and could not inhibit *L. monocytogenes*. At pH 5,5; there was inhibition of the microorganisms tested for 24 to 72 hours of storage at 30°C. The researcher concluded that chitosan acts better at pH 5.5 than at pH 6.5 [14]. Another researcher examined the antibacterial activity of chitosan of different MW at various pH levels (pH 4, 4.5, and 5) on *L. monocytogenes* strains. The results also indicated that, with the exception of two *L. monocytogenes* strains, chitosan with a pH of 5 had the greatest bacterial reduction effect during a 24-hour incubation period [15]. In another study, two pH levels were tested at a concentration of 0.15% (w/v) chitosan. Later an 8-hour incubation, the antibacterial effect was found to be higher at pH 5,0 than pH 6,2 for *S.* Typhimurium, but the opposite for *E. coli* and *Listeria monocytogenes*, where the antimicrobial effect of chitosan at pH 6.2 was stronger than it was at pH 5.0. The effect of chitosan at both pH levels seemed to be dependent on the microorganism. Differences were observed in chitosan at both pH levels of acetic acid compared to control. Chitosan exhibited a pronounced antimicrobial activity at both pH values, particularly on *L. monocytogenes*. Chitosan obtained from both sources, crustaceans and insects, was bacteriostatic or bactericidal for three pathogenic microorganisms at pH 5.0 [6].

Several hypotheses have been proposed about the antimicrobial function of chitosan. Ionic interactions occcuring between the positive charges of amino groups and negative bacterial surface molecules under acid conditions change the membrane permeability which leads to cellular lysis. Interaction with necessary nutrients for bacteria could be another mechanism. Chitosan's bactericidal effect may also be affected by the inoculum size to the bacteria growth [6]. In some studies, all compounds tested after 4 hours of incubation for an inoculum size of 10^3 cells/mL were bactericidal at any concentration of chitosan tested. In contrast, at a higher initial inoculum concentration, 0.1% (w/v) chitosan was only bacteriostatic. Regardless of the inoculum level, any chito-oligosaccharide mixture of 0.25% (w/v) was sufficient to reduce the starting population of *E. coli* by at least 3 log cycles. However, the results regarding the effect of inoculum size are not conclusive because they vary with pH and type of microorganism. Therefore, it is not possible to predict higher antimicrobial activity at a given inoculum size in all cases [16, 17].

Included in the peptidoglycan layer on the cell surface, teichoic acid is vital for the growth of Gram-positive bacteria as well as for cell division. Chitosan and its derivatives can bind to teichoic acid on the surface of Gram-positive bacteria non-covalently. Chitosan's effect on the cell membrane has not been clearly discovered yet; however, it is well-known that it affects the cell membrane because it has a greater hydrodynamic diameter than peptidoglycans' pore size. Strangely, chitosan with a MW of 5 kDa suppresses DNA synthesis and promotes *Bacillus megaterium* filamentation, which suggests that the chitosan's MW plays a role in its potential to affect cell membrane permeability [18]. In addition, the effect of chitosan on teichoic acid has been demonstrated by testing *Staphylococcus aureus* mutant strains lacking the genes needed for teichoic acid biosynthesis [19]. Mutant strains were found to be more resistant to the environmental conditions than the wild-type strain, which indicates that anionic teichoic acid improves chitosan's antibacterial properties against Gram-positive bacteria. Teichoic acid has, interestingly, many functions. It controls activities of enzymes, helps to cope with environmental stress, and manages the cationic concentration in the cell cover by binding to the cell surface and the cell receptor. The mechanism of antimicrobial effect of chitosan on Gram-positive bacteria is due to electrostatic effect with teichoic acid, resulting in disruption and death of cell [18]. Two different mechanisms mediate the interactions between chitosan and the outer membrane of Gram-negative bacteria. The first mechanism involves chelating chitosan with various cations when pH is higher than pKa, resulting in a breakdown in the uptake of essential nutrients and a breakdown in cell wall integrity. The second mechanism involves electrostatic interactions between chitosan and anions associated with lipopolysaccharides in the outer membrane. Chitosan also creates disruptions in the inner membrane, causing intracellular content to leak. In addition, chitosan can pass through the cell membrane of Gram-negative bacteria where it likely interferes with DNA/RNA synthesis and triggers an intracellular response in cells. Thus, the electrostatic interactions between chitosan and the anionic surface are crucial to chitosan's antimicrobial properties against Gram-negative bacteria. Moreover, chitosan can bind non-covalently to the cell membrane of Gram-negative bacteria, suggesting it plays an important role in antimicrobial activity [20, 21]. The difference between Gram-positive and Gram-negative bacteria is more obvious compared to chitosan-resistant fungi and chitosan-sensitive fungi. Chitosan is belived to interact with a phospholipid component of chitosan-sensitive fungi electrostatically, thereby breaking it down and participating the cell, fnally leading to the prevention of DNA/RNA as well as protein synthesis. However, chitosan is unable to make the cell wall of chitosan-resistant fungi permeable due to its variable fluidity, so it remains on the cell surface and forms a polymer to function as a barrier against oxygen and necessary nutrients, ultimately resulting in cell death. The lowered antimicrobial activity of chitosan was also seen in a *Neurospora crassa* mutant strain, explaining the lower levels of unsaturated fatty acids relative to the wild-type strain. Thus, the antibacterial effect of chitosan on fungi is greatly affected by the fluency of the cell membrane and the type of mushrooms [18]. Chitosan inhibited the growth of *Aspergillus flavus* and aflatoxin in liquid culture, pre-harvest corn and peanuts, and increased the production of phytoalexin in germinating peanuts. Chitosan has become the first compound in the list of basic substances approved by the European Union for plant protection in agricultural practices, both for organic agriculture and for integrated pest control (Tes. EU 66 2014/563). Thus, chitosan can be used as a biodegradable fungicide. In addition, chitosan shows antiviral activity against plant viruses. It has been demonstrated that chitosan inhibits productive infection caused by bacteriophages. The efficacy of bacteriophage inhibition is directly dependent on the final concentration in the medium. The main factors by which chitosan suppresses phage infections are phage particle inactivation and inhibition of bacteriophage growth at the cellular level. Chitosan can be used for induction of phagoresistance in industrial microorganism cultures to prevent unwanted phagolysis caused by inoculum contamination with virulent bacteriophages or spontaneous prophage induction in lysogenic culture [22].

As stated previously, pH can play a key role in chitosan's antimicrobial activity, and the pKa of chitosan sequences from 6,3 to 6,5 [23]. Chitosan only dissolves in acidic aqueous environment where it becomes polycationic when the pH value is lower than the pKa amount. Polycationic chitosan molecules react with negatively charged cell wall molecules, including proteins, phospholipids, polysaccharides, and fatty acids because of the high intensity of amino groups found on the polymer surface, ultimately causing intracellular materials to leak. Chitosan exhibits higher antimicrobial activity at low pH values (< 6) because its amino group is ionized at low pH rates. Moreover, the positive charge of chitosan improves at low pH values, increasing the absorption of chitosan at the bacterial cell wall. Moreover, at upper pH values (> 6) the amino group of chitosan becomes aprotic, which may lead to precipitation from solution [24]. One study informed that chitosan's antimicrobial activities against *Klebsiella pneumoniae* partly resulted from the polycationic nature of chitosan, thus being associated with the protonation of the amino group. The protonation of the amino group is related to the degree of polymerization as well as the pH of the environment. For example, chitosan is more effective against *Candida lambica* at pH 4 than pH 6 [18].

In the early 1960s, chitosan's ability to bind to red blood cells was investigated. At that time, it was also seen as a hemostatic agent. Chitosan has been used in water purification for the last 30 years. Since then, numerous studies have been conducted to find ways to use these materials. Today, chitosan is known as a dietary supplement for weight loss. In fact, it has been marketed for this purpose in Japan as well as Europe for about 20 years. Many people even call it "anti-fat" [25]. Chitosan has attracted great attention because of its increasing demand as a highly beneficial biopolymer in recent times. Chitosan, which is obtained by deacetylation of chitin with sodium hydroxide (NaOH), can be extracted from a variety of fungi, insects, and crustaceans. Basically, chitosan is a polymer consisting of randomly distributed units of N-acetyl-D-glucosamine and D-glucosamine with different deacetylation degree, acetylation type, and molecular weight which could be chemically modified to its derivatives. These derivatives affect antibacterial influence of chitosan and its solubility in acidic solutions. Chitosan's three reactive functional groups are: the amino group at the C-6 position, the primary hydroxyl group at the C-6 position, and the secondary hydroxyl group at the C-3 position. The amino group at the C-6 position differs from chitosan obtained from chitin due to its chemical, physical, and biological functions [18]. Chitosan is a very useful and attractive biopolymer due to its diverse chemical structure. Structural diversity can be seen in MW ranging from low (100 kDa) to high (300 kDa) as well as DD ranging from chitin (< 60%) to chitosan (> 60%). The wide range of chitosan samples described in different studies is surprising. Moreover, there are various conflicts regarding the use of chitosan in different biological applications [26].

Speaking of the synthesis of chitosan derivatives, the most beneficial advantage of chitosan is that it can be chemically modified into a wide variety of derivatives. Due to the presence of a primary alcohol group and an amino group, N, O-modified chitosan, as well as O-modified chitosan, can be modified to N-modified chitosan. The main reason for the synthesis of different chitosan derivatives is to improve certain properties. For example, quaternized chitosan derivatives have improved antimicrobial activity and water solubility, while phosphorylated chitosan derivatives have improved solubility, and N-benzyl/N-alkyl chitosan derivatives show improved antimicrobial activity [27]. Today, chitosan can be modified using two methods: Selective and non-selective modifications. The hydroxyl group is less nucleophilic than the amino group; however, both groups can still interact with electrophiles, including isothiocyanates and acids. These reactions lead to the selective O-chitosan derivative to be synthesized by a one-point reaction, while the non-selective N, O-chitosan derivative is synthesized. An acidic solution like sulfuric acid (H_2SO_4) can be used in production of the O-chitosan derivative. The amino group is protonated by using an acidic solution, which makes the alcohol functional group more reactive. This reaction preserves 90–95% of the amino acids; it is also a very effective and easy way of obtaining the O-modified chitosan derivative. On the other hand, the selective chitosan derivative equiped using this method is just limited to electrophiles and can only react with the amino group [28–30].

Due to its low cost, biocompatibility, absence of toxicity, and biodegradability, chitosan has applications in various fields such as tissue engineering, cosmetics, biomedicine, and biotechnology. Chitosan can be used to clarify agent wastewater and remove dye or metal ions due to its potential to protonate the amino group [31]. It can widely be used in the food industry as a browning inhibitor in juices, an antioxidant in sausages, a purifying agent in apple juices, and an antimicrobial agent. Chitosan can also be used to deliver transmucosal proteins and peptides thanks to its ability to adhere to the mucosa and open epithelial cell connections. Finally, it can be used as a carrier of macromolecular drugs. Conventionaly, chitosan has been used in its natural form with some limitations such as low surface area, low porosity, and low solubility at neutral pH. The functionality of chitosan can be increased by producing different derivatives through various chemical and physical processes [18].

Today, while preserving the organoleptic and nutritional properties of food products, great importance is attached to microbiological food safety. To accommodate these processes, the food industry must use special packaging materials that protect the quality and safety of food. Moreover, new generation food packaging materials are expected to have antimicrobial properties which create an environment that delays or completely prevents microbial growth, thus extending the shelf life of food products. Antimicrobial materials can be classified into two broad categories: organic materials and inorganic materials [32, 33]. Of particular interest as inorganic materials are metals, metal phosphates, and metal oxides considered safe for human and/or animal use. Inorganic substances are stable under severe conditions. However, examples of organic antimicrobial materials include halogenated compounds, quaternary ammonium salts, and phenols. Also, recent studies have found that natural polymers like chitosan and its derivatives have antibacterial activities. Thus, chitosan is promising substance that can be used in food packaging due to its ability to prevent gas or aroma in dry status and to form an excellent film [18] and for this purposes chitosan is used in various foods to extend shelf life mentioned in **Table 1**.

The antibacterial function of chitosan and its derivatives can be affected by different food ingredients. Charges and electrostatic forces on chitosan are the key

Food	Impact / Finding
Applications in Fruits and Vegetables	
Apple, banana, citrus, mango, peach, carrot and lettuce coated with chitosan, Strawberry coated with chitosan, Lychee fruit	Decreased respiratory rate and ethylene production, caries control and softening delay were observed.
Strawberry coated with chitosan	It has been observed that shelf life increases due to its antifungal properties and / or its ability to stimulate defense enzymes (chitinase and-1,3-glucanase).
Lychee fruit	The browning is delayed by preventing the increase in polyphenol oxidase activity.
Juices	
Use of soluble chitosan as a purifier in apple, grape, lemon, and orange juice	Fruit juices are purer than bentonite and gelatin, and the acceptance of fruit juices has increased.
To control the acidity of carrots and apple juice	It was observed to cause a significant decrease in titration acidity.
Apple and pear juice	It has also been indicated it prevent enzymatic browning.
Applications in Meat and Meat Products	
Beef	It was determined that the value of thiobarbutyric acid (TBA) decreased by 70% compared to the control sample and had a positive effect on maintaining the red color of the meat during storage.
Beef, fowl	It was determined that the addition of 3% chitosan-glutamate reduced the development of Clostridium perfringens spores.
Pork products	It was determined that chitosanglutamate used at 0.3% level and 0.6% was an effective preservative and the total number of bacteria, yeast, mold and lactic acid bacteria decreased to 3 records as a result of storage at 4° C for 18 days.
Sausage	It has been determined that chitosan reduces the use of sodium nitrite in sausage by half (150 ppm) without affecting quality and storage stability, and has also been found to reduce the amount of residual nitrite.
Applications in Dairy Products	
Cheese	*It has been reported that it inhibits the growth of L. monocytogenes and S. aureus, but does not affect Gram-negative Pseudomonas aeroginosa.*
Mozzarella	It has been determined that when used with the Lysozyme enzyme for film and coating purposes, it inhibits the growth of E. coli, L. monocytogenes, *Pseudomonas fluorescens* and yeast and molds and improves shelf life.
Applications in Eggs	
Coated with chitosan (3% chitosan in 1% acetic acid)	Reported at least 2 weeks longer shelf life of eggs at 25° C according to the control sample.
Coated with chitosan-lysozyme mixture	*Growth inhibition of L. monocytogenes, Salmonella enterica, coliforms, yeast and mold, delayed moisture loss and pH changes have been reported.*

Table 1.
The effect of chitosan on some food groups.

factors enabling its antibacterial property; therefore, any food ingredient that can affect these factors inhibits chitosan's antimicrobial activity. For instance, inorganic cations (Mg2+) inhibit the adhesion of *E. coli* to hexadecane via chitosan as a result of disruption of the electrostatic interaction liable for chitosan adsorption to the organism cell surface. Also, the addition of a metal ion lessened the antimicrobial influence of the chitosan derivative against *Staphylococcus aureus*. It has also been informed that starch, α-lactalbumin and β-lactoglobulin (whey proteins), and sodium chloride (NaCl) have a negative effect on antibacterial function of chitosan; however, fat had no effect [34, 35].

Chitosan is used as a food additive in many countries, including Japan, Korea, and Italy, due to its many properties. Today, customers demand safe and quality food products. The food industry's need to extend the shelf life of food products has pushed research to identify improved preservation strategies [36]. The food industry is an area where important applications of chitosan are widely used. Reducing or preventing the number of chemicals in food is highly demanded in food industry. To meet this growing demand, chitosan can be used as an additive in food products. Chitosan can react with metals and prevent the initiation of lipid oxidation; therefore, it can be used as a secondary antioxidant. What's more, the antioxidant effect of chitosan can be increased by combining it with many other naturally occurring ingredients. For example, combining chitosan with glucose enhances its antioxidant property, but it does not affect its antibacterial influence against *E. coli*, *S. aureus*, *Bacillus subtilis*, and *Pseudomonas*. Chitosan can also be bound to other naturally occurring substances such as xylan to improve their antibacterial and antioxidant properties [37, 38]. In addition, the low oxygen permeability of chitosan can decrease the contact of food with oxygen, thereby reducing the oxidation rate. Chitosan and its derivatives can be used as a promising substance to extend the shelf life of various food products. For example, when a chitosan-based substance is used to coat certain food products, it can decrease bread hardness, retrogradation, weight loss, and bacterial development. The surface of eggs and fruits can be coated with chitosan to create a protective barrier that can decrease respiration and sweating rates, as well as prevent the transfer of gas and moisture from albumin through eggshells. Thus, chitosan can be used to improve the structure and quality of food products as well as prevent microbial growth and color changes [18]. It is known that cattle act as a native reservoir for the *E. coli* O157:H7 agent that causes most foodborne diseases. Unfortunately, the inhibition of *E. coli* O157:H7 contamination on meat and meat products has not been successful. Controlling the contamination of these pathogens is very important during processing level and to reduce the contamination of *E. coli* O157:H7 in cattle to an acceptable value. The effect of chitosan on *E. coli* O157:H7 infected calves was researched and the results defined that the time of fecal contamination was remarkably decreased in chitosan-treated animals compared to untreated animals. Also, chitosan administration did not cause any ration profitability or abnormal behavior [39].

One of the factors affecting the antimicrobial activity of chitosan is the DD. An increase in DD means an increased number of amino groups on chitosan. As a result, chitosan has an increasing number of protonated amino groups in an acidic condition and is fully soluble in water, which increases the likelihood of interaction between chitosan and negatively charged cell walls of microorganisms. A variation of the deacetylation process resulted in the variation of MW as well as significant differences in the % DD of chitosan. It has been proven that chitosans with low MW (< 10 kDa) have more antimicrobial activity than natural chitosans. Low MW fractions have little or no activity. Chitosan with a MW ranging from 10,000 to 100,000 Da will be useful in inhibiting bacterial growth. In addition, chitosan with an average MW of 9300 Da, was effective against *E. coli*. One researcher reported that while D-glucosamine hydrochloride (chitosan monomer) did not exhibit any growth inhibition against several bacteria, chitosan was effective. This suggests that the antimicrobial activity of chitosan is not only related to its cationic nature but also its chain length. However, another researcher found that 10,000 Da chitosan was least effective in bactericidal activities, while 220,000 Da chitosan was most effective [36].

Chitosan is also used as an encapsulation material to improve food processing. Encapsulation is an attractive technology for protecting chemicals to prevent unwanted changes. Encapsulation materials can be formed with one or more compounds, such as chitosan, maltodextrin, acacia gum, hydroxypropyl methylcellulose phthalate gelatin, and starch, which can be used as a mixture or alone, among others. Chitosan has also attracted attention due to its applications in food and pharmacy. The antimicrobial and antifungal activities of chitosan are some of the most intriguing properties for improving food preservation and reducing the use of chemical preservatives. One study reported the use of chitosan in combination with essential oils, using nanoencapsulation processes, which have the potential to be applied in food industries. Due to the fact that essential oils such as thymol, eugenol, and carvacrol found in thyme, clove, and thyme essential oils easily degrade in light, air, and high temperatures, nanoencapsulation has recently been developed as an effective technique to protect them from evaporation and oxidation [40].

The ion binding character of chitosan is another important quality. Chitosan has proven to have the best chelating properties among other natural polymers. Although hydroxyl groups may also be involved in absorption, the amino groups of chitosan are responsible for compound formation, in which nitrogen is a donor of electron pairs. The mechanism for collaborating the reactive groups with metal ions is very different and can link to the ion pattern, pH, and also the key ingredients of the solution. The constitution of compounds can also be reported based on Lewis' acid–base theory: the metal ion (acting as an acid) is the acceptor of the double electron given by the chitosan (acting as the base) [41]. With regard to food applications of chitosan, information on the selective binding of essential metal ions to chitosan is important for its application as a cholesterol-lowering agent and its more controversial use as a weight loss agent [42].

Recently, researchers are increasingly interested in active food packaging materials, and there has been more interest in finding materials that provide biological activity to thin films as well as improving their properties. With the widespread use of non-fragile petroleum-based plastics, environmental pollution has become increasingly apparent. Most countries have placed restrictions on plastics, and there is an increasing demand for biodegradable functional packaging materials. Among the many natural biopolymers, chitosan has gained increasing attention thanks to its non-toxicity, biodegradability, biocompatibility, antibacterial activity, and excellent film-forming ability. Chitosan is a native cationic linear polysaccharide created of D-glucosamine and N-acetyl-D-glucosamine units prepared by partial deacetylation of chitin. Chitosan has excellent features that enable it to be used as wound dressing in the medical area, for tissue engineering, and as food packaging in the industrial area [8]. As a result, chitosan is one of the most important edible films used worldwide, produced by the deacetylation of chitin. Many native biopolymers can be used to compose edible films; however, among them chitosan attracts the attention for its excellent film-forming activity, flexibility, stability, biocompatibility, non-toxicity, biodegradability, and commercial usability. Chitosan, which is a traditionally available polysaccharide with the deacetylation of chitin, was generally accepted as safe by FDA (United States Food and Drug Administration) in 2005 and was confirmed for use as a food supplement suitable for human diets [7].

The most prominent properties of chitosan, as a compound obtained by various methods, can be attributed to its antimicrobial and antioxidant properties. Scientific publications reporting the antimicrobial activity of chitosan are specified in **Tables 2** and **3**. Considering these properties, the use of chitosan as an edible film to extend the shelf life of foods has been studied by many researchers.

Chitosan or its derivatives	Preparation method and/or foods	Target microorganisms and/or findings
Modified chitosan	Chitosan obtained from shrimp chitin in three particle sizes by deacetylating with different concentrations of NaOH (30%, 40%, and 50%) under microwave irradiation for 10 minutes	*Salmonella typhimurium* *Escherichia coli* The inhibitory effect was greater against *S. typhimurium* than *E. coli.*
	In 1% acetic acid, 73.68% classical deacetylated chitosan, and 83.55% ultrasound-assisted deacetylated chitosan	*Staphylococcus aureus* *Escherichia coli* *Pseudomonas aeruginosa* *Klebsiella pneumonia* *Candida albicans* *Candida parapsilosis* Antimicrobial activities are directly proportional to the increasing degree of deacetylation.
	Chitosan obtained by treating chitin with 50% NaOH and dissolved in 1% acetic acid without modification and with modification with ultraviolet or ozone	*Staphylococcus aureus* *Bacillus cereus* *Bacillus subtilis* *Escherichia coli* *Pseudomonas aeruginosa* *Aspergillus niger* *Candida albicans* *Candida tropicalis* ve *Rhizopus* No difference was observed in the antibacterial properties of unmodified and modified chitosan.
Kitooligosaccharides	Chitin (338 kDa MW and 35% deacetylation grade) Kitooligosaccharide (chitin hydrolyzed with HCl) Kitooligosaccharide (HCl hydrolyzed chitosan, 80% deacetylation degree)	*Staphylococcus aureus* *Bacillus subtilis* *Bacillus cereus* *Escherichia coli* *Pseudomonas aeruginosa* *Salmonella typhimurium* *Vibrio cholerae* *Shigella dysenteriae* *Prevotella melaninogenica* *Bacteroides fragilis* Chitin showed a bacteriostatic effect on *E. coli*, *V. cholerae*, *S. dysenteriae,* and *B. fragilis*, while Chitosan showed a bacteriostatic effect on all bacteria tested except *S. typhimurium*.
	Chitosan oligomers hydrolyzed with nitrous acid ($NaNO2$ + $CH3COOH$) and dissolved in 1% acetic acid	*Enterobacter aerogen* *Enterococcus faecalis* *Escherichia coli* *Staphylococcus aureus* Inhibition was observed in the microorganisms tested, but sharp inhibition was detected against *E. faecalis.*

Chitosan or its derivatives	Preparation method and/or foods	Target microorganisms and/or findings
	Chitooligosaccharides of different molecular weights: > 100 kDa, 100 to 10 kDa, 10 to 1 kDa	*Bacillus cereus* *Bacillus subtilis* *Staphylococcus aureus* *Escherichia coli* *Pseudomonas aeruginosa* *Candida albicans* *Saccharomyces chevalieri* *Macrophomina phaseolina* *Aspergillus niger* Antimicrobial effects were attributed to the type of strains. There was no association with MW.
	Chitooligosaccharides using papaya and dissolved in 0.25% acetic acid	*Escherichia coli* *Staphylococcus aureus* *Salmonella typhimurium* *Salmonella enteritidis* All microorganisms tested were inhibited but a higher effect was reported for *E. coli*.
Chitosan (0–2.0% w/w)	Surimi gel made from black catfish (*Clarias gariepinus*)	Bacterial growth is inhibited.
Chitosan solution prepared in 1% acetic acid	Culture tilapia (*Oreochromis niloticus*)	A shelf life of 6 days was observed for the control group, while a shelf life of 12 days was observed for samples treated with chitosan.
Chitosan coating solution (1% and 2% w/v in 1% acetic acid)	Sardine (*Sardinella longiceps*) fillets	Shelf life increased to 7 and 9 days, respectively, for fillets treated with 1% and 2% chitosan compared to the control group, whose shelf life was 5 days.
Chitosan coating solution (2% w/v in 1% acetic acid)	Rainbow trout (*Oncorhynchus mykiss*)	The shelf life of hot smoked fillets with a shelf life of 14–16 days, vacuum-packed and stored at +4 °C was extended to 24 days for fillets treated with chitosan.
Chitosan coating solution prepared with 2% (w/v) chitosan in 1% acetic acid	Carp (*Cyprinus carpio*)	A decrease was determined in the total number of aerobic organisms, psychrophilic bacteria, lactic acid bacteria, and *Enterobacteriaceae* bacteria.
1% (w/v) chitosan coating solution in 1% v/v acetic acid and 0.2% (w/v) bamboo leaves	Silver carp (*Hypophthalmichthys molitrix*)	The total number of living beings was higher in the control group stored at 4 °C for 24 days.
Chitosan, deacetylated 2% (w/v) in acetic acid at 1% v/v Chitosan coating solutions with 1.5% cinnamon oil added	Rainbow trout (*Oncorhynchus mykiss*) fillets	When chitosan only and chitosan with essential oil were added, the shelf life with chitosan was doubled compared to the control group.
Chitosan-based edible coatings	Deepwater pink shrimp (*Parapenaeus longirostris*)	The shelf life of shrimp treated with chitosan was extended by 3 days.
Chitosan (2% w/v) prepared in 1% acetic acid added with thyme oil (1% w/v)	Butterfly-shaped rainbow trout (*Oncorhynchus mykiss*)	Compared to the control group, the shelf life of fillets treated with chitosan was extended by more than 15 days.

Chitosan or its derivatives	Preparation method and/or foods	Target microorganisms and/or findings
(0.125% and 0.25% w/v) carvacrol added chitosan (2% w/v)	Tilapia (*Oreochromis niloticus*)	During the storage period of 21 days, total viable *Vibrio parahaemolyticus, Vibrio cholerae, Vibrio alginolyticus,* and the total coliform inhibitory effect were observed in fillets. Increasing the carvacrol concentration increased this effect.

Table 2.
Studies revealing the antibacterial properties of chitosan, accordind to Olatunde et al. [43].

Microorganism Bacteria / Yeast / Mold	Foods
Aeromonas hydrophila	Sausage, Seafood
Bacillus cereus	Meat, Seafood
Bacillus licheniformis Bacillus subtilis	Bread, Meat, Sausage
Brochothrix thermosphacta	Milk, Fruits and vegetables, Meat
Clostridium historyticum Clostridium perfringens Coliform	Sausage, Meat, Soybean Sprouts
Enterobacter aeromonas Enterococcus faecalis Escherichia coli	Fruits and vegetables, Bread, Meat
Listeria monocytogenes	Fruits and vegetables, Meat, Sausage, Seafood
Pseudomonas aeruginosa	Meat, Sausage, Seafood
Salmonella Enteritidis	Mayonnaise, Meat, Sausage
Salmonella Typhimurium	Bread, Meat, Sausage, Seafood
Staphylococcus aureus	Bread, Meat, Sausage, Seafood
Vibrio cholerae	Seafood
Vibrio parahaemolyticus	Seafood
Candida albicans	Seafood
Saccharomyces cerevisiae	Bread
Zygosaccharomyces baili	Juice
Aspergillus niger	Bread
Aspergillus parasiticus	Seafood
Fusarium oxysporum	Seafood
Rhizopus nigricans	Bread

Table 3.
Antimicrobial activity of chitosan against some organisms in foods.

3. Conclusion and results

Chitosan is a versatile biopolymer that has a variety of commercial applications. However, individual research reports have used chitosans from various sources with varying physicochemical properties. Hence, the question arises as to how to globally produce chitosans with consistent properties. Each batch of chitosan produced from the same manufacturer may differ in its quality. Functional properties of chitosan vary with molecular weight and degree of deacetylation. With proper modification of chitosan, its functional properties and biological activities can be further enhanced, and more applications are being developed. Chitosan with different structures shows different biological activities and not all the biological activities are found in one kind of chitosan. Each special type of bioactive chitosan should be developed for its potential application. Moreover, many studies carried out on chitosan and chitooligosaccharide bioactivity have not provided detailed molecular mechanisms. Hence, it is difficult to explain exactly how these molecules exert their activities. Therefore, future research should be directed toward understanding their molecular-level details, which may provide insights into the unknown biochemical functions of chitosan. One major drawback of chitosan film is its high sensitivity to humidity, and thus, it may not be appropriate for use when it is in direct contact with moist foods. More research is needed to develop antimicrobial chitosan films that are less sensitive to humidity. Numerous researches conducted on food applications of chitosans have been done at a small or laboratory scale. Further research on quality and shelf life of foods, containing or coated with chitosan, should be conducted on scale-up with large volumes typical of commercial conditions.

Chitosan is a polysaccharide-based film applied to the outer surface of foods and is effective in controlling physiological, morphological, and physiochemical changes in foods. Chitosan films can control oxygen and moisture permeability and have antioxidant and antimicrobial effects on food. The most widely accepted hypotheses about the antimicrobial effect of chitosan are: 1) ionic surface interaction resulting in cell wall leakage; 2) inhibition of mRNA and protein synthesis by the penetration of chitosan into the nuclei of microorganisms; and 3) creating an external barrier, chelating metals and triggering suppression of microbial growth in essential nutrients. All of these situations are likely to occur at the same time but at different densities. The MW and DD are also important factors in determining such activity. Generally, the lower the MW and DD, the higher the effectiveness in reducing microorganism growth and proliferation. Despite the many advantages of chitosan, there are also various restrictions related to its use. The most important limitation of chitosan is its low solubility at neutral pH. To compensate for this deficiency, various chemical and physical processes have been used to increase its solubility.

Author details

Sadik Büyükyörük
Department of Food Hygiene and Technology, Faculty of Veterinary Medicine, Adnan Menderes University, Işikli, Aydin, Turkey

*Address all correspondence to: sbuyukyoruk@adu.edu.tr

References

[1] Badawy MEI and Rabea EI, (2011). A biopolymer chitosan and its derivatives as promising antimicrobial agents against plant pathogens and their applications in crop protection. International Journal of Carbohydrate Chemistry, 1-29, doi:10.1155/2011/460381.

[2] Beaney P, Lizardi-Mendoza J and Healy M, (2005). Comparison of chitins produced by chemical and bioprocessing methods. Journal of Chemical Technology and Biotechnology, 80:2, 145-150.

[3] Yıldırım Z, Öncül N and Yıldırım M, (2016). Chitosan and antimicrobial properties. Nigde University Journal of Engineering Sciences, 5:1, 19-36.

[4] Bostan K, Aldemir T and Aydin A, (2007). Chitosan and its antimicrobial activity. Journal of Turkish Society of Microbiology, 37:2, 118-127.

[5] Ibañez-Peinado D, Ubeda-Manzanaro M, Martinez A, Rodrigo D, (2020). Antimicrobial effect of insect chitosan on *Salmonella* typhimurium, *Escherichia coli* O157:H7 and *listeria monocytogenes* survival. PLoS ONE, 15:12, e0244153.https://doi.org/10.1371/journal. pone.0244153

[6] Barrera-Ruiz DG, Cuestas-Rosas GC, Sánchez-Marıñez RI, Álvarez-Aınza ML, Moreno-Ibarra GM, López-Meneses AK, Plascencıa-Jatomea M, Cortez-Rocha MO, (2020). Antibacterial activity of essential oils encapsulated in chitosan nanoparticles, Food Science and Technology, Campinas. 40:2, 568-573.

[7] Demircan B and Özdestan-Ocak Ö, (2020). Effects of lemon essential oil and ethyl lauroyl arginate on the physicochemical and mechanical properties of chitosan films for mackerel fillet coating application. Journal of Food Measurement and Characterization, https://doi.org/10.1007/s11694-020-00745-1.

[8] Cui R, Yan J, Cao J, Qin Y, Yuan M and Li L, (2020). Release properties of cinnamaldehyde loaded by montmorillonite in chitosan-based antibacterial food packaging. International Journal of Food Science and Technology, doi:10.1111/ijfs.14912

[9] Bourakadi KE, Merghoub N, Fardioui M, El Mehdi MM, Kadmiri IM, El Mokhtar E, El Kacem AQ, Bouhfid R, (2019). Chitosan/polyvinyl alcohol/thiabendazoluim-montmorillonite bionanocomposite films: Mechanical, morphological and antimicrobial properties. Composites Part B: Engineering, 172, 103-110.

[10] Huang YK, Mei L, Chen XG and Wang Q, (2018). Recent developments in food packaging based on nanomaterials. Nanomaterials, 8:10, 830-859.

[11] Qin YY, Zhang ZH, Li L, Yuan ML, Fan J and Zhao TR, (2015). Physio-mechanical properties of an active chitosan film incorporated with montmorillonite and natural antioxidants extracted from pomegranate rind. Journal of Food Science and Technology-Mysore, 52, 1471-1479.

[12] No HK, Park NY, Lee SH, Meyers SP, (2002). Antibacterial activity of chitosans and chitosan oligomers with different molecular weights. International Journal of Food Microbiology, 74:1, 65-72.

[13] Jumaa M, Furkert FH, Muller BW, (2002). A new lipid emulsion formulation with high antimicrobial efficacy using chitosan. European Journal of Pharmaceutics and Biopharmaceutics, 53:1,115-123.

[14] Wang GH, (1992). Inhibition and inactivation of five species of foodborne pathogens by chitosan. Journal of Food Protection, 55:11, 916-919.

[15] Gücükoğlu A, Yildirim Y, Terzi G, Erdem, S, (2016). In vitro effects of chitosan on the survival of *listeria monocytogenes*. Veterinary Medicine, 13, 11-18.

[16] Goy RC and Assis OBG (2014). Antimicrobial analysis of films processed from chitosan and N,N,N-trimethylchitosan. Brazilian Journal of Chemical Engineering, 31, 643-648.

[17] Jia Z, shen D and Xu W, (2001). Synthesis and antibacterial activities of quaternary ammonium salt of chitosan. Carbohydrate Research, 333:1, 1-6.

[18] Rajoka MSR, Mehwish HM, Wu Y, Zhao L, Arfat Y, Majeed K and Anwaar S, (2020). Chitin/chitosan derivatives and their interactions with microorganisms: A comprehensive review and future perspectives. Critical Reviews in Biotechnology, 40:3, 365-379.

[19] Costa EM, Silva S, Tavaria FK, Pintato MM, (2017). Insights into chitosan antibiofilm activity against methicillin-resistant *Staphylococcus aureus*. Journal of Applied Microbiology, 122:6, 1547-1557.

[20] Tang H, Zhang P, Kieft TL, Ryan SJ, Baker SM, Wiesmann WP and Rogelj S, (2010). Antibacterial action of a novel functionalized chitosan-arginine against gram-negative bacteria. Acta Biomaterialia, 6:7, 2562-2571.

[21] Helander IM, Nurmiaho-Lassila EL, Ahvenainen R, Rhoadesc J, Roller S, (2001). Chitosan disrupts the barrier properties of the outer membrane of gram-negative bacteria. International Journal of Food Microbiology,71:2-3, 235-244.

[22] Raafat D and Sahl HG, (2009). Chitosan and its antimicrobial potential – A critical literature survey. Microbial Biotechnology, 2:2, 186-200.

[23] Kiang T, Wen J, Lim HW, Leong KW, (2004). The effect of the degree of chitosan deacetylation on the efficiency of gene transfection. Biomaterials, 25:22, 5293-5301.

[24] Younes I, Sellimi S, Rinaudo M, Jellouli K, Nasri M, (2014). Influence of acetylation degree and molecular weight of homogeneous chitosans on antibacterial and antifungal activities. International Journal of Food Microbiology, 185, 57-63.

[25] Guo X, Sun T, Zhong R, Ma L, You C, Tian M, Li H and Wang C, (2018). Effects of chitosan oligosaccharides on human blood components. Frontiers in Pharmacology, 9:1412, 1-10.

[26] Verlee A, Mincke S, Stevens CV, (2017). Recent developments in antibacterial and antifungal chitosan and its derivatives. Carbohydrate Polymers, 164, 268-283.

[27] Kurita Y and Isogai A, (2012). N-Alkylations of chitosan promoted with sodium hydrogen carbonate under aqueous conditions. International Journal of Biological Macromolecules, 50:3, 741-746.

[28] Sahariah P, Arnadottir B and Masson M, (2016). Synthetic strategy for selective N-modified and O-modified PEGylated chitosan derivatives. European Polymer Journal, 81, 53-63.

[29] Carvalho LCR, Queda F, Santos CVA and Marques MMB, (2016). Selective modification of chitin and chitosan: An route to tailored oligosaccharides. Chemistry an Asian Journal, 11:24, 3468-3481.

[30] Rabea EI, Badawy ME, Rogge TM, Stevens CV, Steurbaut W, Höfte M and Smagghe G, (2006). Enhancement of fungicidal and insecticidal activity by reductive alkylation of chitosan. Pest Management Science, 62:9, 890-897.

[31] Vakili M, Rafatullah M, Salamatinia B, Abdullah AZ,

Ibrahim MH, Tan KB, Gholami Z and Amouzgar P, (2014). Application of chitosan and its derivatives as adsorbents for dye removal from water and wastewater: A review. Carbohydrate Polymers, 113, 115-130.

[32] Higazy A, Hashem M, ElShafei A, Shaker N and Hady MA, (2010). Development of antimicrobial jute packaging using chitosan and chitosan–metal complex. Carbohydrate Polymers, 79:4, 867-874.

[33] Lee SY, Lee SJ, Choi DS and Hur SJ, (2015). Current topics in active and intelligent food packaging for preservation of fresh foods. Journal of the Science of Food and Agriculture, 95:14, 2799-2810.

[34] Devlieghere F, Vermeulen A and Debevere J, (2004). Chitosan: Antimicrobial activity, interactions with food components and applicability as a coating on fruit and vegetables. Food Microbiology, 21:6, 703-714.

[35] Chung YC, Yeh JY and Tsai CF, (2011). Antibacterial characteristics and activity of water soluble chitosan derivatives prepared by the maillard reaction. Molecules, 16:10, 8504-8514.

[36] Kong M, Chen XG, Xing K and Park HJ, (2010). Antimicrobial properties of chitosan and mode of action: A state of the art review. International Journal of Food Microbiology, 144:1, 51-63.

[37] Schreiber SB, Bozell JJ, Hayes DG and Zivanovic S, (2013). Introduction of primary antioxidant activity to chitosan for application as a multifunctional food packaging material. Food Hydrocolloids, 33:2, 207-214.

[38] Kanatt SR, Chander R and Sharma A, (2008). Chitosan glucose complex – A novel food preservative. Food Chemistry, 106:2, 521-528.

[39] Jeong KC, Kang MY, Kang J, Baumler DJ and kaspar CW, (2011). Reduction of *Escherichia coli* O157:H7 shedding in cattle by addition of chitosan microparticles to feed. Applied and Environmental Microbiology, 77:8, 2611-2616.

[40] No HK, Meyers SP, Prınyawıwatkul W and Xu Z, (2007). Applications of chitosan for improvement of quality and shelf life of foods: A review. Journal of Food Science, 72:5, 87-100.

[41] Tong T, Li Y, Hou R, Wang X and Wang S, (2017). Decoration of chitosan microspheres with Brønsted hetero-polyacids and Lewis ion Ti: Tri-functional catalysts for esterification to biodiesel. RSC Advances, 7, 42422-42429.

[42] Bokura H and Kobayashi S, (2003). Chitosan decreases total cholesterol in women: A randomized, double-blind, placebo-controlled trial. European Journal of Clinical Nutrition, 57, 721-725.

[43] Olatunde OO, Benjakul S and Yesilsu AF, (2020). Antimicrobial compounds from crustaceans and their applications for extending shelf-life of marine-based foods. Turkish Journal of Fisheries and Aquatic Science, 20:8, 629-646.